Urban Air Pollution in Megacities of the World

EARTHWATCH

Global Environment Monitoring System

Urban Air Pollution in Megacities of the World

Published on behalf of
the World Health Organization
and the United Nations Environment Programme
by

BLACKWELL
Reference

UNEP

For bibliographic and reference purposes the publication should be referred to as: WHO/UNEP 1992 Urban Air Pollution in Megacities of the World, World Health Organization, United Nations Environment Programme, Blackwell, Oxford.

First published 1992
Reprinted 1993

Blackwell Publishers
108 Cowley Road
Oxford OX4 1JF
UK

238 Main Street
Cambridge, Massachusetts 02142
USA

British Library Cataloguing in Publication Data

A CIP catalogue record for this book is available from the British Library.

Library of Congress Cataloging-in-Publication Data

Urban air pollution in megacities of the world: earthwatch: global environment monitoring system.
p. cm.
ISBN 0–631–18404–X (pbk.)
1. Air–Pollution. 2. Urbanization—Environmental aspects.
3. Metropolitan areas. I. World Health Organization. II. United Nations Environment Programme.
TD883.U729 1992
363.73'92'091732—dc20 92–19968

DISCLAIMER
The designations employed and the presentations do not imply the expression of any opinion whatsoever on the part of WHO or UNEP concerning the legal status of any country, territory, city or area or its authority, or concerning the delimitation of its frontiers or boundaries.

Typeset in Palatino
by Imogen Bertin, Cork
Printed in Great Britain by The Alden Press, Oxford

This book is printed on acid-free paper

Contents

PART II – DETAILED MEGACITY REPORTS

Appendices

Foreword

Throughout the world, urban centres are developing at an unprecedented rate. Increasing urban populations and growing levels of industrialization have inevitably led to a series of environment-related problems, not the least of which is worsening air quality.

By the end of the decade around half of the world's population will be living in urban areas, exposing millions of dwellers in urban areas to poor shelter and lack of safe water, sanitation and drainage. These problems are often compounded by harmful levels of urban air pollutants caused mainly by emissions from industries, motor vehicles and the domestic combustion of fossil fuels.

To address the issue of urban air pollution, the World Health Organization (WHO) and the United Nations Environment Programme (UNEP) have, since 1974, been collaborating in an Urban Air Quality Monitoring Programme, commonly referred to as GEMS/Air. The GEMS/Air programme forms part of the United Nations system-wide Global Environment Monitoring System (GEMS), which in turn represents a collective effort of the world community to acquire, through monitoring and assessment, the scientific data and information that are essential for the rational and sustainable management of the environment.

The present report was initiated in order to focus more attention onto the air quality situation in megacities, particularly in tropical developing countries. It is already emerging that the lessons to be learned from this investigation will apply in large measure to other cities not covered in this report.

The report illustrates the urgent need for the more effective planning of energy requirements and transport to reduce human exposure to pollutants and to decrease risks to health and the environment. The need for cleaner technology for energy production from fossil fuels, and for the use of alternative, less-polluting fuels, is highlighted. Better public transport should be introduced to lessen reliance on private motor vehicles. The results show that promotion of human health, environmental protection and sustainable development are intimately linked.

The collection of accurate and reliable data necessary for the evaluation of urban air quality in the twenty megacities covered by this report has been a collective effort of scientists, managers and monitoring specialists in each city. Many of the basic data have come from the GEMS/Air programme, supplemented with information from national reports and scientific publications.

By studying past and present air pollution problems and air quality management strategies, UNEP and WHO believe that many of the problems currently faced by megacities can be avoided by the megacities of the future.

Hiroshi Nakajima
Director-General
World Health Organization

Mostafa K. Tolba
Executive Director
United Nations Environment Programme

Acknowledgements

The preparation of this Report was a collective undertaking of the Division of Environmental Health of WHO in Geneva (Dr D. Mage and Mr G. Ozolins), the GEMS Monitoring and Assessment Research Centre in London (Mr A. D. Webster and Dr P. J. Peterson), and the UNEP/GEMS Programme Activity Centre in Nairobi (Dr M. D. Gwynne, Dr R. Orthofer and Dr V. Vandeweerd). They were supported in this task by two advisers, Dr J. G. Kretzschmar, Belgium and Dr M. Raizenne, Canada.

The basic information for this report was obtained from the appropriate national or local authorities for the 20 megacities and from the GEMS/Air Quality Data Bank operated by the US Environmental Protection Agency (Mr E. G. Evans).

The collection of national and city data was facilitated by WHO Regional Offices: American (Dr H. W. de Koning), Eastern Mediterranean (Dr M. I. Sheikh), South-East Asia (Mr M. L. Gupta) and Western Pacific (Mr S. Tamplin).

The co-sponsoring organizations wish to thank individuals and organizations listed below for assisting with data collection or initial city analysis and for commenting upon particular portions of the text in its various stages:

Dr A. L. Aggarwal and Dr R. Thakre, National Environmental Engineering Research Institute, Nagpur, India.

Dr A. Beg, Pakistan Agro Chemicals, Karachi, Pakistan.

Dr B. Bierck, Stevens Institute of Technology, Hoboken NJ, USA.

Mr J. Bower, Warren Spring Laboratory, Stevenage, UK.

Dr J. R. Brook, Health and Welfare Canada, Ottawa, Canada.

Dr Chen Changjie, Institute of Environmental Health Monitoring, Beijing, China.

Mr A. Fernández-Bremauntz, London, UK.

Dr R. Gonzalez-Garcia, Secretariat of Urban Development and Ecology, Mexico DF, Mexico.

Professor Liang Xiyan, Beijing Municipal Environmental Monitoring Centre, Beijing, China.

Ing. Maria Dolores Mazzola, Direccion de Relaciones Sanitarias Internacionales, Buenos Aires, Argentina.

Mr J. O. E. Moore, Health and Welfare Canada, Ottawa, Canada.

Dr A. B. Murray, UNEP-Harmonization of Environmental Measurement Office, Munich, Germany.

Professor Mahmoud M. Nasralla, Cairo National Research Centre, Cairo, Egypt.

Dr D.-G. Rhee, Ministry of the Environment, Seoul, Korea.

Professor F. Ya. Rovinsky, Institute of Global Climate and Ecology, Moscow, Russia.

Mr J. Siriswasdi, Office of the National Environment Board, Bangkok, Thailand.

Dr Enrique Toliva, Tecnoconsult, Mexico DF, Mexico.

Thanks are due to the staff of the GEMS Monitoring and Assessment Research Centre where editing and production took place: Mr J. A. Jackson and Ms C. J. Meads for editing, and Ms F. Preston for secretarial services. Finally, thanks are due to Ms I. Bertin for typesetting, design and layout.

The cover photograph was provided by Mark Edwards of Still Pictures.

1

Introduction

Many of the major cities of the world are beset by a series of environmentally-related problems, not the least of which is their deteriorating air quality. Exposure to air pollution is now an almost inescapable part of urban life throughout the world. The available information shows that World Health Organization (WHO) guidelines are being regularly exceeded in many major urban centres, in some cases to a great extent. Given the rate at which these cities are growing, and the general absence of pollution control measures in many of them, the air pollution situation will probably worsen and the quality of life of many urban residents will continue to deteriorate. Although good progress has been made in controlling air pollution problems in many industrialized countries over the last two decades, air quality - particularly in the larger cities of the developing countries - is worsening. The WHO Commission on Health and Environment (WHO, 1992a), which recently concluded its work, identified urban air pollution as a major environmental health problem deserving high priority for action.

Of foremost concern is the health and well-being of urban residents. The concentrations of ambient air pollutants which prevail in many urban areas are sufficiently high to cause increased mortality, morbidity, deficits in pulmonary function and cardiovascular and neurobehavioural effects. Indoor sources of air pollution, such as cooking fires and smoking tobacco add to these ambient concentrations and can result in even more severe exposures for people in their homes. In addition to health, there are other concerns. Air pollution is seriously damaging material resources of the cities, such as buildings and various works of art. Its impact on vegetation is also of concern. Finally, urban agglomerations are also the major sources of regional and global atmospheric pollution and emissions of greenhouse gases.

In order to assess the problems of urban air pollution in a global context, WHO and the United Nations Environment Programme (UNEP) initiated a detailed study of air quality in 20 of the 24 megacities of the world (see Chapter 4, Section 4.2). For the purposes of this study, megacities were defined as urban agglomerations with current or projected populations of 10 million or more by the year 2000 (Figure 1.1) (UN, 1989). The urban areas chosen included cities in all parts of the world - two in North America, four in Central and South America, one in Africa, eleven in Asia and two in Europe. The megacities are not necessarily the world's most polluted cities. The primary reasons for singling out the megacities are that they all have serious air pollution problems; they encompass large land areas and many people (the total population of the 20 megacities in 1990 was estimated to be 234 million); and many other cities are heading for megacity status. This last point is of particular importance. A review of the air pollution situation in the present-day megacities and identification of their difficulties in finding solutions can serve as a warning of the problems facing the rapidly growing urban areas, and act as a guide for solving and preventing some of them.

The study was carried out within the framework of the WHO/UNEP long-standing urban air quality monitoring and assessment programme, known as GEMS/Air, which is a component of the Global Environment Monitoring System. This programme has been operating since 1974 and its activities include collecting, handling and assessing certain air quality data from over 50 cities in 35 countries. A recent review of the programme has proposed that additional pollutants should be included in the GEMS/Air monitoring programme, that emissions inventories should be compiled and that additional countries should be encouraged to join in order to strengthen the world-wide coverage (WHO, 1992b). WHO and UNEP have issued two earlier comprehensive assessments based on the results from the programme (WHO, 1984; UNEP/WHO, 1988). For the present report, the GEMS/Air data were supplemented by other air quality data, information on sources and emissions of air pollutants, and other factors of importance in evaluating air pollution. Such data and information were supplied by national and municipal authorities directly to WHO/UNEP secretariats or obtained through WHO/UNEP staff, consultant missions, and in some instances from the scientific literature. The draft assessments were

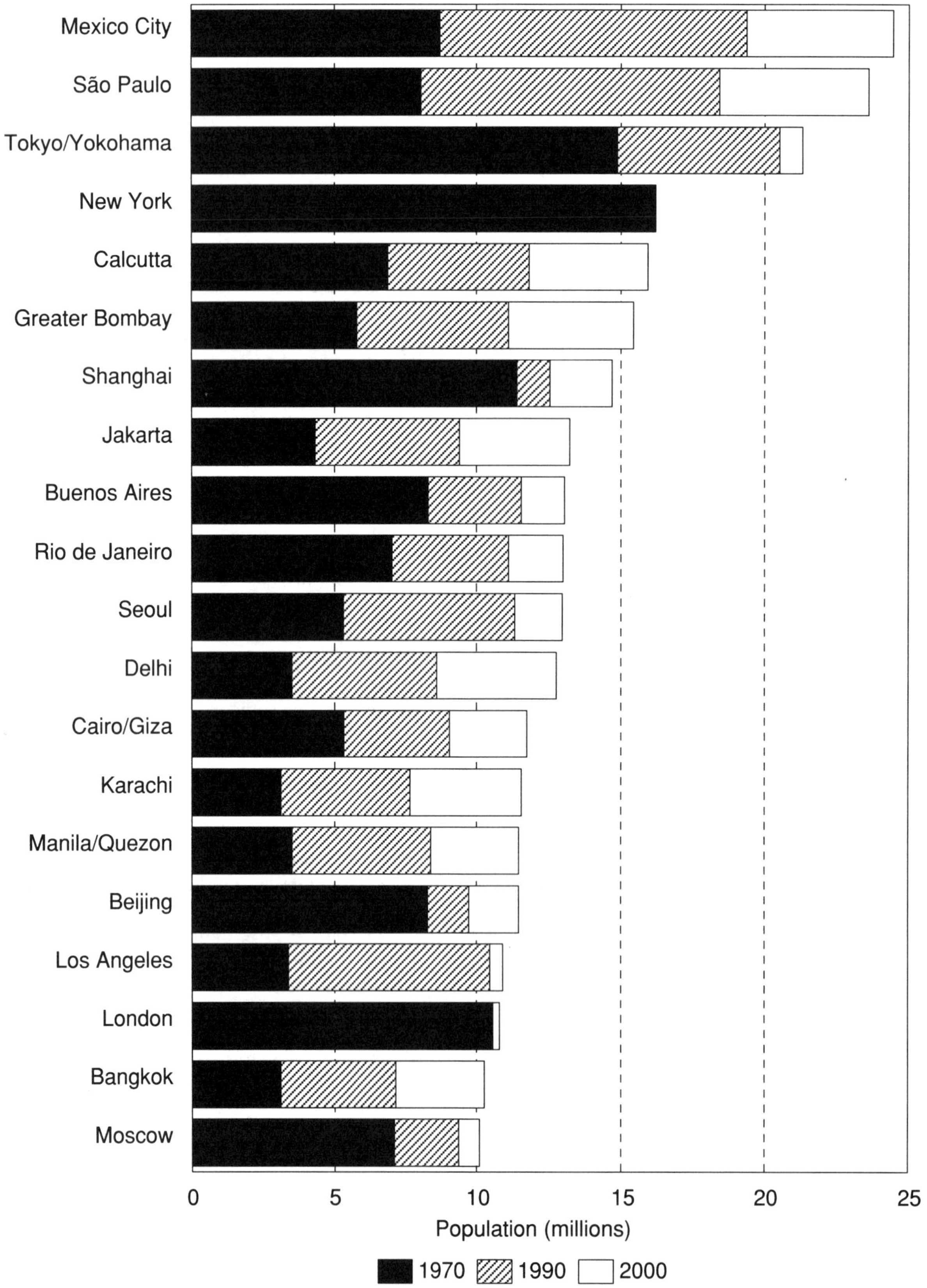

Figure 1.1 *Estimated population of 20 megacities in 1970 and 1990 and projected population in 2000*
Source: UN, 1989

reviewed by a WHO/UNEP Government-designated Expert Group meeting on Urban Air Pollution Monitoring which was convened in Geneva from 5–8 November 1991 (WHO, 1992b). This group also made recommendations for the conclusions and suggested actions which the final report should contain. In order to ensure that the report contains accurate information on the air pollution situation in different cities, the relevant sections of the draft report were submitted to the respective authorities in each country/city for verification.

The Preparatory Committee of the UN Conference on Environment and Development (UNCED) was informed of WHO and UNEP's study at its First Substantive Session in August 1990. The Committee welcomed this undertaking and requested WHO and UNEP to keep it informed of progress. A report was consequently presented to the Committee at its Third Session in August 1991. Publication of the present report follows the convening of UNCED in Brazil in 1992.

The compilation of a global assessment of urban air pollution in megacities is difficult. The available data sets on the pollutants and their health and environmental impacts are often incomplete and sometimes out of date. There are difficulties in compiling the data for a report such as this because of differences in methodology and reporting between countries and even within countries and cities. Shortcomings in the data which were used, including problems of representativeness and comparability of different data sets, are noted in the report where appropriate. Nevertheless, the data sets which have been collected and analysed provide the first valid and comprehensive overview of the air pollution situation and trends in the megacities. Where appropriate, these data sets have been compared with health guidelines established by WHO and with the respective national standards set by governments. This assessment should contribute substantially to the formulation of national, regional and global policies for the prevention and control of urban air pollution world-wide.

It is clear that air pollution in many of the world's megacities, as well as in other cities, is a major health and environmental concern. This concern is increasing and should command high priority for action. National efforts to deal with this increasing problem are under way in many of these cities, but such efforts must be strengthened. Global and regional initiatives aimed at counteracting global warming and transboundary air pollution, through measures such as cleaner production and energy conservation, will also result in reducing the emissions of many of the air pollutants of local concern. Additional strategies will, however, be needed to deal with air pollution within the cities. The WHO Commission on Health and Environment, Panel on Energy (WHO, 1992c) identified the following as possible strategic elements "... urban and transportation planning, particularly in fast-growing cities, provision of safe and convenient mass transport, control of emissions from vehicles (especially automobiles) and industries, a switch to cleaner fuels, emphasis on district heating with co-generation plants, stringent controls on new industrial operations, and appropriate siting of power generating plants in relation to residential areas".

Following this general introduction, the remainder of the report is divided into two major parts. Part I provides an overview of the findings, including a chapter containing basic background information on urban air pollution and its impacts. It also includes a chapter containing single sheet summaries on the air pollution situation in the 20 megacities. This is followed by a comparative overview of air pollution in the 20 megacities. The final chapter of Part I presents summary conclusions and recommendations based on the data presented.

Part II provides more detailed information and comprises individual chapters on each of the 20 megacities summarizing demographic and meteorological situations, sources and emissions of air pollutants, air quality and its adverse impacts on human health and the environment. Where appropriate these chapters also contain specific recommendations for action. Air quality guidelines and standards are listed in Appendix I. Full country names, their abbreviations and city names used throughout this report are listed in Appendix II.

References

UN 1989 *Prospects of World Urbanization 1988*, Population Studies No. 112, United Nations, New York.

UNEP/WHO 1988 *Assessment of Urban Air Quality*, United Nations Environment Programme and World Health Organization, Nairobi.

WHO 1984 *Urban Air Pollution 1973-1980*, World Health Organization, Geneva.

WHO 1992a *Our Plant Our Health: Report of the WHO Commission on Health and Environment*, World Health Organization, Geneva.

WHO 1992b *Urban Air Pollution Monitoring: Report of a Meeting of UNEP/WHO Government-Designated Experts, Geneva, 5-8 November 1991*, WHO/PEP/92.2, UNEP/GEMS/ 92.A.1, World Health Organization, Geneva.

WHO 1992c *WHO Commission on Health and Environment, Panel on Energy*, World Health Organization, Geneva.

PART I –
AIR POLLUTION REVIEW

2

Understanding Urban Air Pollution Problems

2.1 Introduction

The air pollution problems of the megacities differ greatly and are influenced by a number of factors, including topography, demography, meteorology, the level and rate of industrialization and socio-economic development. These problems are of increasing importance because the projected growth in the urban population world-wide increases the number of people exposed to urban air pollution. United Nations (UN, 1989) estimates indicate that by the year 2000, 47 per cent of the global population will be living in urban areas. Figure 1.1 (Chapter 1) shows the projected population growth in the 20 megacities discussed in this report. In 1990, 69 cities had populations of 3 million or more and it is projected that by the year 2000 85 cities will be in this category (Table 2.1).

2.2 Sources of Urban Air Pollution

The production and consumption of energy influences most aspects of urban life. Energy is required for cooking, heating and lighting, for motorized transport and for industrial processes. Fossil fuels meet most of these energy demands in cities throughout the world, either directly or via conversion to electrical energy. Growing urban populations and levels of industrialization inevitably lead to greater energy demand which is usually reflected in increasing pollutant emissions.

The combustion of fossil fuels for domestic heating, for power generation, in motor vehicles, in industrial processes and in disposal of solid wastes by incineration, are generally the principal sources of air pollutant emissions to the atmosphere in urban areas. Traditionally the most common air pollutants in urban environments include sulphur dioxide (SO_2), the nitrogen oxides (NO and NO_2, collectively termed NO_x), carbon monoxide (CO), ozone (O_3), suspended particulate matter (SPM) and lead (Pb). Combustion is the principal man-made source of the traditional air pollutants; combustion of fossil fuels in stationary sources leads to the production of SO_2, NO_x and particulates – both primary particulates in the form of fly ash and soot and secondary particulates, sulphate (SO_4^{2-}) and nitrate (NO_3^-) aerosols formed in the atmosphere following gas to particle conversion. Domestic solid fuel use, mainly coal and wood, also represents a significant source of these pollutants in some cities, particularly those in the developing world. Petrol-fuelled motor vehicles are the principal sources of NO_x, CO and Pb, whereas diesel-fuelled engines emit significant quantities of particulates and SO_2 in addition to NO_x.

Ozone, a photochemical oxidant and the main constituent of photochemical smog, is not emitted directly from combustion sources, but is formed in the lower atmosphere in the presence of sunlight from NO_x and volatile organic compounds (VOCs). The VOCs may be emitted from a variety of man-made sources including road traffic, production and use of organic chemicals (e.g., solvents), transport and use of crude oil, the use and distribution of natural gas and, to a lesser extent, from waste disposal sites and waste-water treatment plants. Cities in warmer, sunny locations with high traffic densities tend to be especially prone to the net formation of O_3 and other photochemical oxidants from precursor emissions.

Although detailed emissions inventories are not widely available for individual cities, on the basis of observed trends in national emissions and recent increases in vehicle registrations, it may be concluded that motor vehicles now constitute the main source of air pollutants in the majority of cities in industrialized countries. This is particularly true of the pollutants CO, NO_x and, to a lesser extent, SPM.

The cities of developing countries, in contrast, exhibit a greater variety in the nature of air pollution sources. The relative contributions of mobile and stationary sources to air pollutant emissions differs markedly between cities and is dependent on the level of motorization and the level, density, and type

Table 2.1 *Population in urban agglomerations of more than 3 million*

Rank[a]	Agglomeration	Country	Population (millions)					
			1950	1960	1970	1980	1990	2000
1	Tokyo/Yokohama	Japan	6.74	10.69	14.87	17.67	20.52	21.32
2	Mexico City	Mexico	2.88	4.93	8.74	13.97	19.37	24.44
3	São Paulo	Brazil	2.75	4.71	8.06	12.50	18.42	23.60
4	New York	U.S.A.	12.34	14.16	16.19	15.58	15.65	16.10
5	Shanghai	China	10.26	10.67	11.41	11.80	12.55	14.69
6	Calcutta	India	4.45	5.50	6.91	9.00	11.83	15.94
7	Buenos Aires	Argentina	5.13	6.69	8.31	9.88	11.58	13.05
8	Seoul	Korea	1.02	2.36	5.31	8.28	11.33	12.97
9	Greater Bombay	India	2.90	4.06	5.81	8.05	11.13	15.43
10	Rio de Janeiro	Brazil	3.45	4.93	7.04	8.98	11.12	13.00
11	London	United Kingdom	10.25	10.72	10.55	10.44	10.57	10.79
12	Osaka/Kobe	Japan	3.83	5.75	7.60	8.71	10.49	11.18
13	Los Angeles	U.S.A.	4.05	6.53	8.38	9.51	10.47	10.91
14	Beijing	China	6.64	7.30	8.29	9.10	9.74	11.47
15	Jakarta	Indonesia	1.73	2.73	4.32	6.42	9.42	13.23
16	Moscow	Russia	4.84	6.29	7.11	8.21	9.39	10.11
17	Tehran	Iran	1.04	1.87	3.29	5.55	9.21	13.78
18	Cairo/Giza	Egypt	2.41	3.71	5.33	6.93	9.08	11.77
19	Paris	France	5.44	7.23	8.33	8.68	8.75	8.76
20	Delhi	India	1.39	2.28	3.53	5.54	8.62	12.77
21	Manila/Quezon	Philippines	1.54	2.27	3.53	5.96	8.40	11.48
22	Tianjin	China	5.36	5.97	6.87	7.68	8.38	9.96
23	Milan	Italy	3.63	4.50	5.53	6.77	7.90	8.74
24	Karachi	Pakistan	1.03	1.85	3.13	4.95	7.67	11.57
25	Lagos	Nigeria	0.29	0.76	2.02	4.45	7.60	12.45
26	Bangkok	Thailand	1.36	2.15	3.11	4.75	7.16	10.26
27	Chicago	U.S.A.	4.94	5.98	6.72	6.78	6.89	6.98
28	Lima-Callao	Peru	1.01	1.69	2.84	4.41	6.50	8.78
29	Dacca	Bangladesh	0.42	0.65	1.50	3.29	6.40	11.26
30	Madras	India	1.40	1.71	3.03	4.19	5.69	7.85
31	Bogota	Colombia	0.68	1.30	2.37	3.91	5.59	6.94
32	Hong Kong	Hong Kong	1.75	2.59	3.40	4.48	5.44	6.09
33	St Petersburg	Russia	2.62	3.46	3.98	4.70	5.39	5.84
34	Baghdad	Iraq	0.58	1.02	2.11	3.54	5.35	7.66
35	Madrid	Spain	1.55	2.23	3.37	4.30	5.06	5.42
36	Bangalore	India	0.76	1.17	1.62	2.81	4.86	7.67
37	Pusan	Korea	0.95	1.15	1.81	3.12	4.75	5.82
38	Santiago	Chile	1.33	2.03	2.84	3.70	4.70	5.58
39	Shenyang	China	2.22	4.47	3.14	3.82	4.46	5.50
40	Naples	Italy	2.75	3.19	3.59	4.02	4.33	4.46
41	Philadelphia	U.S.A.	2.94	3.64	4.02	4.12	4.23	4.33
42	Lahore	Pakistan	0.83	1.26	1.97	2.85	4.08	5.93
43	Caracas	Venezuela	0.68	1.28	2.05	2.94	3.96	4.79
44	Detroit	U.S.A.	2.77	3.55	3.97	3.81	3.86	3.92
45	Belo Horizonte	Brazil	0.47	0.87	1.59	2.52	3.81	5.01
46	Sydney	Australia	1.70	2.13	2.67	3.28	3.79	4.06
47	Rome	Italy	1.57	2.33	3.07	3.48	3.72	3.82

Continued

Table 2.1 *Continued*

Rank[a]	Agglomeration	Country	Population (millions) 1950	1960	1970	1980	1990	2000
48	Wuhan	China	1.25	2.17	2.73	3.21	3.66	4.47
49	Guangzhou	China	1.43	1.93	2.50	3.07	3.62	4.49
50	Ankara	Turkey	0.28	0.64	1.27	2.20	3.62	5.19
51	Katowice	Poland	1.69	2.41	2.76	3.15	3.56	3.88
52	Ahmedabad	India	0.86	1.18	1.69	2.45	3.55	5.09
53	Hyderabad	India	1.12	1.24	1.75	2.47	3.49	4.94
54	Algiers	Algeria	0.44	0.87	1.20	2.11	3.43	5.16
55	San Francisco	U.S.A.	2.03	2.44	2.99	3.20	3.39	3.53
56	Toronto	Canada	1.07	1.74	2.53	2.96	3.32	3.61
57	Barcelona	Spain	1.56	1.95	2.66	3.07	3.31	3.38
58	Alexandria	Egypt	1.04	1.50	1.09	2.53	3.28	4.29
59	Casablanca	Morocco	0.71	1.10	1.51	2.21	3.26	4.63
60	Houston	U.S.A.	0.71	1.15	1.69	2.43	3.19	3.62
61	Yangon	Myanmar	0.67	0.98	1.43	2.18	3.18	4.45
62	Ho Chi Minh City	Viet Nam	0.87	1.32	2.00	2.48	3.17	4.42
63	Porto Alegre	Brazil	0.67	1.01	1.52	2.22	3.11	3.94
64	Melbourne	Australia	1.54	1.87	2.33	2.79	3.10	3.27
65	Chengdu	China	0.70	1.12	1.58	2.35	3.06	3.98
66	Guadalajara	Mexico	0.40	0.88	1.51	2.28	3.06	3.89
67	Washington D .C.	U.S.A.	1.30	1.82	2.49	2.77	3.03	3.19
68	Birmingham	United Kingdom	2.50	2.67	2.80	2.92	3.02	3.10
69	Medan	Indonesia	0.35	0.46	0.61	1.34	3.00	5.36

[a] Ranked by estimated population in 1990

After UN ,1989

of industry present. Cities in Latin America, for example, tend to have higher vehicle densities than those in other developing regions and therefore are likely to experience a higher contribution from motor vehicles to the total urban pollution load. Motor vehicle contributions are less important in cities with lower levels of motorization (e.g., cities in Africa) and in cities located in temperate regions that are dependent on coal or biomass fuels for space heating and other domestic purposes (e.g., cities in China, and parts of eastern Europe). It is also worth noting that in many developing countries, vehicle fleets tend to be older and poorly maintained – a factor which will increase the significance of motor vehicles as a pollutant source.

At the present time the vehicle fleet of the world is concentrated in the high income economies of the world. In 1988, the Organisation for Economic Co-operation and Development (OECD) countries alone accounted for 80 per cent of the world's cars, 70 per cent of trucks and buses and over 50 per cent of two- or three-wheeled vehicles. Since 1950, the global vehicle fleet has grown tenfold and is expected to double from the present total of 630 million vehicles within the next 20–30 years (Faiz et al., 1990). The rate of growth of the world's vehicle fleet is projected to out-pace that of both the total and urban population (see Figure 2.1).

Much of the expected growth in vehicle numbers is likely to occur in developing countries and in eastern Europe. In contrast, much of the demand for motor vehicles in the developed countries will be for vehicle replacement. The contribution of motor vehicles to the pollution load is thus set to increase in the developing countries; in the absence of the introduction of stringent control measures for traffic-related pollutants, it is expected that air quality will deteriorate in these regions.

In addition to the more common or 'traditional' air pollutants, a large number of toxic and carcinogenic chemicals are increasingly being detected in urban air, albeit at low concentrations. Examples include selected heavy metals (e.g., beryllium, cadmium and mercury), trace organics (e.g., benzene, polychlorodibenzo-dioxins and -furans, formaldehyde, vinylchloride and polyaromatic hydrocarbons

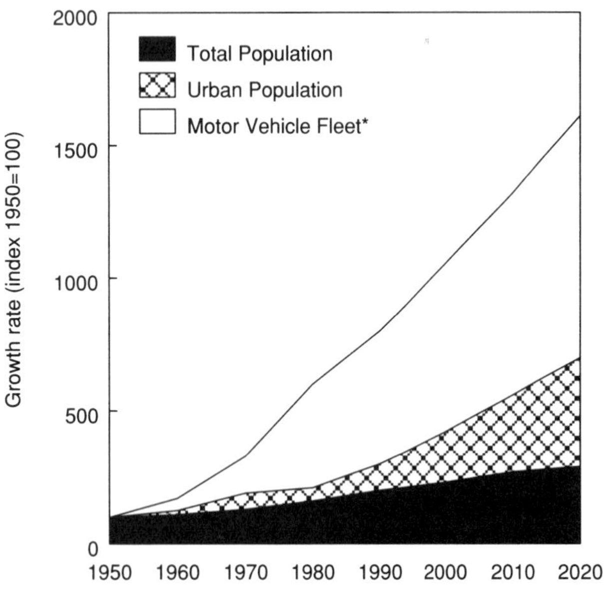

***Figure 2.1** Estimated and projected rate of increase of the total population, the urban population and motor vehicle numbers, 1950–2020*

*Excluding motorized two- and three-wheelers
After Faiz et al., 1990

(PAHs)), radionuclides (e.g., radon) and fibres (e.g., asbestos). Such chemicals are emitted from a wide range of sources including waste incinerators, sewage-treatment plants, industrial and manufacturing processes, solvent use (e.g., in dry-cleaning establishments), building materials and motor vehicles. Although emissions of these chemicals are generally low compared with those of the traditional pollutants, these pollutants may pose significant risks to health in view of their extremely high toxicity or carcinogenic potential or a combination of both. As the measurement of low concentrations of toxic chemical contaminants in air presents analytical difficulties, very little monitoring, let alone routine monitoring, is currently conducted.

2.3 Dispersion and Transport

Very often urban air pollution problems are aggravated by meteorological and topographical factors which often concentrate pollutants in the city and prevent proper dispersion and dilution. Many cities are surrounded by hills which may act as a downwind barrier trapping pollution close to the city. Thermal inversions are a particular problem in temperate and cold climates. Under normal dispersive conditions hot pollutant gases rise as they come into contact with colder air masses with increasing altitude. However, under certain circumstances the temperature may increase with altitude and an inversion layer forms a few tens or hundreds of metres above the ground. This inversion layer may then trap pollutants close to the emissions source and act as a heat cover prolonging the inversion. These conditions are of greatest concern when wind speeds are low. Isothermal conditions, that is when there is no change in temperature with altitude, may have a similar effect. Another meteorological phenomenon which greatly influences air quality is the "urban heat island". The heat generated by a city causes the air to rise which draws in colder, and possibly more polluted, air from surrounding areas.

On a more local scale, buildings and other structures can have a great effect upon pollutant dispersion. The "street canyon" effect occurs when the dispersion of low level emissions by the prevailing wind is prevented by tall buildings on each side of a busy road.

It is apparent that the long-range transport of air pollution from megacities may have national and regional impacts. Oxides of nitrogen and sulphur in the "urban plume" may contribute to acid deposition at great distances from the city. Ozone concentrations are often elevated downwind of urban areas due to the time lag involved in photochemical processes and NO scavenging in polluted atmospheres.

2.4 Air Quality Monitoring

In the 1960s, recognition of the ubiquitous nature of pollutants, such as SO_2, NO_x, CO, SPM, Pb and O_3 in urban air and concern for potential adverse impacts on human health prompted many national institutions to set up systematic monitoring networks for the routine measurement of urban air quality. National air quality standards and other forms of legislation were also introduced in order to protect human health. In many of the more developed countries, early legislation and monitoring efforts focused on SO_2 and SPM. Since the late 1970s, however, and as motor vehicles became an increasingly important source of air pollutants, networks have typically expanded to incorporate the routine monitoring of the traffic-related pollutants such as CO, NO_x and Pb. During the 1980s, urban air quality monitoring for the traditional air pollutants has also been established in the lesser developed countries, especially those in Asia and South America.

In more recent years, greater attention has been paid to the need for the monitoring of the photochemical oxidants, notably O_3, and their VOC precursors. Although the instrumentation is now well developed, relatively few countries routinely monitor O_3 as an indicator of photochemical pollution. In the case of VOCs, reliable monitoring instrumentation has only recently been developed and consequently urban VOC data are scarce. Furthermore, greater attention is being paid to characterizing the nature of SPM. The heterogeneous nature of SPM means that total SPM data have only limited value for health effect assessments. Monitoring objectives in some countries have therefore narrowed to determine specific size fractions of particulate matter (e.g., matter less than 10 μm - PM_{10} - which is in the respirable size range) and chemical characteristics (e.g., heavy metals, such as mercury, cadmium and lead; organic matter, such as PAHs).

Despite advances in the scale and scope of urban air quality monitoring in recent years, major difficulties in acquiring comprehensive and reliable air quality data for assessment purposes still exist. Monitoring capabilities vary markedly between countries, depending largely on national priorities and objectives and availability of economic resources and manpower. Typically, resources are not available to support the monitoring of all air pollutants in a routine programme and a selection of priority or key pollutants is made, taking into account the ease of measurement of the pollutant and its relative significance in terms of risks to human health. Furthermore, fixed site monitoring stations only provide data about a city's air pollution levels at the specific time and place of sampling. A sparse set of observations at a few place–time co-ordinates does not necessarily give an accurate picture of the extent and severity of the air pollution problem over a larger area. Even in the highly industrialized countries, there is generally a limit to the number of routine observation sites that can be established owing to the high cost of the equipment required. Thus, routine monitoring is often supplemented by a combination of spot surveys and short-term studies in order to characterize a particular problem. Increasingly sophisticated modelling techniques are being developed in order to complement monitoring data by providing estimates of ambient air pollution levels over wider areas.

Monitoring data may be expressed as mass per unit mass (such as parts per million (ppm)) or as mass per unit volume (such as μg m^{-3} or mg m^{-3}). A conversion factor is used when converting from ppm to μg m^{-3} - different conversion factors are used to allow for gas volume changes with temperature (usually 0°C or 25°C, both at 1 atm). Individual countries express data in ppm or as μg m^{-3} or both. Where national data are routinely expressed as ppm, national practice on conversion factors was followed for this report. Where no information on national practice was available, standard conversion factors at 0°C, 1 atm were used in this report, as in WHO (1979a).

2.5 Air Pollution Impacts

Air pollution can adversely affect human health, not only by direct inhalation but also indirectly by other exposure routes, such as drinking water contamination, food contamination and skin transfer.

Most of the traditional air pollutants directly affect the respiratory and cardiovascular systems. Increased mortality, morbidity and impaired pulmonary function have been associated with elevated levels of SO_2 and SPM. Nitrogen dioxide and O_3 also affect the respiratory system; acute exposure can cause inflammatory and permeability responses, lung function decrements and increases in airway reactivity. Ozone is also known to irritate the eyes, nose and throat and to cause headaches. Carbon monoxide has a high affinity for haemoglobin and is able to displace oxygen in the blood, which in turn can lead to cardiovascular and neurobehavioural effects. Lead inhibits haemoglobin synthesis in red blood cells in bone marrow, impairs liver and kidney function and causes neurological damage.

The direct human health effects of air pollution vary according to both the intensity and the duration of exposure and also with the health status of the population exposed. Certain sectors of the population may be at greater risk, for example, the young and the elderly, those already suffering from respiratory and cardiopulmonary disease, hyper-responders and people exercising.

At the present time, assessment of air quality for public health purposes consists essentially of examining ambient air quality against established guidelines. The WHO has recommended air quality guidelines for a wide range of pollutants; these guidelines indicate the level and exposure time at which no adverse effects on human health are expected to occur. Many countries set their own national air quality standards which are primarily designed to protect human health and are legally enforced. The WHO guidelines are presented in Table 2.2 and the national standards applicable to the 20 cities are featured in city summaries.

In addition to human health impacts, a number of the pollutants considered in detail in this report have additional or indirect impacts on the environment.

Table 2.2 *A Summary of WHO Air Quality Guidelines*

Pollutant	Time-weighted average	Units	Averaging time
Sulphur dioxide[a,b]	500	$\mu g\ m^{-3}$	10 minutes
	350	$\mu g\ m^{-3}$	1 hour
	100–150[1]	$\mu g\ m^{-3}$	24 hours
	40–60[1]	$\mu g\ m^{-3}$	1 year
Carbon monoxide[a,c]	30	$mg\ m^{-3}$	1 hour
	10	$mg\ m^{-3}$	8 hours
Nitrogen dioxide[a,d]	400	$\mu g\ m^{-3}$	1 hour
	150	$\mu g\ m^{-3}$	24 hours
Ozone[a,e]	150–200	$\mu g\ m^{-3}$	1 hour
	100–120	$\mu g\ m^{-3}$	8 hours
Suspended particulate matter[a,b]			
Measurement method			
Black smoke	100–150[1]	$\mu g\ m^{-3}$	24 hours
	40–60[1]	$\mu g\ m^{-3}$	1 year
Total suspended particulates	150–230[1]	$\mu g\ m^{-3}$	24 hours
	60–90[1]	$\mu g\ m^{-3}$	1 year
Thoracic particles (PM_{10})	70[1]	$\mu g\ m^{-3}$	24 hours
Lead[a,f]	0.5–1	$\mu g\ m^{-3}$	1 year

[1] Guideline values for combined exposure to sulphur dioxide and suspended particulate matter (they may not apply to situations where only one of the components is present).
Note: a fuller presentation of WHO and national guidelines is given in Appendix I.
Sources: [a] (WHO, 1987) [b] (WHO, 1979a) [c] (WHO, 1979b) [d] (WHO,1977b) [e] (WHO,1978) [f] (WHO,1977a)

The sulphur and nitrogen oxides are the principal precursors of acidic deposition; long-range transport of SO_2, NO_x and their corresponding acidic transformation products has been linked to soil and freshwater acidification with consequent adverse impacts on aquatic and terrestrial ecosystems. Sulphur dioxide, NO_2 and O_3 are phytotoxic; O_3, in particular, has been implicated in crop losses and forest damage. Impaired visibility and damage to materials (e.g., nylon and rubber), buildings and works of art are also attributed to SO_2 (and acid sulphate aerosols) and O_3.

2.6 Control Strategies

The basis of pollution control strategies relies upon local, national and regional authorities setting air quality and emissions standards. By requiring adherence to such standards governments can encourage industry to develop new and better technologies. In theory this approach has been adopted by most of the megacities discussed in this report. However, in many cases such strategies are not applied because some countries lack the financial and human resources to enforce compliance with standards.

Control technologies generally centre around modification of the fuel or the combustion technique or, alternatively, removing pollutants from flue gases. The type and location of the source, i.e., mobile or static, indoor or outdoor, and the overall cost effectiveness of the different techniques influence the ultimate choice of control method.

Pre-combustion control techniques involve the use of low pollutant fuels. Fuel cleaning is widely used to reduce the sulphur, dust and ash content of coal. Distillate oils are often used in urban areas due to their reduced sulphur content. Many nations have reduced atmospheric emissions through the use of natural gas and nuclear power. In addition the use of certain petrol additives such as tetraethyl lead has been reduced and even eliminated in several countries. Pre-combustion control techniques are often the simplest and most cost-effective method of reducing emissions.

Combustion modification techniques such as low NO_x burners and fluidized bed combustion have been employed in some countries to reduce SO_2 and NO_x emissions. However, modern combustion technology is relatively expensive and therefore beyond the scope of many developing countries.

Post-combustion control involves the removal of pollutants from flue gases and vehicle exhausts. There are a number of methods for removing SPM from flue gases including gravity settling chambers, cyclones, spray chambers, bag filters and electrostatic precipitators. A combination of these techniques is usually employed on larger coal-fired boilers. Flue gas desulphurization is being increasingly adopted by many industrialized nations to meet emissions standards and to achieve internationally agreed targets. Catalytic reduction is at present the best available technique for reducing NO_x emissions. The introduction of simple chimneys and vents to domestic stoves and heaters has greatly improved indoor air quality in developing regions. The use of three-way catalytic converters (TWCs) to control motor vehicle pollutants has resulted in significant reductions of NO_x, CO and hydrocarbons emissions from new vehicles.

The introduction of new manufacturing processes has led to significant reductions in industrial emissions. An obvious example is the use of low temperature hydrometallurgical techniques which reduce the SO_2 emissions associated with traditional metal smelting methods. Many countries have adopted energy conservation measures, mainly on economic grounds, and have effectively limited energy consumption and demand and also improved generation and distribution efficiency.

2.7 References

Faiz, A., Sinha, K., Walsh, M. and Valma, A. 1990 *Automotive Air Pollution: Issues and Options for Developing Countries*, WPS 492, The World Bank, Washington DC.

UN 1989 *Prospects of World Urbanization 1988*, Population Studies No. 112, United Nations, New York.

WHO 1977a *Lead*, Environmental Health Criteria 3, World Health Organization, Geneva.

WHO 1977b *Oxides of nitrogen*, Environmental Health Criteria 4, World Health Organization, Geneva.

WHO 1978 *Photochemical oxidants*, Environmental Health Criteria 7, World Health Organization, Geneva.

WHO 1979a *Sulfur oxides and suspended particulate matter*, Environmental Health Criteria 8, World Health Organization, Geneva.

WHO 1979b *Carbon monoxide*, Environmental Health Criteria 13, World Health Organization, Geneva.

WHO 1987 *Air quality guidelines for Europe*, WHO Regional Publications, European Series No. 23, World Health Organization, Regional Office for Europe, Copenhagen.

3

Twenty Megacity Air Quality Summaries

Introduction

The following 20 one-page summaries synthesize the most important data and information from the main city chapters (Part II).

General geographical, demographic and meteorological data are summarized at the head of each page, together with factors which greatly affect emissions in the city, such as motor vehicle registrations (total number of licensed vehicles), the predominant industries and the main energy sources. The estimated population in 1990 and the projected population in 2000 are taken from published UN sources. Climate is classified according to the Köppen system (See Table 4.1).

A brief situation analysis describes the main air quality problems together with contributory factors. Where possible ambient concentrations of pollutants are compared with WHO guidelines.

The text also refers to major epidemiological studies undertaken in the cities. Brief mention may also be made of control programmes and where appropriate specific recommendations are made.

On the right-hand side of each page is a graphical representation of annual trend data. Only those data sets with a significant time trend are presented. Where applicable, WHO guidelines are also shown.

Gross emissions estimates are tabulated (for the most recent year available) at the foot of the page. These data have been rounded to the nearest 100 or 1,000 tonnes.

Finally, data sources are not given in these summaries. However, this information can be found in the main city chapters in Part II.

Bangkok

Estimated Population 1990: *7.16 million*
Projected Population 2000: *10.26 million*
Map Reference: *Latitude 13°44'N, Longitude 100°34'E*
Altitude: *7 m*
Area: *1,565 km^2*
Climate: *Tropical savanna*

Annual Mean Precipitation: *1,500 mm*
Temperature Range: *25 – 30°C*
Motor Vehicle Registrations: *1,760,000 (1988)*
Industry: *Food processing, textiles, building materials, electronic assembly*
Energy: *Two natural gas-fired power plants*

Situation Analysis

Bangkok, the capital of Thailand, is growing rapidly. Large numbers of motor vehicles congest the city and create noise and air pollution problems. Many small-scale factories are located in the area and new industry is being moved to the suburbs.

Sulphur dioxide (SO_2) is not a problem as there is no winter heating season and WHO guidelines are being met. Suspended particulate matter (SPM) exceeds the WHO annual guideline at all three GEMS monitoring sites. The maximum WHO daily guideline of 230 μg m^{-3} is exceeded at two of the three stations.

Lead (Pb) meets WHO guidelines at ambient monitoring stations but not at kerbside. Carbon monoxide (CO) meets WHO guidelines at ambient air monitoring stations. Nitrogen dioxide (NO_2) and ozone (O_3) meet WHO guidelines where measured in Bangkok, although higher values could occur in downwind areas not presently monitored.

Epidemiological data for air pollutant effects are not available. The population exposed to these levels of automotive pollutants are likely to suffer pulmonary impairments leading to some restrictions on activity. Susceptible individuals will have increased rates of respiratory illness. The World Bank estimates that vehicle emissions will double by the year 2000. Unless strict motor vehicle controls are put in place the air quality will worsen significantly in the next decade.

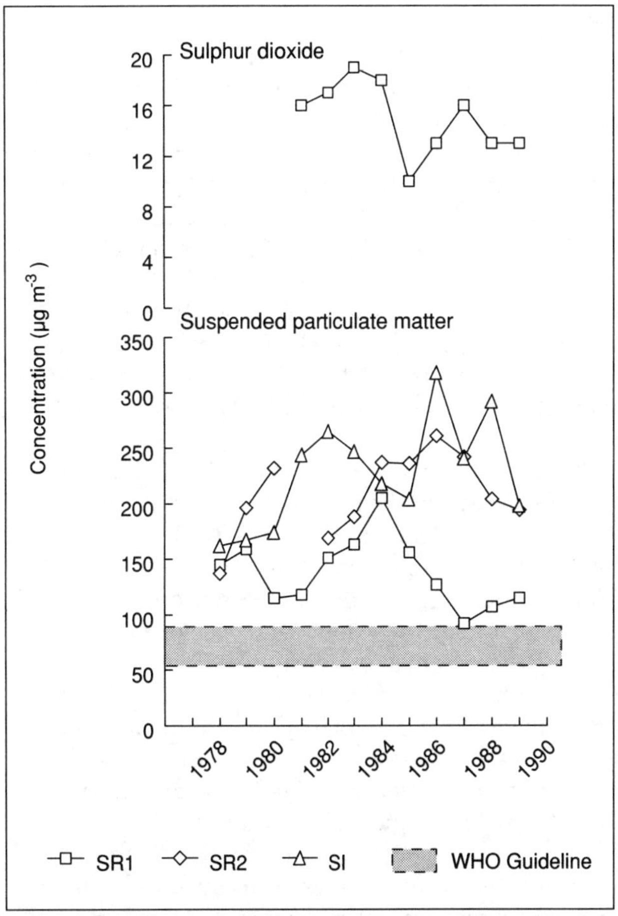

Annual mean SO2 and SPM (TSP) concentrations

Estimated emissions 1980 and 1990 (tonnes per annum)

	1980	1990
SO_2	120,000	n.a.
SPM	40,000	80,000
Pb	20,000*	1,000
CO	120,000	280,000
NO_x	30,000	n.a.

* Doubtful value
n.a. No data available

Beijing

Estimated Population 1990: *9.74 million*
Projected Population 2000: *11.47 million*
Map Reference: *Latitude 39°48'N, Longitude 116°28'E*
Altitude: *44 m*
Area: *16,800 km²*
Climate: *Continental warm summer*

Annual Mean Precipitation: *584 mm*
Temperature Range: *–5 – 26°C*
Motor Vehicle Registrations: *308,000 (1991)*
Industry: *Chemicals, smelting, textiles, metal products*
Energy: *24 thermal power plants. Coal is the major industrial fuel.*

Situation Analysis

Beijing, the capital of China, is growing rapidly. The city is crowded, but with relatively little motor vehicle traffic as bicycles and rapid transit systems are major modes of transport. Dust storms in spring contribute to the suspended particulate matter (SPM) loading. Small industries are interspersed throughout the city; their emissions and domestic smoke from coal stoves at low elevations produce serious sulphur dioxide (SO_2) and SPM problems during the winter. Carbon monoxide (CO) is also a problem indoors during winter months where coal is a cooking fuel.

Annual mean SO_2 meets WHO guidelines in the suburban areas. In the city centre the WHO guideline is exceeded by a factor of two and peak values reach 700 μg m^{-3}. Suspended particulate matter exceeds the WHO guideline at all four GEMS stations by up to a factor of more than four and peak values over 1,000 μg m^{-3} are reached during the winter months. These high values of SO_2 and SPM, often occurring simultaneously at over 500 μg m^{-3} are likely to be associated with increases in mortality of the aged and morbidity in children and other sensitive groups although epidemiological data are sparse.

Lead (Pb), CO and oxides of nitrogen (NO_x) do not currently pose a problem; ozone (O_3) data may exceed WHO guidelines, although lack of routine data and emission inventories hamper detailed evaluations.

Beijing is in the process of converting from coal to liquid petroleum gas (LPG) and natural gas, replacing multiple small units with central heating plants with emission controls and converting from raw coal to briquettes in domestic stoves. Flue gas desulphurization may be necessary to bring SO_2 concentrations down below WHO guidelines. With the planned increase in industrial productivity and in motor vehicles, Beijing faces a difficult period of pollution control.

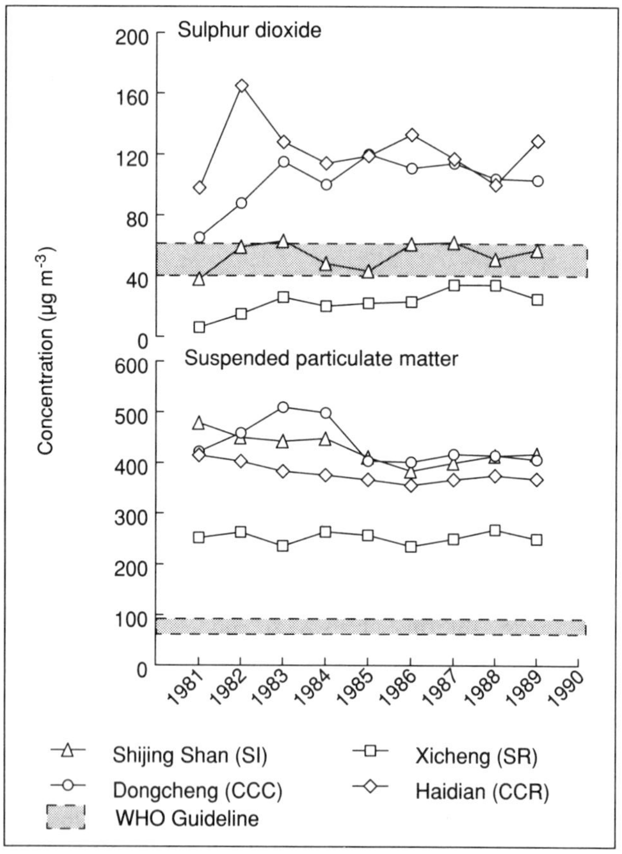

Annual mean SO_2 and SPM (TSP) concentrations

Estimated emissions, 1985 (tonnes per annum)

Source	SO_2	TSP
Industry	203,000	68,000
Power plants	126,000	n.a.
Coking	124,000	n.a.
Household stoves	73,000	6,000
Heating boilers	n.a.	34,000
Commercial stoves	n.a.	7,600
Total	526,000	115,600

n.a. No data available

Bombay

Estimated Population 1990: *11.13 million*
Projected Population 2000: *15.43 million*
Map Reference: *Latitude 18°54'N, Longitude 72°49'E*
Altitude: *11 m*
Area: *603 km^2*
Climate: *Tropical savanna*

Annual Mean Precipitation: *2,078 mm*
Temperature Range: *14 – 34°C*
Motor Vehicle Registrations: *588,000 (1989)*
Industry: *Textiles (cotton), chemical and engineering*
Energy: *One thermal power plant. Natural gas is a major industrial fuel.*

Situation Analysis

Bombay is the financial and commercial centre of India and is a major industrialized port. Urban spread is restricted because of the city's island location. Acute overcrowding has led to strict planning controls on industries. Bombay's motor vehicle density is high considering the restricted area. The main air quality problems in Bombay are attributable to increasing industrial productivity and motor vehicle traffic.

Planning control measures and the introduction of natural gas have reduced sulphur dioxide (SO_2) emissions in Bombay while ambient concentrations are now well below WHO annual guideline levels. Suspended particulate matter (SPM) annual mean and 98 percentile concentrations exceed guidelines at all stations in all the years monitored.

Airborne lead levels have decreased over two decades to below WHO guidelines. Emissions of oxides of nitrogen (NO_x) and carbon monoxide (CO) are increasing in line with motor vehicle traffic. Since 1985 the WHO daily guideline for NO_2 has been regularly exceeded. Ozone (O_3) is not monitored but monitoring should be initiated to identify whether O_3 may be a health hazard.

An epidemiological study indicated deleterious health effects in areas of high air pollution, but more recent data are required to examine the current situation. Bombay's authorities should continue to encourage industrial relocation out of the city and to promote further use of natural gas. Urbanization should be slowed to allow infrastructure development designed to prevent worsening air quality.

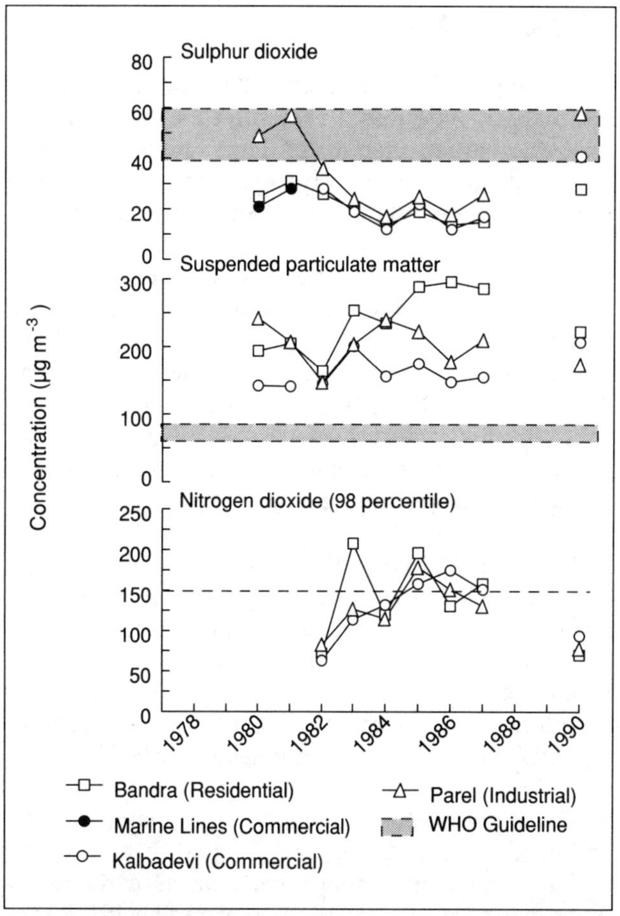

Annual mean SO_2 and SPM concentrations and annual 98 percentile concentrations

Estimated emissions, 1990 (tonnes per annum)

SO_2	157,000
SPM	50.000
CO	188,000
NO_x	58,000

Buenos Aires

Estimated Population 1990: *11.58 million*
Projected Population 2000: *13.05 million*
Map Reference: *Latitude 34°35'S, Longitude 58°29'W*
Altitude: *25 m*
Area: *7,000 km^2*
Climate: *Humid sub-tropical*
Annual Mean Precipitation: *1,000 mm*
Temperature Range: *10 – 25°C*
Motor Vehicle Registrations: *Data not available; estimated at 1 million*
Industry: *Heavy engineering to small-scale industries*
Energy: *Three power plants burning fuel oil and natural gas*

Situation Analysis

Population in the Metropolitan Area of Buenos Aires has more than doubled in recent years; increasing urbanization is expected to continue. The main air pollution problems are poorly characterized in view of the absence of substantial monitoring programmes.

No recent long-term data on sulphur dioxide (SO_2) were made available; data from 1968–1973 showed that values were within or lower than the WHO guideline range. Emissions data for the Metropolitan Area are sparse. Suspended particulate matter (SPM) concentrations at some sites exceed WHO guidelines but detailed assessments were not possible in view of the absence of supporting data.

The lead (Pb) content of petrol is relatively high which gives rise for concern, especially in view of increasing motorization. Emissions of carbon monoxide (CO) and oxides of nitrogen (NO_x) are mainly attributable to motor vehicles, but no systematic monitoring data were available. Ozone (O_3) monitoring data, if collected, were not made available.

The authorities in Buenos Aires should be encouraged to initiate regular air quality monitoring and to gather epidemiological data as a matter of priority to ascertain environmental and health risks.

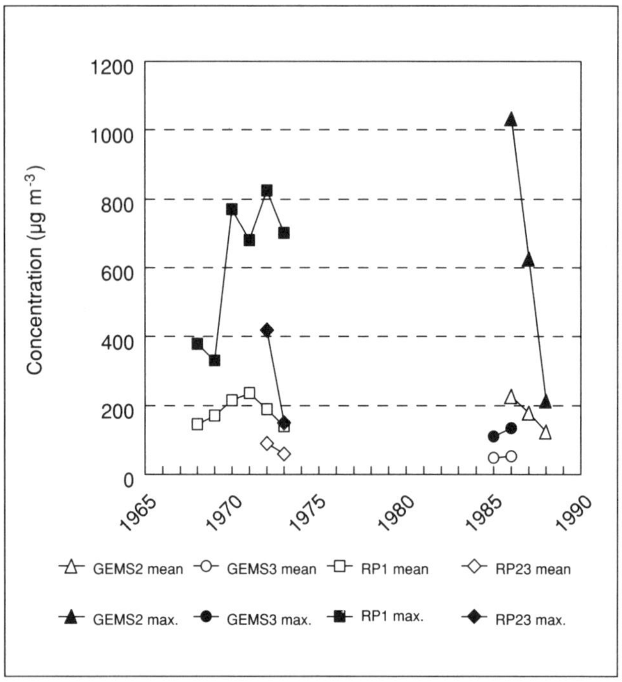

Annual mean and maximum SPM concentrations

Estimated emissions, 1989* (tonnes per annum)

SO_2	34,600
SPM	3,900
CO	240,000
NO_x	27,100

*City of Buenos Aires only

Cairo

Estimated Population 1990: *9.08 million*
Projected Population 2000: *11.77 million*
Map Reference: *Latitude 30°08'N, Longitude 31°34'E*
Altitude: *74 m*
Area: *214 km² (Governorate of Cairo only)*
Climate: *Desert*

Annual Mean Precipitation: *22 mm*
Temperature Range: *14 – 29°C*
Motor Vehicle Registrations: *939,000 (1990)*
Industry: *Textiles, iron and steel, motor vehicles and cement*
Energy: *Ten thermal power plants, mainly oil and natural gas*

Situation Analysis

Cairo is the most populous city in Africa and is the site of increasing industrial development and of vehicular traffic causing noise and air pollution problems.

Sulphur dioxide (SO_2) data are scant because of problems associated with monitoring techniques. However, short-term studies have shown that exceedences of WHO guidelines are a common event as most of Cairo's power stations burn heavy fuel oil. WHO annual mean guidelines are exceeded by a factor of five in the city centre and by a factor of two in some residential areas.

Annual mean smoke and total suspended particulates (TSP) concentrations exceeded WHO guidelines at most monitoring stations. In 1989, only two out of eight stations monitoring smoke were within the WHO annual guideline. A high proportion of the suspended particulate fraction is natural dust because of the dry desert climate. However, fugitive emissions from cement factories and from other industries are major sources of particulates.

The lead (Pb) content of petrol is relatively high which is reflected in high Pb levels in ambient air and in blood lead concentrations. A recent survey of tropospheric ozone (O_3) levels at a residential site indicated that concentrations were generally below WHO guidelines, but ozone is not routinely monitored.

Overall, the lack of adequate air quality data, emissions inventories and epidemiological data make any assessment difficult. It can be expected that air pollution problems are typical of a rapidly growing industrialized city.

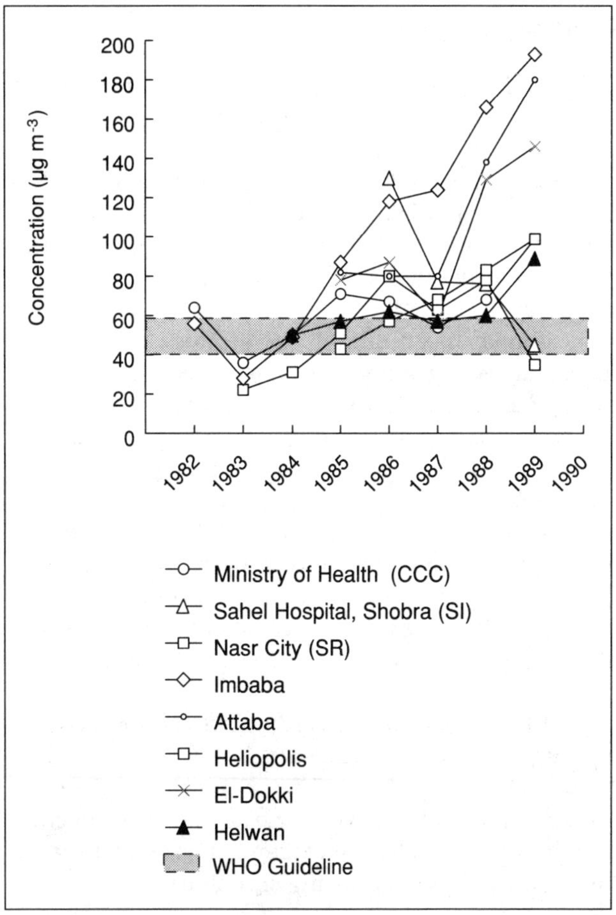

Annual mean SPM (smoke) concentrations

Estimated Emissions

No emissions estimates available

Calcutta

Estimated Population 1990: *11.83 million*
Projected Population 2000: *15.94 million*
Map Reference: *Latitude 22°32'N, Longitude 88°20'E*
Altitude: *6 m*
Area: *1,295 km^2*
Climate: *Tropical savanna*

Annual Mean Precipitation: *1,582 mm*
Temperature Range: *20 – 31°C*
Motor Vehicle Registrations: *500,000 (1989)*
Industry: *Heavy engineering, chemicals, jute, textiles*
Energy: *Two thermal power plants. Coal is the major fuel source.*

Situation Analysis

Calcutta has the largest urban population in India and overcrowding is a major problem.

Sulphur dioxide (SO_2) levels are surprisingly low - below WHO guidelines - owing to the low sulphur content of local coal. Suspended particulate matter (SPM) from the industrial burning of coal represents the greatest air pollution problem in Calcutta. Ambient annual mean SPM concentrations of over five times the WHO guideline have been recorded. Industrial emissions have stabilized in recent years and are projected to remain constant, or even to decline, over the next 10 years.

Lead (Pb), carbon monoxide (CO) and ozone (O_3) are not monitored on a routine basis. Annual 98 percentile nitrogen dioxide (NO_2) concentrations exceeded the WHO guideline at two out of three sites in 1987, but by 1990 were well below guidelines.

Motor vehicle emissions are of increasing concern although, at present, Calcutta has a relatively low motor vehicle population. Motor vehicle registrations are doubling every six years. Strict vehicle inspection and improved traffic management are required to help to slow down increasing emissions.

Cleaner coal burning is a necessity in order to reduce SPM emissions from domestic and industrial sources. The apparent high incidence of respiratory disease associated with air pollution should be urgently examined.

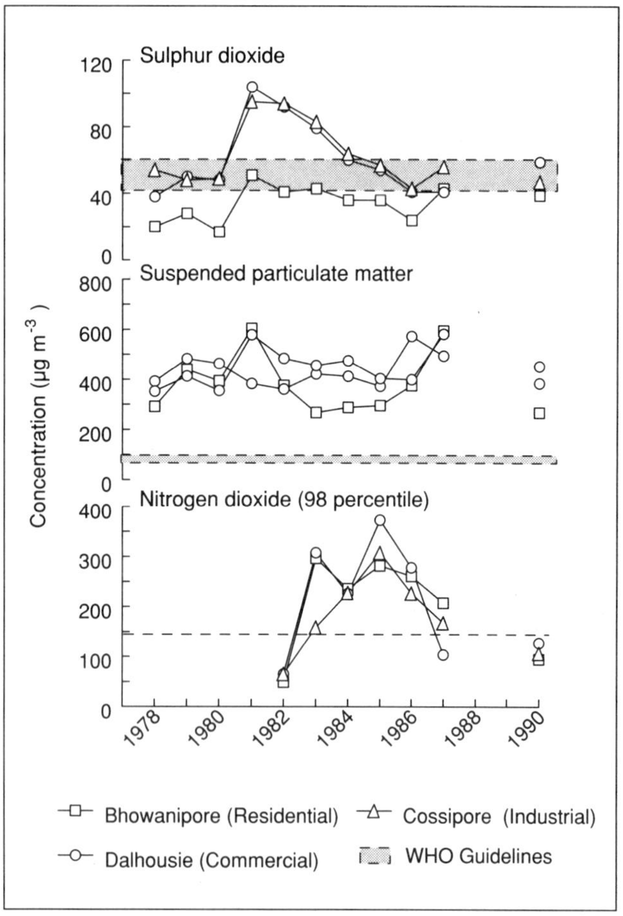

Annual mean SO_2 and SPM concentrations and annual 98 percentile NO_2 concentrations

Estimated Emissions, 1990 (tonnes per annum)

SO_2	25,500
SPM	200,000
CO	177,000
NO_x	40,200

Delhi

Estimated Population 1990: *8.62 million*
Projected Population 2000: *12.77 million*
Map Reference: *Latitude 28°35'N, Longitude 77°12'E*
Altitude: *216 m*
Area: *591 km^2 (1981)*
Climate: *Tropical steppe*

Annual Mean Precipitation: *715 mm*
Temperature Range: *14 – 34°C*
Motor Vehicle Registrations: *1,660,000 (1989)*
Industry: *Service industries, engineering, clothing, chemicals*
Energy: *Two thermal power plants. Coal is a major fuel source, two power stations in city.*

Situation Analysis

Delhi is the capital city of India and is a rapidly expanding centre of government, trade, commerce and industry. Dust storms are a regular climatic feature; these deposit large amounts of suspended particulate matter (SPM) in the city's atmosphere. Delhi is a major transport hub and has a high motor vehicle population compared with other Indian cities.

Sulphur dioxide (SO_2) concentrations regularly exceed annual and daily WHO guidelines at commercial and industrial stations. A significant upward trend in SO_2 concentrations has been observed between 1978 and 1987, but isolated 1990 data suggest that concentrations have decreased to below the WHO annual guideline. Suspended particulate matter concentrations are well above WHO annual guidelines and 98 percentile concentrations show an increasing trend between 1982 and 1987.

Lead (Pb) in air spot samples were variable but generally below the WHO guideline value. Despite Delhi's large motor vehicle population, ambient oxides of nitrogen (NO_x) concentrations are relatively low, although nitrogen dioxide (NO_2) levels are increasing each year. This trend will continue as a result of increasing motor vehicle numbers. Carbon monoxide (CO) emissions are likely to follow a similar trend. Ozone (O_3) and CO are not routinely monitored so their risks are difficult to evaluate.

Changes in fuel use have helped to stabilize emissions from commercial and domestic sources. However, the rapid rate of industrial expansion means that emissions from this source are likely to rise. Overall, Delhi's air quality is likely to deteriorate rapidly over the next decade. Epidemiological data are urgently needed.

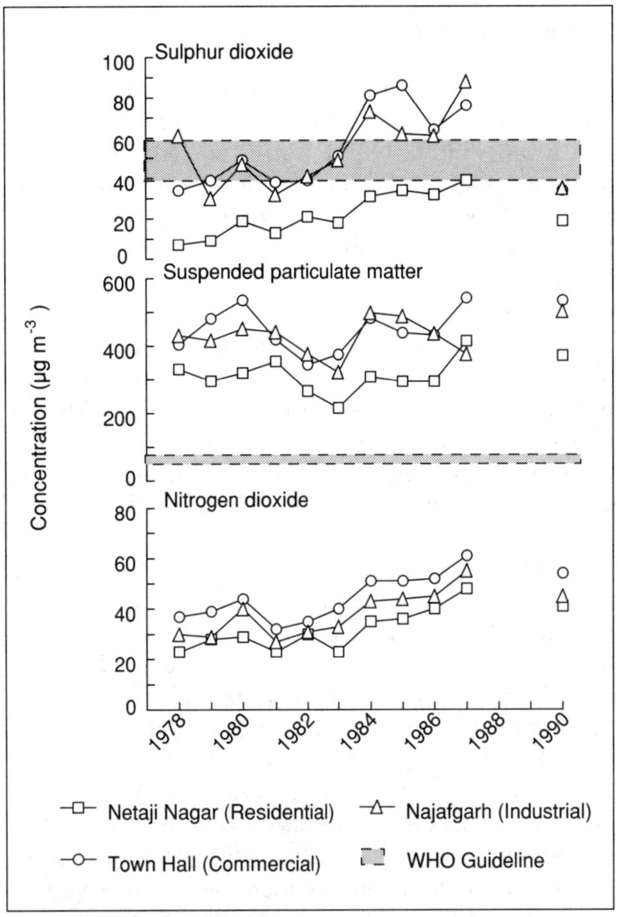

Annual mean SO_2, SPM (TSP) and NO_2 concentrations

Estimated emissions, 1990 (tonnes per annum)

SO_2	46,000
SPM	116,000
CO	265,000
NO_x	73,000

Jakarta

Estimated Population 1990: *9.42 million*
Projected Population 2000: *13.23 million*
Map Reference: *Latitude 6°08'S, Longitude 106°45'E*
Altitude: *7 m*
Area: *590 km^2 (Metropolitan Jakarta)*
Climate: *Tropical rain forest*

Annual Mean Precipitation: *1,760 mm*
Temperature Range: *22 – 33°C*
Motor Vehicle Registrations: *1,380,000 (1987)*
Industry: *Chemicals, textiles, shipping, food products, metal products*
Energy: *Three thermal power plants. Oil is the major fuel source.*

Situation Analysis

Jakarta, the capital of Indonesia, is growing rapidly. The city is often congested by motor vehicle traffic as hardly any public rapid transit system exists. The major air quality problems arise from motor vehicle traffic with maxima of suspended particulate matter (SPM) in the city centre, although insufficient data hinder the evaluation.

Sulphur dioxide (SO_2) presents no problem in Jakarta as both GEMS and BMG Jakarta stations now record SO_2 annual means below the WHO guideline. However, SPM values are reported by BMG Jakarta to be more than 400 μg m^{-3} annual mean in the city centre indicating a potential for health problems.

High lead (Pb) concentrations are reported in short-term studies, an indication of the serious motor vehicle pollution in the central areas. The concentrations of carbon monoxide (CO) exceed WHO guidelines in areas congested by traffic. Nitrogen dioxide (NO_2) and ozone (O_3) annual data for Jakarta meet the WHO guidelines, and although O_3 short-term concentrations have been high in the past, the most recent data indicate concentrations below WHO eight-hour guidelines. These pollutants may become more of a problem with the increase in motor vehicle traffic.

It is expected that the growth of the population and the number of motor vehicles in Jakarta will increase the air pollution levels above the present concentrations. Motor vehicle emission controls will be required to prevent the health of the population from being severely impacted. Epidemiological data are lacking and this should be corrected.

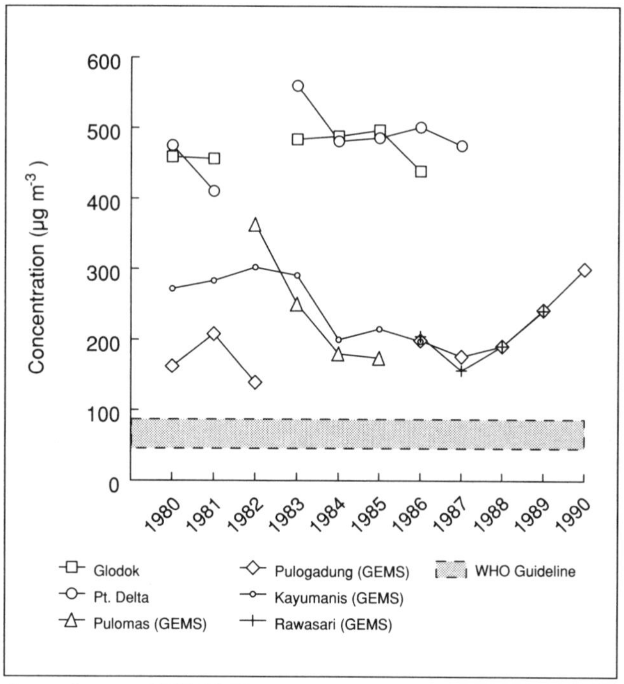

Annual mean SPM (TSP) concentrations

Estimated emissions, 1989 (tonnes per annum)

SO_2	24,700
SPM	n.a.
CO	325,000
NO_X	20,500

n.a. No data available

Karachi

Estimated Population 1990: *7.67 million*
Projected Population 2000: *11.57 million*
Map Reference: *Latitude 24°51'N, Longitude 67°02'E*
Altitude: *1.5 – 37 m*
Area: *3,530 km^2 (1982)*
Climate: *Desert*

Annual Mean Precipitation: *200 mm*
Temperature Range: *20 – 30°C*
Motor Vehicle Registrations: *650,000 (1989)*
Industry: *Textiles, footwear, metal, food processing*
Energy: *Oil and natural gas in industry and power generation*

Situation Analysis

Karachi is the largest city in Pakistan with a rapidly increasing population and concomitant urban spread. Power stations burn mainly heavy fuel oil and are also a significant source of pollutant emissions. In the absence of long-term monitoring, the air pollution situation cannot be adequately characterized.

Karachi is situated on the coast and therefore sea breezes aid pollutant dispersion during the summer. Despite this, maximum sulphur dioxide (SO_2) concentrations exceed WHO guidelines in some areas. The region's desert climate means that a high proportion of natural particulates are present in the atmosphere which affects suspended particulate matter (SPM) concentrations. Values exceed WHO annual mean guidelines and increasing trends are apparent.

Open burning of solid refuse is a major source of SPM, carbon monoxide (CO) and oxides of nitrogen (NO_x). As monitoring data are sparse, data on air quality in general, and on lead (Pb), CO, nitrogen dioxide (NO_2) and ozone (O_3) concentrations in particular, are difficult to evaluate. Ambient concentrations of Pb in air exceed WHO guidelines in areas of heavy traffic.

Karachi's motor vehicle population is growing 2.5 times faster than the human population and therefore will become an increasingly important source of pollution. Control measures aimed at improving the city's refuse disposal and at reducing motor vehicle emissions would help to reduce pollutant concentrations significantly, but in the absence of adequate data air quality management options cannot be evaluated. Epidemiological studies should be initiated.

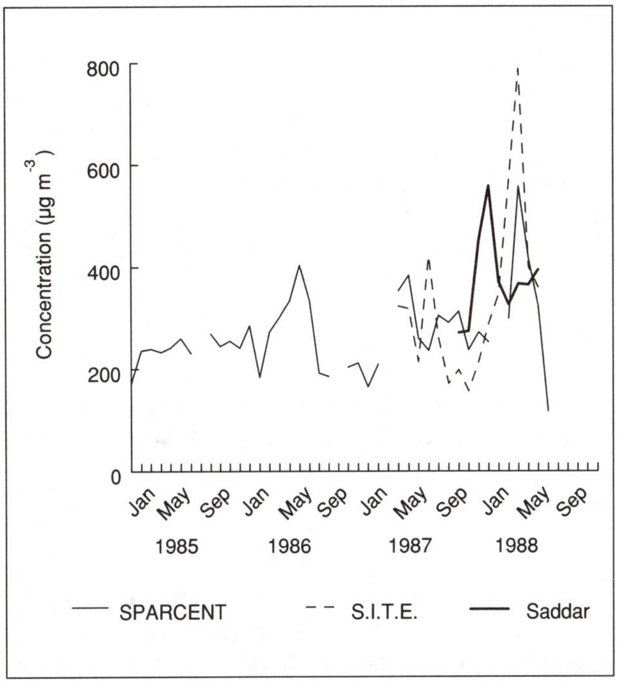

Monthly mean SPM (TSP) concentrations

Estimated emissions, 1989 (tonnes per annum)

SO_2	77,000
SPM	106,000
CO	271,000
NO_x*	50,000

* not including industry and domestic

London

Estimated Population 1990: *10.57 million (urban agglomeration)*
Projected Population 2000: *10.79 million (urban agglomeration)*
Map Reference: *Latitude 51°30'N, Longitude 0°10'W*
Altitude: *5 m*
Area: *1,579 km^2 (Greater London)*
Climate: *Marine west coast*

Annual Mean Precipitation: *597 mm*
Temperature Range: *5.5 – 18.1°C*
Motor Vehicle Registrations: *2,700,000 (1988) (Greater London)*
Industry: *Commerce and banking, consumer services, manufacturing industries*
Energy: *Natural gas – major industrial/commercial fuel*

Situation Analysis

London is the capital of the United Kingdom and is one of the world's major financial centres. The enforcement of clean air legislation in London and major changes in fuel use have brought about a dramatic reduction in smoke (SPM) and sulphur dioxide (SO_2) concentrations over the past 30 years. Major epidemiological studies have been undertaken in earlier years.

Annual mean smoke and SO_2 concentrations are now well below WHO annual guidelines. However, short-term high pollution episodes caused by temperature inversions are still a problem during the winter.

Emissions from motor vehicles now present the most significant threat to air quality. The increasing motor vehicle traffic is likely to offset any reduction in oxides of nitrogen (NO_x), and possibly carbon monoxide (CO), emissions brought about by the introduction of exhaust emission tests and catalytic converters. High localized concentrations of CO are of major concern.

London is a major source of photochemical precursors (NOx and volatile organic compounds (VOCs)) and is believed to be responsible for high tropospheric ozone concentrations 'downwind' of the city. Ozone (O_3) concentrations monitored in the city are not as high as at rural sites, probably as a result of NO scavenging processes.

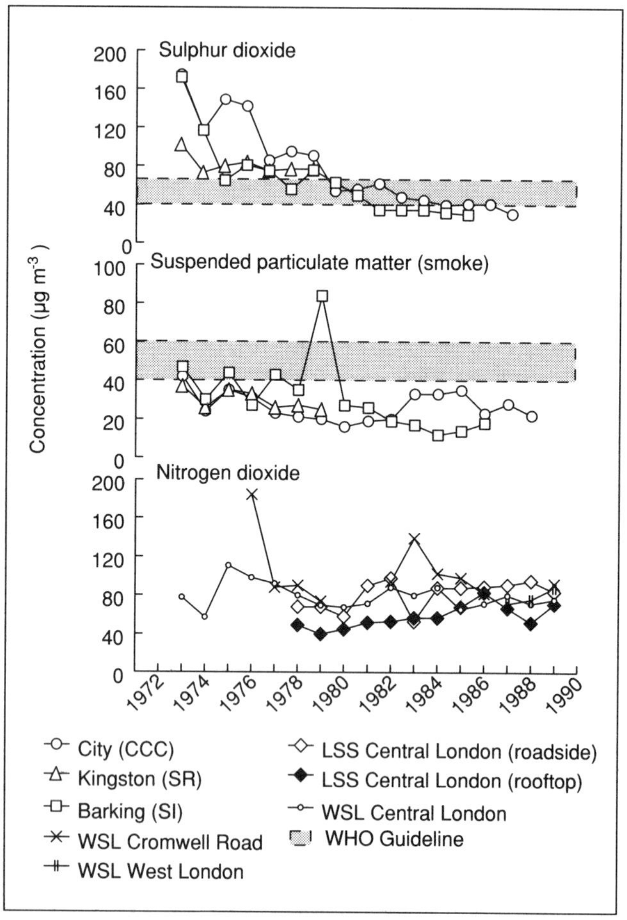

Annual mean SO_2, SPM (smoke) and NO_2 concentrations

Estimated emissions, 1983 (tonnes per annum)

SO_2	49,000	
Smoke	11,000	
Pb	350–700	(no date)
CO	950,000	(1978)
NO_x	79,000	

Los Angeles

Estimated Population 1990: *10.47 million*
Projected Population 2000: *10.91 million*
Map Reference: *Latitude 34°00'N, Longitude 118°15'W*
Altitude: *113 m*
Area: *16,600 km^2*
Climate: *Mediterranean*

Annual Mean Precipitation: *367 mm*
Temperature Range: *12 – 21°C*
Motor Vehicle Registrations: *8,000,000 (1989)*
Industry: *Mainly light industry, aerospace, electronics, consumer services*
Energy: *Mainly hydroelectric, some natural gas power plants.*

Situation Analysis

Considerable efforts have been made to establish well-planned monitoring methods and to control air pollution in the Los Angeles Basin. The situation is still the worst in the USA despite the introduction of stringent emission controls. The main air quality problems relate to increasing motor vehicle use.

Sulphur dioxide (SO_2) annual arithmetic mean concentrations are well below the WHO guidelines, although the hourly maxima are rarely exceeded. Total suspended particulates (TSP), on the other hand, continue to exceed State/Federal and WHO guidelines, the greatest proportion being from anthropogenic sources.

With the introduction of lead-free petrol, the ambient air exposure route is no longer of health significance except in the vicinity of secondary lead smelters. Despite early successes in reducing carbon monoxide (CO) pollution with very strict emission controls, levels and the number of exceedences are rising. Los Angeles has failed to meet the national air quality standards for nitrogen dioxide (NO_2), the WHO one-hour NO_2 guidelines are often exceeded, and it has the most serious ozone (O_3) problem in the USA.

Epidemiological studies consistently show that the health of the population of Los Angeles has been significantly affected by ambient air pollution. Continuing plans to reduce air pollution, particularly from motor vehicle use, are considerable and will place increasing constraints on the Los Angeles Basin residents. The continued increases in urban population and in motor vehicle use will continue to exacerbate the situation.

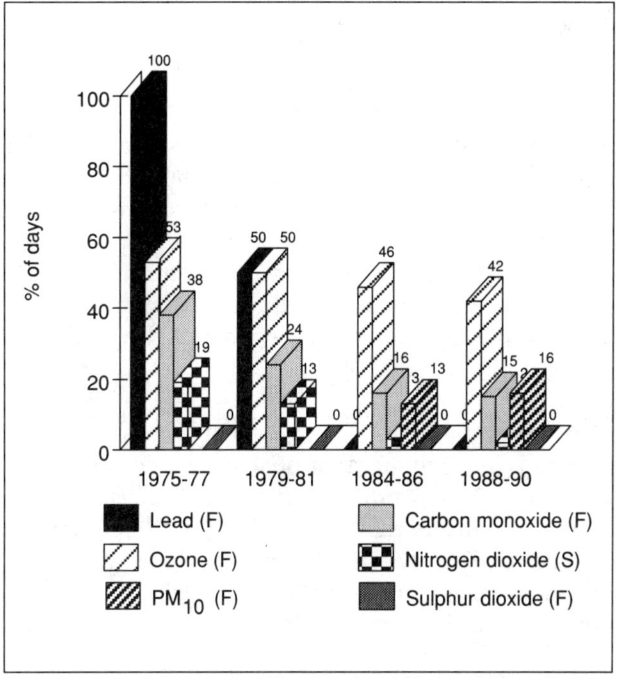

Percentage of days exceeding levels of Federal (F) or State (S) standards, 1975–1990

Estimated emissions, 1987 (tonnes per annum)

SO_x	50,000
PM_{10}	400,000
CO	1,800,000
NO_x	440,000

Manila

Estimated Population 1990: *8.40 million*
Projected Population 2000: *11.48 million*
Map Reference: *Latitude 14°36'N, Longitude 120°59'E*
Altitude: *5 m*
Area: *636 km^2 (Manila City)*
Climate: *Tropical rain forest*

Annual Mean Precipitation: *2,159 mm*
Temperature Range: *25 – 27.8°C*
Motor Vehicle Registrations: *510,000 (1988)*
Industry: *Chemicals, textiles, shipping, food products, metal products*
Energy: *Three thermal power plants. Oil is major industrial fuel.*

Siituation Analysis

Metropolitan Manila, the national capital region of the Philippines, is growing rapidly. While a rapid transport system known as the Light Rail Transit transports some 350,000 commuters daily through 30 km of rail system, the metropolis is still very often congested by motor vehicle traffic. Motor vehicles, light industry and open burning contribute major portions to the urban air pollution. Monitoring data are incomplete, thus hindering detailed analysis of the air pollution situation.

Sulphur dioxide (SO_2) does not appear to be a problem in Manila as annual means meet WHO guidelines. Suspended particulate matter (SPM), however, substantially exceeds the WHO annual mean guideline. These levels are produced primarily by automotive emissions with contributions from power plants, industry and open burning.

Lead (Pb) has been measured at over three times the WHO guideline so it may be a problem for children who live near heavy traffic areas. No recent data are available on nitrogen dioxide (NO_2), carbon monoxide (CO) and ozone (O_3) which are expected to be high in relation to the dense motor vehicle traffic.

The chronic exposure to motor vehicle exhaust emissions by commuters and residents is likely to lead to increased morbidity in the most highly exposed individuals and to the development of chronic pulmonary conditions. Epidemiological data are required to verify the position. Motor vehicle emission controls, especially on the diesel-engine jeepneys, are a necessary measure for the protection of public health. As Manila develops industrially, strict emission controls will be needed to keep the air quality to the present levels.

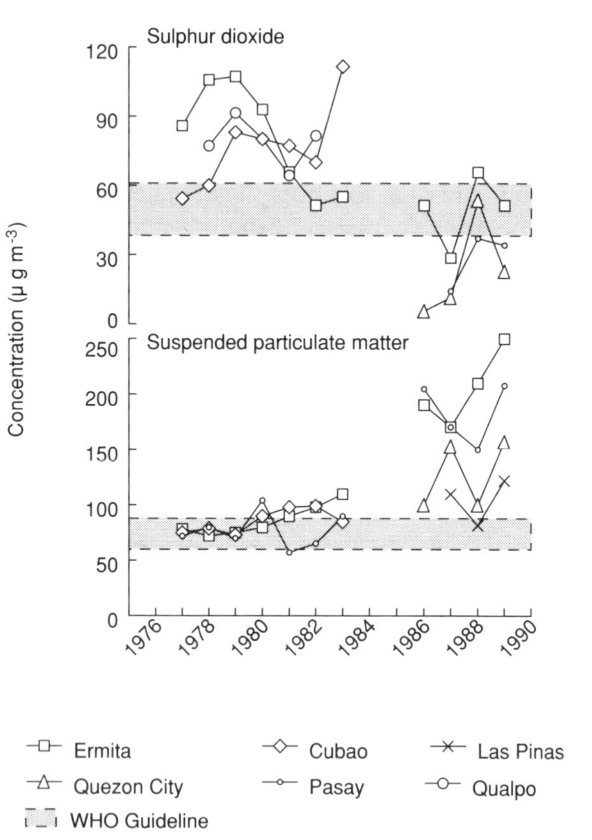

Annual mean SO_2 and SPM concentrations

Estimated emissions, 1987 (tonnes per annum)

SO_2	148,000
SPM	69,000
CO	557,000
NO_x	119,000

Mexico City

Estimated Population 1990: *19.37 million*
Projected Population 2000: *24.44 million*
Map Reference: *Latitude 19°26' N, Longitude 99°07'W*
Altitude: *2,240 m*
Area: *2,500 km^2*
Climate: *Marine west coast*

Annual Mean Precipitation: *725 mm*
Temperature Range: *12 – 17.4°C*
Motor Vehicle Registrations: *2,500,000 (1989)*
Industry: *Mixed, heavy to light industry*
Energy: *Two power generation plants, switched from fuel oil to natural gas in 1991*

Situation Analysis

The Metropolitan Area of Mexico City, with a very large and rapidly developing urban population, is the major economic centre of Mexico. Air pollution is serious and is primarily caused by energy consumption. Monitoring capabilities at both manual and automatic stations are now good.

Ambient annual mean sulphur dioxide (SO_2) levels substantially exceed WHO guidelines while annual trends have not changed over the past 18 years. Closure of a refinery and the switch from fuel oil to natural gas have reduced emissions. WHO and national guidelines for suspended particulate matter (SPM) are frequently exceeded at monitoring sites by subtantial amounts with no annual trends apparent over an 18-year period.

Yearly mean lead (Pb) levels in ambient air have decreased substantially from 1981 to 1986, associated with a lowering of lead in petrol, but concentrations have risen since then at some sites. Carbon monoxide (CO) pollution is very serious. Concentrations exceed WHO guidelines on a regular basis at all stations. High altitude greatly exacerbates the CO problem.

Nitrogen dioxide (NO_2) concentrations again show no yearly time trends, but substantially exceed WHO guidelines. In view of the high oxides of nitrogen (NO_x) concentrations and local meteorological and topographic factors. Extremely high ozone (O_3) levels affect the whole city. The greatest frequency of O_3 excess levels occurs in the south-west zone and on around 254 days of the year. Basic environmental health data related to air quality measurements are lacking.

The serious air pollution problems in Mexico City are now being addressed through technological improvements, use of better fuels, emission control strategies, and replacement of old industries.

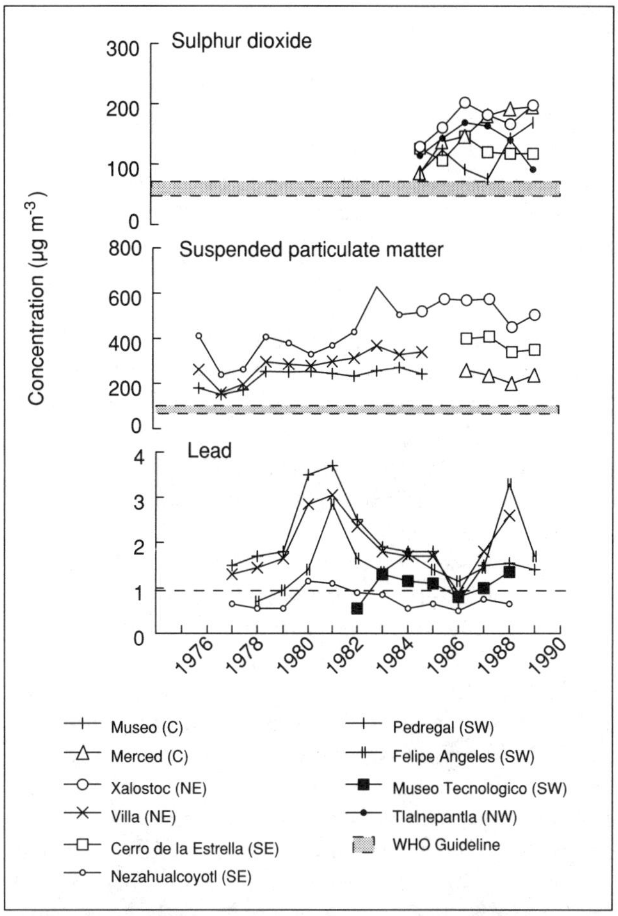

Annual mean SO_2, SPM (TSP) and Pb concentrations

Estimated emissions, 1989 (tonnes per annum)

SO_2	206,000
SPM	451,000
CO	2,951,000
NO_x	177,000

Moscow

Estimated Population 1990: *9.39 million*
Projected Population 2000: *10.11 million*
Map Reference: *Latitude 55°45'N, Longitude 37°34'E*
Altitude: *156 m*
Area: *994 km^2*
Climate: *Continental cool summer*

Annual Mean Precipitation: *575 mm*
Temperature Range: *–10 – 20°C*
Motor Vehicle Registrations: *665,000 (1989)*
Industry: *Engineering, textile, motor vehicle manufacture, steel and chemicals*
Energy: *Natural gas power plants*

Situation Analysis

Moscow is the capital of the Russian Federation and is the State's principal industrial centre. Motor vehicles and power generation account for a large proportion of total emissions. Changes in fuel consumption from fuel oil to natural gas have had a significant effect on emissions and on ambient air quality over the past 10 years.

Problems have been encountered with sulphur dioxide (SO_2) measurements in the past and reported low ambient concentrations are not consistent with emissions figures. Suspended particulate matter (SPM) has decreased in recent years but mean daily concentrations still exceed WHO guidelines.

Lead (Pb) concentrations in air are relatively low, but concentrations of carbon monoxide (CO) exceed both WHO and national guideline values throughout the year. However, emissions estimates show some reduction in overall CO emissions between 1975 and 1990. Emissions of oxides of nitrogen (NO_x) are considered to be the most important air quality problem in Moscow. Levels of nitrogen dioxide (NO_2) have increased significantly between 1986 and 1990 which suggests that motor vehicle traffic and/or power generation have increased over the same period. Tropospheric ozone (O_3) is not considered to be a problem in Moscow.

With the data available it has not been possible to evaluate health effects arising from air pollution. Current monitoring methodology reviews and automated measurements would improve management capabilities.

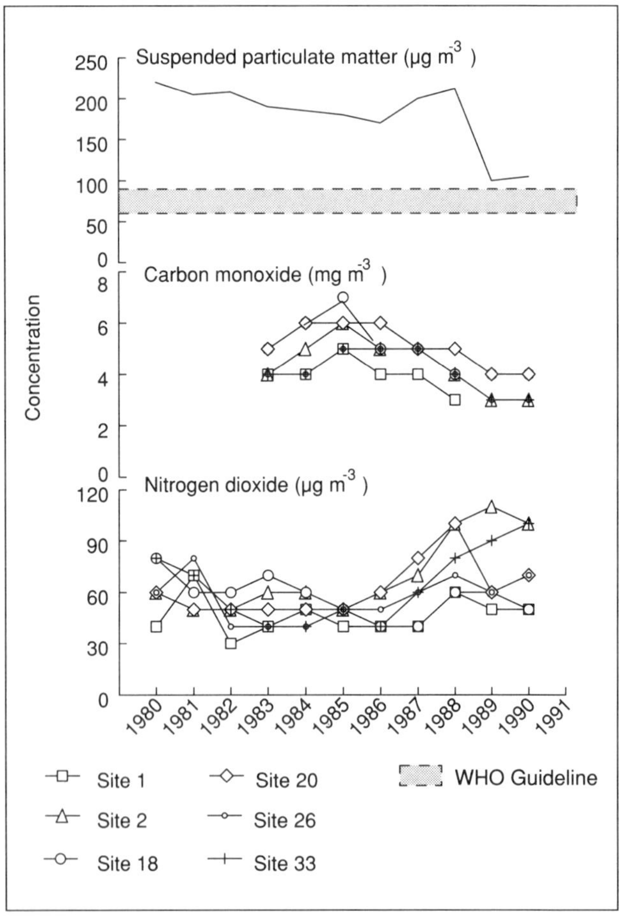

Annual mean SPM, CO and NO_2 concentrations

Estimated emissions, 1990 (tonnes per annum)

SO_2	130,000
SPM	60,000
CO	900,000
NO_x	210,000

New York

Estimated Population 1990: *15.65 million*
Projected Population 2000: *16.10 million*
Map Reference: *Latitude 40°45'N, Longitude 74°00'W*
Altitude: *10 m*
Area: *3,585 km^2 (metropolitan area)*
Climate: *Continental warm summer*

Annual Mean Precipitation: *1,120 mm*
Temperature Range: *0 – 25°C*
Motor Vehicle Registrations: *1,780,000 (1990, city only)*
Industry: *Chemicals, consumer services, electronics, transporation*
Energy: *Oil- and gas-fired power plants*

Situation Analysis

The New York Metropolitan Area is one of high population density. The surrounding metropolitan areas of Connecticut and New Jersey have heavy industry, refineries and populations travelling by motor vehicle into the city, all of which contribute to the air quality problems.

Air quality in New York City (NYC) has improved greatly since the 1950s and early 1960s when sulphur dioxide (SO_2) and smoke (TSP) pollution reached episodic levels which caused excess mortality. Emission controls and elimination of coal burning have brought the levels of TSP and SO_2 below the WHO guideline. Particulate matter < 10 µm (PM_{10}) and lead (Pb) now meet USA air quality standards except in limited street canyon situations where PM_{10} exceeds the standards because of diesel buses and trucks.

The major problems now for NYC are the pollutants generated from automotive traffic. Carbon monoxide (CO) concentrations have decreased significantly as the number of days which exceeded the WHO guideline dropped from 98 in 1980 to 4 in 1989. Nitrogen dioxide (NO_2) meets the USEPA standard. Ozone (O_3) levels have also dropped rapidly, from 119 days exceeding WHO guidelines in 1980 to 9 in 1989, illustrating the effectiveness of controls of hydrocarbon sources and motor vehicles. The maximum effect of emissions from NYC traffic occurs in the downwind areas such as suburban Connecticut.

Air quality in NYC has improved greatly and the main health problem remains high ozone during the warmest summer days. It is a good example of the success of strict emission controls which eliminated SO_2 and TSP as hazards to human health.

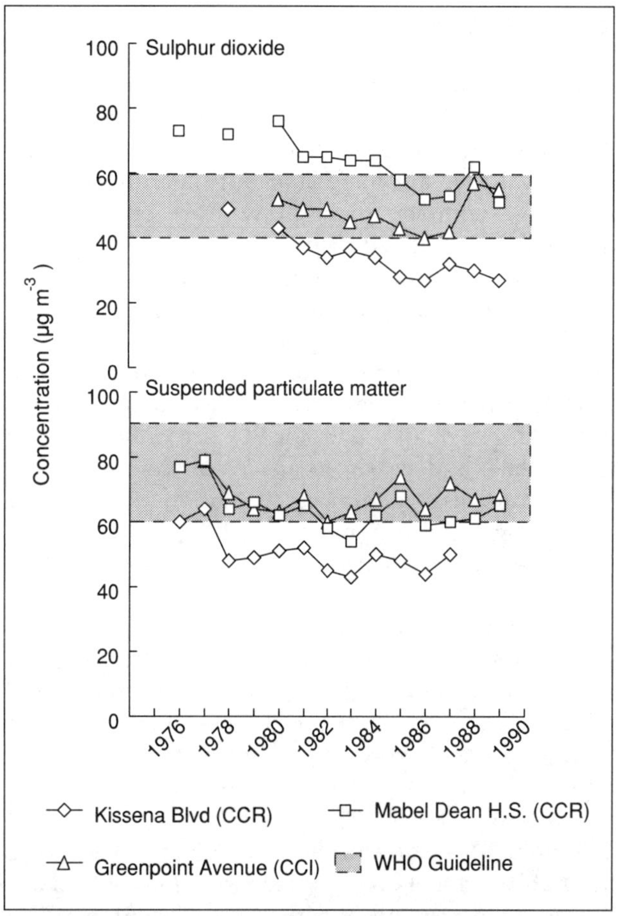

Annual mean SO_2 and SPM (TSP) concentrations

Estimated emissions, 1985 (tonnes per annum)

	New York City	Metropolitan New York
SO_x	55,000	349,000
PM_{10}	56,000	289,000
CO	368,000	1,752,000
NO_x	121,000	513,000
VOC	217,000	872,000

Rio de Janeiro

Estimated Population 1990: *11.12 million*
Projected Population 2000: *13.00 million*
Map Reference: *Latitude 22°54'S, Longitude 43°10'W*
Altitude: *Sea level to 885 m*
Area: *6,500 km²*
Climate: *Tropical rain forest*

Annual Mean Precipitation: *1,000 mm*
Temperature Range: *21 – 26°C*
Motor Vehicle Registrations: *Not available*
Industry: *Metallurgy, engineering, textiles, minerals*
Energy: *Thermal power plants, oil- and coal-fired*

Situation Analysis

Rio de Janeiro, the second largest city in Brazil, is growing rapidly without adequate land use planning. Air pollution sources are complex and uncontrolled. Motor vehicles are increasing but data are unavailable. The complex topography from sea level to mountains over 1,000 metres leads to air stagnation and pollutant build-up. Air quality monitoring in Rio is incomplete and often data are unobtainable.

Sulphur dioxide (SO_2) sources are significant as a result of the use of high-sulphur fuel oil and coal for industry. Air quality data for SO_2 are only available up to 1984; and yearly means were well above the WHO guideline at almost all stations although a decreasing trend is apparent for 1980–1984. Maxima for SO_2 exceeded 400 $\mu g\ m^{-3}$.

Suspended particulate matter (SPM)/total suspended particulates data up to 1989 are almost all above the WHO guideline and one-in-six day maxima over 400 $\mu g\ m^{-3}$ occur. Such days of 400 $\mu g\ m^{-3}$ of both SO_2 and TSP are associated with excess mortality.

Lead (Pb) is not a problem as motor vehicles use alcohol-based fuels and phased reductions of lead additives have resulted in annual means within WHO guidelines. The fragmentary carbon monoxide (CO) data available give rise to concern as hourly means above the WHO guideline have been measured.

Rio has a potential for serious air pollution problems although epidemiological data are scant. Air quality management procedures involving emission controls and improved monitoring are undoubtedly urgently required.

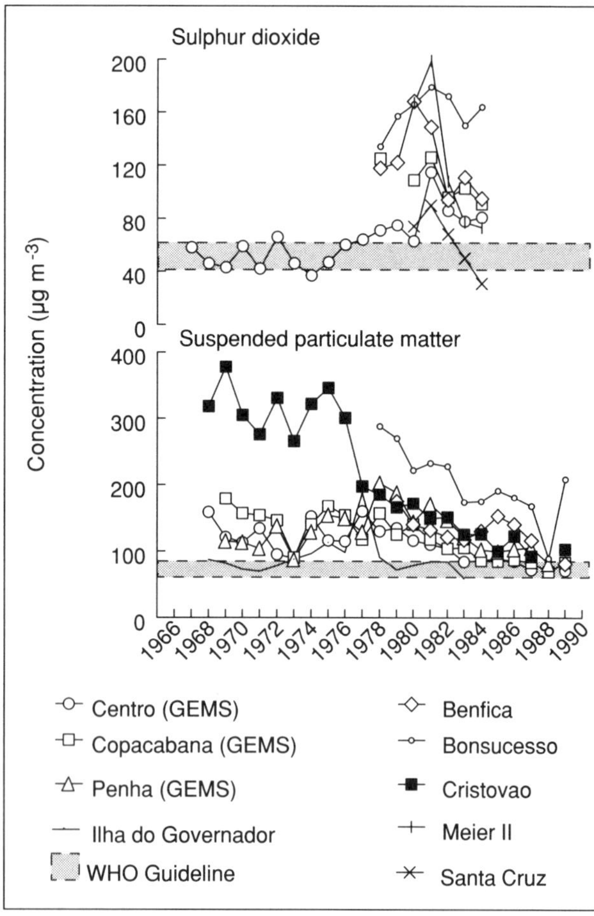

Annual mean SO_2 and SPM (TSP) concentrations

Estimated emissions, 1978 (tonnes per annum)

SO_x	188,000
SPM	194,000
CO	642,000
NO_x	63,000
HC	162,000

São Paulo

Estimated Population 1990: *18.42 million*
Projected Population 2000: *23.60 million*
Map Reference: *Latitude 23°37'S, Longitude 46°39'W*
Altitude: *Approximately 800 m*
Area: *8,000 km^2*
Climate: *Tropical rain forest*

Annual Mean Precipitation: *1,930 mm*
Temperature Range: *15 – 22°C*
Motor Vehicle Registrations: *4,000,000 (1990)*
Industry: *Mixed, economically important region*
Energy: *Power generation largely hydroelectric*

Situation Analysis

The Greater São Paulo Area contains more than 10 per cent of the total Brazilian population and is rapidly expanding. Manual, automatic and mobile monitoring is undertaken but not apparently across all the geographical area. Air quality has been improving in recent years.

Sulphur dioxide (SO_2) emissions and ambient concentrations have decreased substantially during the 1970s as a result of emission controls, and have stabilized since 1984 at levels below or within WHO guidelines. National air quality standards are now met. For total suspended particulates (TSP), the situation is less clear; the manual network data show no trends but exceed WHO guidelines, while the automatic network data illustrate a decreasing trend and are usually within or below WHO standards. High smoke levels are still recorded although values have decreased.

Lead (Pb) is not a major problem because of the use of alcohol as a motor vehicle fuel. In view of the large-scale increase in motor vehicles, traffic-related emissions give cause for concern. Despite an emission control programme, concentrations of carbon monoxide (CO) exceed WHO guidelines, nitrogen dioxide (NO_2) concentrations exceed WHO guidelines at some sites, while ozone (O_3) concentrations exceed the attention level of 400 µg m^{-3} on many days of the year.

Health effect studies of air pollution are urgently needed, particularly for motor vehicle-related pollutants. Although much has been done to reduce SO_2 and Pb levels, SPM, CO, NO_2 and O_3 are of great concern.

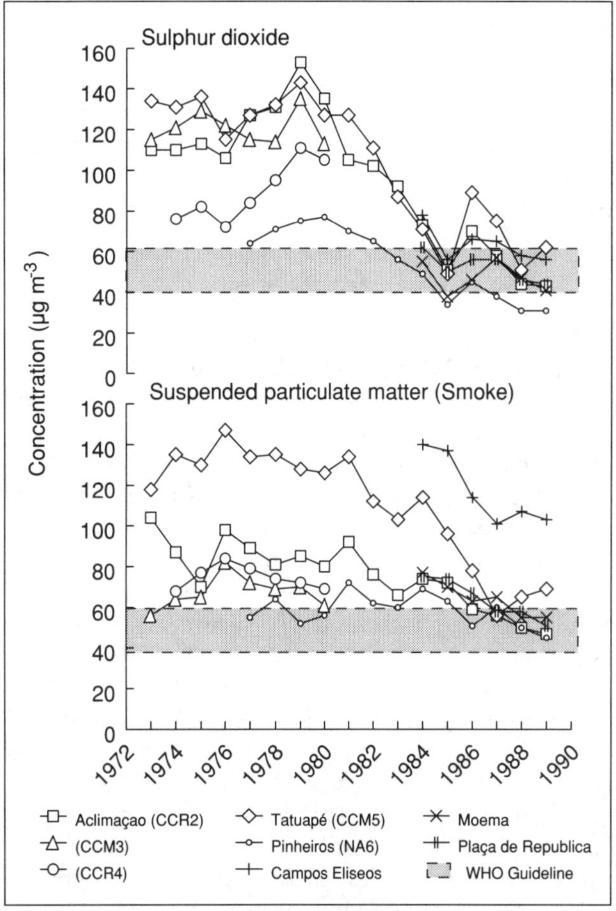

Annual mean SO_2 and SPM (smoke) concentrations

Estimated emissions, 1990 (tonnes per annum)

SO_2	122,000
SPM	77,000
CO	1,391,000
NO_x	217,000
HC	233,000

Seoul

Estimated Population 1990: *11.33 million*
Projected Population 2000: *12.97 million*
Map Reference: *Latitude 37°33'N, Longitude 127°00'E*
Altitude: *Sea level to 100m (up to 800 m)*
Area: *1,650 km^2 (Greater Seoul)*
Climate: *Continental warm summer*
Annual Mean Precipitation: *1,200 mm*
Temperature Range: *-5 – 25°C*
Motor Vehicle Registrations: *2,660,000 (1989)*
Industry: *Electronics, textiles, chemicals, motor vehicles, food processing*
Energy: *Coal is the main fuel (oil is also used).*

Situation Analysis

Seoul is the capital of the Republic of Korea and is the centre of intense industrial expansion and urbanization. A rapid rise in the standard of living has been seen in recent years with a concomitant increase in fuel use by the industry and transport sectors. Monitoring capabilities have been recently improved.

Domestic heating during the winter and industry (including power generation) are the main anthropogenic sources of sulphur dioxide (SO_2) and of suspended particulate matter (SPM). The widespread use of coal (and anthracite briquettes) causes extensive SO_2 and SPM pollution and WHO annual mean guidelines for both pollutants are consistently exceeded. Highest concentrations are reported in the winter heating period. Therefore, long-term exposure poses a significant risk to health resulting in an increased prevalence of respiratory symptoms throughout the population.

The rapid growth of motor vehicle traffic is also of concern. Data on lead are sparse. A high proportion (50 per cent) of vehicles are diesel-fuelled and this is reflected in the large amount of total suspended particulates (TSP) attributable to this source. Concentrations of other automotive pollutants, carbon monoxide (CO) and oxides of nitrogen (NO_x), remain relatively stable despite increasing vehicle numbers. This is through strict enforcement of vehicular emissions standards. Ozone (O_3) data are incomplete although some elevated values could cause health problems.

High pollution episodes sometimes occur due to the surrounding mountains which act as a barrier containing pollution over the city; however, data are not complete. Progressive national actions have improved the air quality although epidemiological data are not readily available.

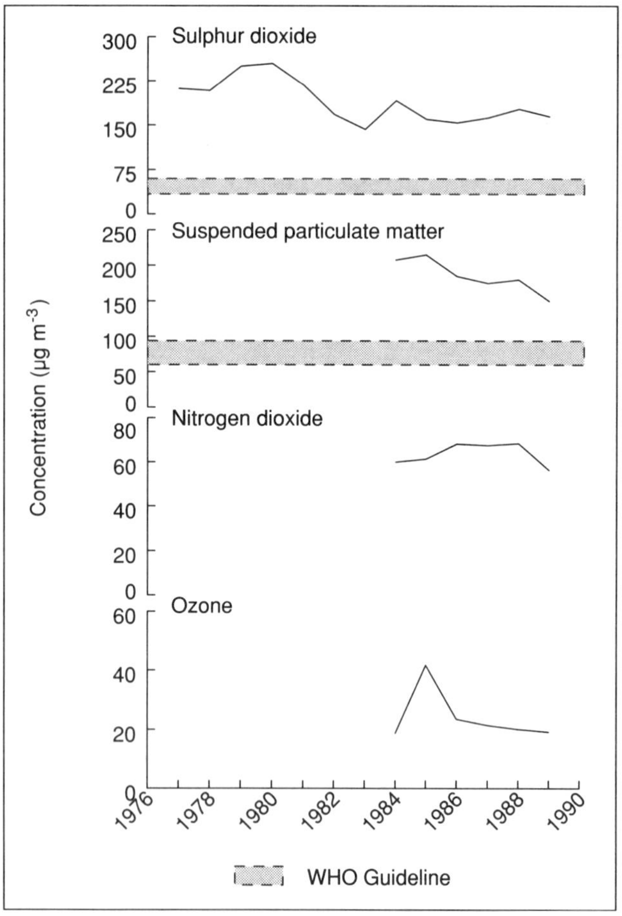

Annual mean SO_2, SPM, NO_2 and O_3 concentrations

Estimated Emissions, 1989 (tonnes per annum)

	Greater Seoul Area	Seoul
SO_2	380,000	138,000
SPM	90,000	49,000
CO	931,000	640,000
NO_x	271,000	130,000
HC	81,000	52,000

Shanghai

Census Population 1990: *13.3 million*
Projected Population 2000: *14.69 million*
Map Reference: *Latitude 31°10'N, Longitude 122°26'E*
Altitude: *4.5 m*
Area: *6,300 km^2*
Climate: *Humid sub-tropical*

Annual Mean Precipitation: *1,040 mm*
Temperature Range: *3 – 28°C*
Motor Vehicle Registrations: *147,700 (1990)*
Industry: *Machine tools, iron and steel, chemicals, textiles*
Energy: *Thermal power plants. Coal is the major industrial fuel.*

Situation Analysis

Shanghai, the commercial centre of China, is one of the most congested cities of the world and the most populous city in China. Housing is cramped and with no zoning restrictions people live next to the many industries throughout the area. The main air quality problems are from industries and the coal smoke emissions from some 91,000 boilers and stoves burning 25 million tonnes of coal annually.

The reported air quality from GEMS and Environment Protection Bureau (EPB) show that suspended particulate matter (SPM) and sulphur dioxide (SO_2) concentrations are well above WHO guidelines. The SO_2 annual means measured by GEMS are at and above the WHO guideline, but the SO_2 concentration may be twice as high in areas not covered by GEMS. The SPM annual means range from 200–300 μg m^{-3} with annual maxima ranging from 600–1,400 μg m^{-3}. These values are associated with morbidity for susceptible individuals and chronic exposure may result in decrements in lung function for children. Shanghai has the highest cancer mortality rate in China and the male lung cancer mortality rate has doubled from 21 to 44/100,000 from 1963 to 1985. No data on ambient carbon monoxide (CO), nitrogen dioxide (NO_2), lead (Pb), and ozone (O_3) were available for this report.

Programmes are in place to reduce emissions but reliance on coal, the great cost of controls, the growth of industry and deterioration of existing facilities may make it difficult to meet air quality goals set for the year 2000.

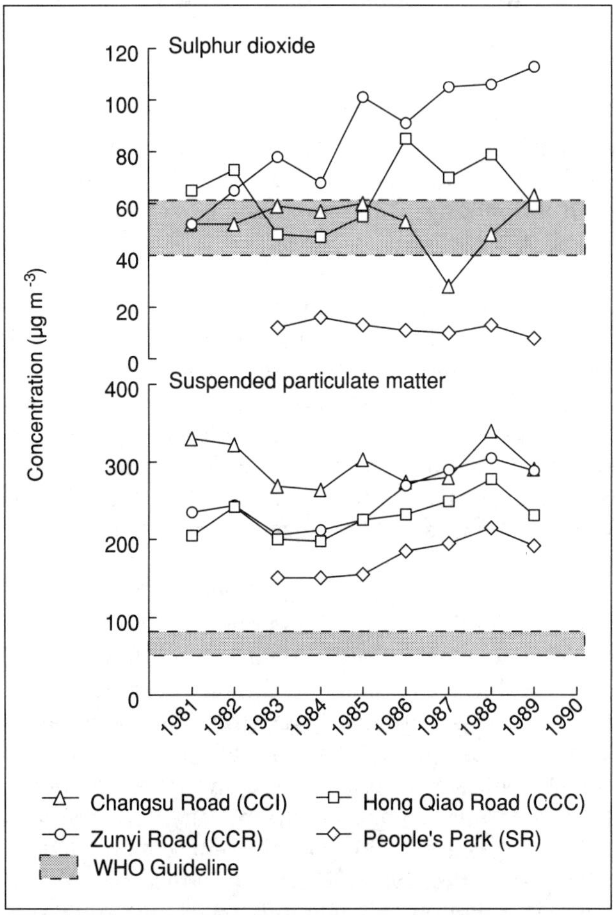

Annual mean SO_2 and SPM (TSP) concentrations

Estimated emissions*, 1983 (tonnes per annum)

SO_2	267,000
SPM	324,000
CO	229,000
NO_x	127,000

* There are additional emissions resulting from motor vehicle exhaust (CO, NO_x, SPM) and from industrial process (SO_2, SPM)

Tokyo

Estimated Population 1990: *20.52 million*
Projected Population 2000: *21.32 million*
Map Reference: *Latitude 35°40'N, Longitude 139°45'E*
Altitude: *Sea level to 220 m*
Area: *2,162 km^2*
Climate: *Humid sub-tropical*

Annual Mean Precipitation: *1,460 mm*
Temperature Range: *5 – 26°C*
Motor Vehicle Registrations: *4,400,000 (1990)*
Industry: *Electronics, electrical goods, textiles, printing, chemicals and petrochemicals, iron and steel, heavy engineering, shipbuilding*
Energy: *Coal, oil, gas, hydroelectric and nuclear*

Situation Analysis

Tokyo is the capital of Japan and in 1990 was the most populous urban agglomeration in the world. Rapid urban and industrial growth has taken place with a doubling of the population over the past 30 years.

No city specific emissions estimates were available for this report. Increasing industrialization and motorization have had the greatest impact upon air quality. Stringent emissions controls have been implemented to control sulphur dioxide (SO_2) and suspended particulate matter (SPM). Annual mean SO_2 and SPM concentrations are now well below WHO guidelines and have been for most of the 1980s.

No regular monitoring data for lead (Pb) were available for this report. Emissions are expected to be small in view of the low lead concentration in petrol. Motor vehicle emissions of carbon monoxide (CO) and oxides of nitrogen (NO_x) remain the greatest threat to air quality. Despite the introduction of vehicle emissions controls, the increases in motor vehicle population and in trip mileage have resulted only in a stabilization of ambient concentrations throughout the 1980s. Ozone (O_3) exceeds WHO criteria although no annual trend is apparent. "Oxidant warning days" are reported where adverse health effects can be expected. Health data are available.

Tokyo is a recognized example of how air pollution can be controlled in an industrial megacity.

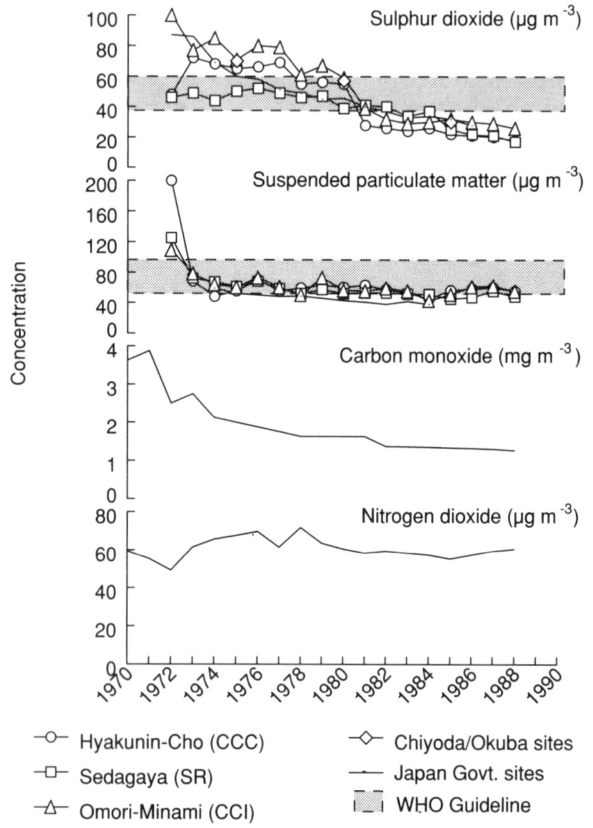

Annual mean SO_2, SPM, CO and NO_2 concentrations

Estimated emissions, 1985 (tonnes per annum)

NO_x	52,700

4

Overview of Air Pollution in theTwenty Megacities

4.1 Introduction

This chapter provides a brief summary of the results of this study. It includes a section on current populations, their growth and the relationship between urbanization and air pollution in the megacities. A brief review is also included on the status of air quality management in these cities, particularly with regard to monitoring and collection and analysis of data, including emissions and sources. A summary of the air pollution situation is presented in the following sections and this chapter is concluded with a brief overview of the trends in air pollution in the megacities.

4.2 Urbanization and Air Pollution

For the purposes of this study megacities were defined as those urban agglomerations with current or projected populations of 10 million or more by the year 2000 (see Chapter 2, Section 2.1). While there are 24 such megacities, the lack of resources and time required that these be reduced to 20. Dacca, Lagos and Tehran were excluded because of general lack of information and Osaka because of its similarity to Tokyo. They will be included in a future update.

The estimated populations of the 20 megacities over the last 30 years are shown in Figure 1.1 (Chapter 1) which also includes their projected populations for the year 2000. Between 1970 and 2000 their combined populations will have doubled, with Manila exhibiting the most rapid growth (4.65 average annual rate of growth, 1970–1985) and New York the lowest (– 0.24 average annual rate of growth, 1970–1985). Currently the most rapidly growing megacities are Lagos and Dacca with a population growth in the period from 1990 to 2000 estimated at 64 per cent and 76 per cent respectively (UN, 1989).

In the coming century, the number of megacities will continue to increase. Between 1970 and 2000 the number of cities with populations of over 10 million will have increased sixfold, from four in 1970 to 24 in 2000. With regard to cities of over three million people, in 1970 there were 35 of them, in 2000 the UN estimates there will be 85 and by 2025 this number could well have doubled (UN, 1989). Table 2.1 (Chapter 2) shows the current and projected populations of the 69 largest cities in the world.

Megacities represent extreme examples of massive urban growth and development. The associated traffic problems, domestic emissions and industrial sources present special problems due to the large scale of the area and are exacerbated by the often fragmented controls on land use and development. Megacities usually have extremely high population densities, especially in central districts, and experience motor vehicle traffic congestion with high-density air pollution emissions as a result. The presence of built-up areas which extend for tens of kilometres alters the normal wind flow and urban "heat islands" may exist due to high energy dissipation. Urban agglomerations form a continuum of developed land without green belts or natural barriers. Air pollution levels can steadily increase as the air masses pass over such areas.

4.3 Status of National Programmes

National programmes for environmental pollution control, including those concerned with air pollution, range from highly developed programmes in some countries to no activities at all in others. A recent survey by WHO showed that of the 60 newly industrializing countries in the world, 34 had limited national programmes, while 15 had essentially no programmes (Schaefer, 1992). These include the countries where the majority of the megacities are located.

A basic requirement for effective air pollution control programmes is the ability of the authorities to monitor air quality, assess human exposures and conduct emissions inventories. Figure 4.1 summarizes the air quality monitoring capabilities in the 20 megacities.

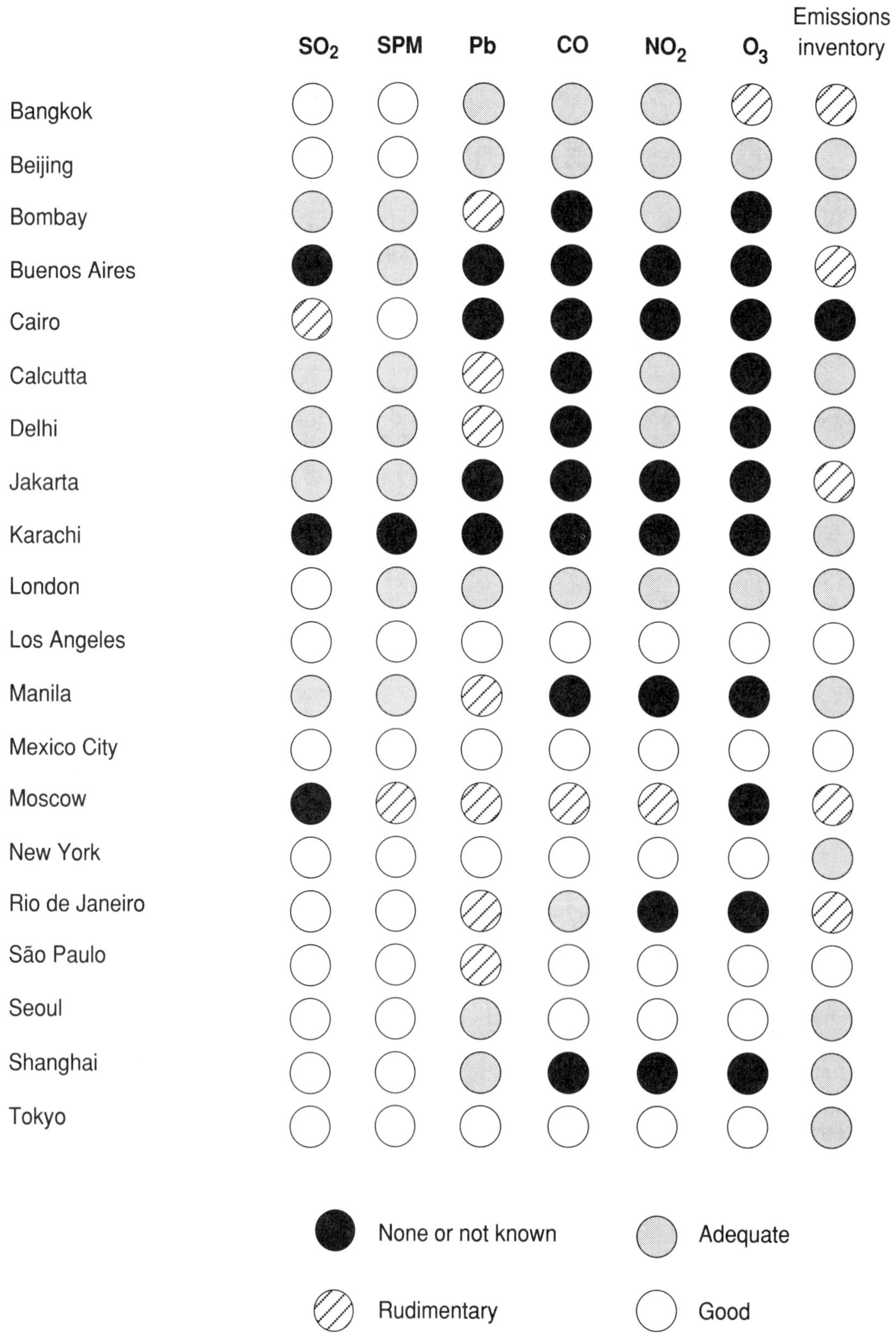

Figure 4.1 *Subjective assessment of the status of monitoring capabilities and availablility of emissions inventories in 20 megacities*

As this present study shows, these functions are less than satisfactorily carried out in many of the countries surveyed. The capabilities of countries fall into three general categories. Cities such as Los Angeles, Mexico City, New York, São Paulo, Seoul and Tokyo maintain comprehensive air quality monitoring networks that often provide real-time data on all major air pollutants. These networks incorporate adequate quality control procedures to ensure that the data are demonstrably valid. In the second category are cities with mostly marginal to adequate air monitoring networks which measure only a few pollutants and usually at fewer sites than desirable. Included in this group are cities such as Bangkok, Beijing, Bombay, London and Rio de Janeiro. The third category are the megacities with inadequate air monitoring capabilities which produce data of unknown quality on a few pollutants. They include such cities as Buenos Aires, Cairo, Karachi, Manila and Moscow. The latter two categories of cities also often lack sufficient quality assurance oversight and there are questions raised about the reliability of the data. It is evident that air quality monitoring capabilities must be improved in many of these megacities.

The maintenance of a current emissions inventory is also a fundamental tool in air quality management programmes. Without emissions inventories it is essentially impossible to design appropriate and cost-effective control strategies. Based on the information collected as part of this study, an estimate could be obtained of the availability and adequacy of emission inventory data. Figure 4.1 shows the present-day status of the availability and inadequacy of such information in many of the 20 megacities.

4.4 Air Pollution and its Impacts

4.4.1 Meteorology and topography

The two major influences on the transport and dispersion of air pollutant emissions are the meteorology of the city (including microclimate effects such as a "heat island") and the topography of the area in relation to the population distribution.

The climates of the megacities range from cooler humid continental (with severe cold seasons such as Beijing) to desert (Cairo) to tropical (no cold season and high temperatures and humidity such as Bangkok). Table 4.1 shows the climate zones according to the Köppen system (Köppen and Geiger, 1935) of the areas of the world in which the 20 megacities are located.

The severity of the cold season determines the amount of residential heating that is required, increasing emissions of, for example, sulphur dioxide (SO_2) in winter and reducing them in the summer. Figure 7.2 (Chapter 7) shows an example of this variability for SO_2 in Beijing for 1989. In the cities with a moderate climate the pollution load tends to be distributed more evenly over the year. The conditions of high solar intensity and temperature in the summer can, however, lead to severe photochemical reactions and production of ozone (O_3) (e.g., Los Angeles and Mexico City).

***Table 4.1** Major climatic regions (Köppen system)*

Climatic regions		Cities
Cooler humid	Continental cool summer	Moscow
	Continental warm summer	Beijing, New York, Seoul
Warmer humid	Mediterranean	Los Angeles
	Humid sub-tropical	Tokyo, Buenos Aires, Shanghai
	Marine west coast	London, Mexico City
Dry	Steppe	Delhi
	Desert	Cairo, Karachi
Tropical humid	Savanna	Bangkok, Bombay, Calcutta
	Tropical rain forest	Jakarta, Manila, Rio de Janeiro, São Paulo

Source: Köppen and Geiger, 1935

The topography of the megacity also influences the manner in which pollutants are transported and dispersed. The megacities fall into the following three groupings:

- Relatively level topography, no influence on climate by a body of water (Beijing, Cairo, Delhi, Moscow)
- Relatively level topography, influence on climate by a body of water (Bangkok, Bombay, Buenos Aires, Calcutta, Jakarta, Karachi, London, Manila, New York, Shanghai, Tokyo)
- Variable topography, influenced by surrounding mountains (Los Angeles, Mexico City, Rio de Janeiro, São Paulo, Seoul)

The presence of a significant body of water can lead to microclimatological effects and to on-shore and off-shore diurnal wind patterns. The presence of internal topographical features, such as mountain ridges, can serve as barriers to transport within the megacity. When the city is surrounded by high mountains (Los Angeles, Mexico City) the pollutants may be trapped within the air shed (i.e., the area in which, owing to meteorological factors and topographic features, the air is more or less homogeneous and across whose boundaries pollutant transport is minimal) over the megacity for several days.

4.4.2 Sources and emissions

All the major air pollutants and their sources are present in the 20 megacities, although to varying degrees. Some sources are common to all of them, while others are of importance in only certain megacities. For the megacities motor vehicle traffic is a major source of air pollution; in nearly half of them it is the single most important source. In addition to the megacities in the developed countries, other megacities such as Bangkok, Jakarta, Manila, Mexico City, São Paulo and Seoul have overwhelming traffic-created air pollution. As a result, the emissions of carbon monoxide (CO), hydrocarbons, oxides of nitrogen (NO and NO_2, collectively termed NO_x) and lead (Pb) (where it is still used as a petrol additive), are high. In Bangkok, Manila and Seoul a substantial proportion of the motor vehicle fleet is diesel-powered, a characteristic which brings its own special problems, such as greater suspended particulate matter (SPM), SO_2 and NO_x. In comparison, in Beijing, Karachi, Moscow and Shanghai motor vehicle emissions are still relatively low but are increasing.

Emissions from power generation and industry are likewise a problem in almost all the megacities. With the possible exception of Jakarta, New York and Shanghai, basic heavy industry is generally not located in or near the megacities. The multitude of other industries (including power generation) are, however, dominant sources of air pollution, especially of SPM and sulphur and nitrogen oxides. Beijing, Bombay, Buenos Aires, Cairo, Calcutta, Seoul and Shanghai are cities with high industrial emissions. In cities such as Bangkok, Beijing, Manila and Shanghai, the problem of industrial emissions is compounded by the bulk of the industrial sources being interspersed with residential areas, thus creating the potential for particularly high human exposure.

Coal and high-sulphur oil have been, or are being, phased out in many megacities to be replaced by cleaner fuels containing less sulphur, such as natural gas. At present, the principal fuel in the 20 megacities is split about equally among coal, oil and natural gas. Coal is still the predominant fuel for industry and energy in Delhi, Beijing, Calcutta, Shanghai and Seoul.

A high degree of domestic coal or biomass fuel use is a very serious problem for health as there may be substantial indoor and outdoor exposures. The concentrations of SPM, hydrocarbons and, in the case of coal, SO_2 are typically very high. Areas of Beijing, Calcutta, Delhi, Shanghai and Seoul are prime examples of where this situation persists.

The open burning of refuse is a major source of urban air pollution in many developing countries but information is scarce. In several of the megacities, most notably Karachi, Mexico City and Manila, pollution from this source is recognized.

High levels of natural wind-blown particulates are another feature which complicates air pollution problems. The proximity to desert areas or barren lands leads to high natural loadings of SPM in cities like Beijing, Cairo, Delhi, Karachi and Mexico City.

4.4.3 Comparative air quality

Figure 4.2 presents a subjective overview of the relative air quality situation in the 20 megacities. In order to compare the cities it was necessary to group the cities into broad categories, those with:

- Serious problems, WHO guidelines exceeded by more than a factor of two;
- Moderate to heavy pollution, WHO guidelines exceeded by up to a factor of two (short-term guidelines exceeded on a regular basis at certain locations);
- Low pollution, WHO guidelines are normally met (short-term guidelines may be exceeded occasionally);
- No data available or insufficient data for assessment.

These assessments of air pollution (Figures 4.2–4.9) are based on the most recent data presented in the city chapters (Part II) and compared with WHO guidelines (Table 2.2). Obviously such comparisons are very difficult owing to differences in monitoring

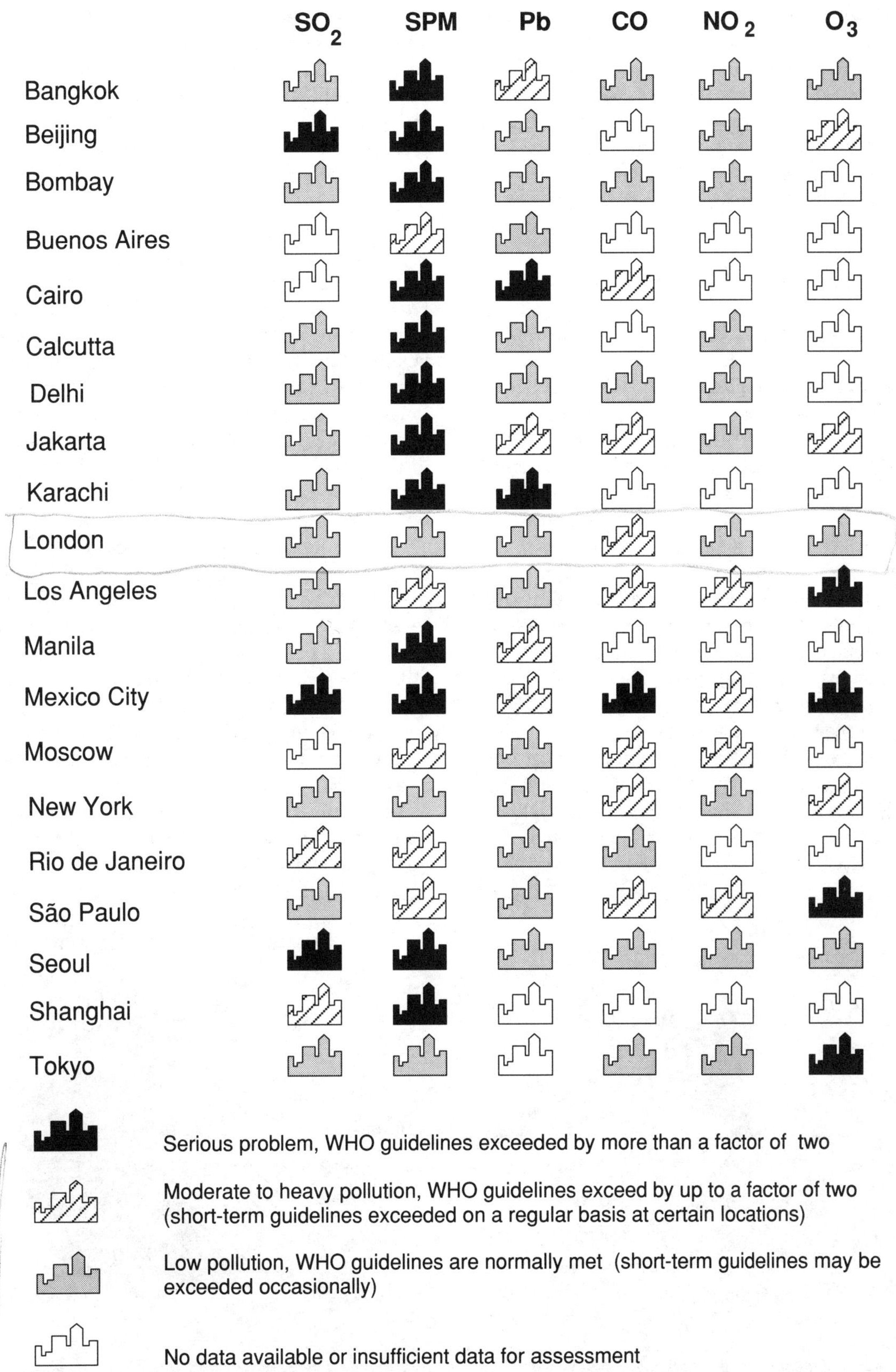

Figure 4.2 *Overview of air quality in 20 megacities based on a subjective assessment of monitoring data and emissions inventories*

methods, reporting procedures, and so forth.

No assessment of the air quality situation has been made where insufficient data were available or where the quality of the data was questionable. Not all of the data was provided by recognized monitoring networks and some assessments have been based upon the results of recent studies and/or surveys.

The following sections provide a brief overview on a pollutant-by-pollutant basis.

4.4.4 Sulphur dioxide

The concentrations of SO_2 have decreased dramatically in a number of the megacities. At present, however, the three cities of Beijing, Mexico City and Seoul can be considered to have serious SO_2 pollution (Figure 4.3). The ambient levels in these cities are in excess of WHO guidelines by a factor of nearly three for annual average concentrations; peak daily concentrations exceed 700 $\mu g\ m^{-3}$. These concentrations are well within the range at which, in the presence of particulate matter, there are observable effects on health such as increased mortality and morbidity of the aged and infirm. It should be noted that a large refinery in Mexico City has been recently closed and the concentrations of SO_2 should decrease markedly.

There are two cities where the concentrations of SO_2 are near or somewhat in excess of the WHO health guidelines. In 12 of the megacities SO_2 does not appear to be a major problem. The concentrations are typically within WHO guidelines, although there may well be some days or even months when the levels may be higher at certain locations in some of these cities. For example, one area in São Paulo has been slightly exceeding the guideline value and there are periodic excursions above the guideline value during the non-monsoon seasons in Calcutta. While London now meets the WHO annual guidelines, it still exceeds hourly and daily guideline maxima. No SO_2 data for Moscow are presented as their validity is questionable. No recent SO_2 data for Buenos Aires have been made available.

4.4.5 Suspended particulate matter

Particulate air pollution is, as illustrated in Figure 4.4, a very serious problem in 12 of the megacities surveyed. The ambient concentrations in these cities are persistently above the WHO guidelines, both for long-term averages and for peak concentrations. The annual average levels across these cities are typically in the range of 200–600 $\mu g\ m^{-3}$ and peak concentrations are frequently above 1000 $\mu g\ m^{-3}$. In combination with high exposure to SO_2, there are notable effects on health as already mentioned. In some of these cities, including Beijing and Seoul, there is a notable decrease in SPM during the non-heating season; in others, such as Shanghai, the high concentrations persist all the year round. In Cairo, and other cities mentioned in Section 4.4.2, wind-blown dust greatly influences the SPM.

In the second group of cities, the concentrations of SPM can be in excess of WHO guidelines by a factor of as much as two. However, in these cities there are some monitoring stations which record concentrations within or near the WHO guidelines, indicating relatively clean areas.

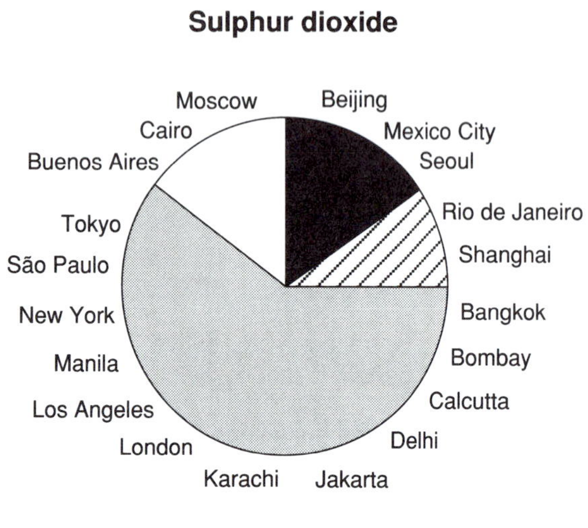

Figure 4.3 *Levels of sulphur dioxide in the 20 megacities (see key in Figure 4.2)*

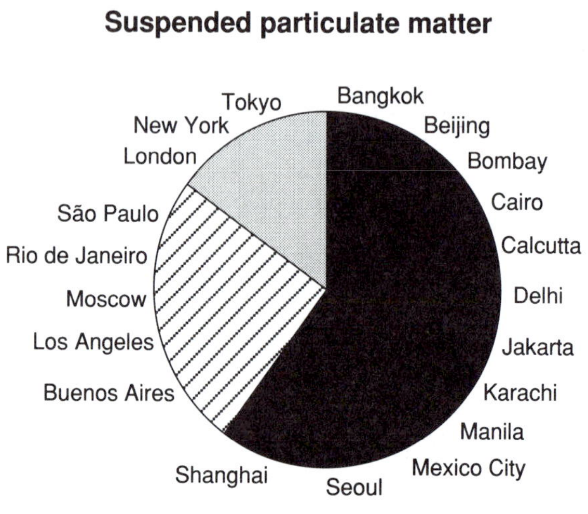

Figure 4.4 *Levels of suspended particulate matter in the 20 megacities (see key in Figure 4.2)*

Of the 20 megacities there are three - London, New York and Tokyo - where, by and large, the WHO guidelines are being met. All of these have undergone massive and costly air pollution reduction programmes and, where necessary, sources are equipped with effective controls.

4.4.6 Lead

Airborne Pb is closely associated with the density of motor-vehicle traffic using leaded fuel and the concentration of Pb additives in the fuel (Figure 4.5). Table 4.2 presents a summary of the latter showing that the concentrations range from 0 -2 g l^{-1}. It is therefore not surprising to see that airborne Pb concentrations in Cairo and Karachi are high - well above the WHO guideline. It should be noted that with a few exceptions there are no routine network data on Pb in ambient air; most information comes from special investigations. High exposure to Pb is associated with reduced mental development of infants and children.

The available information on Pb in air in Bangkok, Jakarta, Manila and Mexico City shows that it is somewhat above the 1 $\mu g\ m^{-3}$ guideline of WHO. In these cities programmes are under way or planned to reduce the amount of Pb in petrol.

Twelve other cities, for which some data on ambient levels of Pb were available, show that such levels are low and well within the WHO guidelines. Among these, London, Los Angeles, New York, Seoul and Tokyo are moving towards totally lead-free petrol. In Rio de Janeiro and São Paulo the levels are low because of reliance on alcohol-powered vehicles coupled with reduced Pb levels in petrol. The relatively low traffic density in Beijing and Moscow results in levels less than 0.5 $\mu g\ m^{-3}$. In Bombay, Calcutta and Delhi, the majority of the spot samples were below WHO guidelines, although some samples were higher.

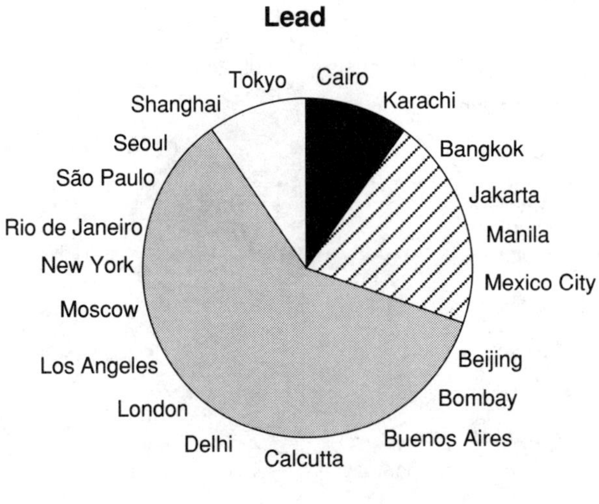

Figure 4.5 *Levels of lead in the 20 megacities (see key in Figure 4.2)*

Table 4.2 *Lead content of petrol*

City	Lead content of petrol* (g l^{-1})	Comments
Bangkok	0.15	
Beijing	0.4–0.8	80% unleaded petrol
Bombay	0.15	
Buenos Aires	0.6 –1.00	
Cairo	0.80	
Calcutta	0.10	
Delhi	0.18	
Jakarta	0.6–0.73	1987 levels
Karachi	1.5–2.00	
London	0.15	>33% unleaded petrol
Los Angeles	0.026	>95% unleaded petrol
Manila	1.16	
Mexico City	0.54	
Moscow	0	No leaded fuel sales in city
New York	0.026	95% unleaded petrol
Rio de Janeiro	0.45	ethanol and gasohol used
São Paulo	0.45	40% ethanol, 60% petrol/ gasohol
Seoul	0.15	
Shanghai	0.40	
Tokyo	0.15	>95% unleaded petrol

* Lead content may be average lead content of petrol or limit value
Note: Many countries have recently introduced (or will soon introduce) sales of unleaded petrol

Sources: Walsh, 1992; NEERI, 1991; Morley ,1991

4.4.7 Carbon monoxide

Concentrations of CO depend particularly on the site of sampling near main traffic corridors. Comparisons between monitoring results for the 20 megacities are therefore difficult to make.

The measurements in Mexico City indicate very high concentrations of CO at different monitoring sites (Figure 4.6). Recent data on exposures of commuters show one-hour concentrations up to 67 mg m^{-3} which is well above both the national standard and the WHO guidelines. In London, Los Angeles, and several other

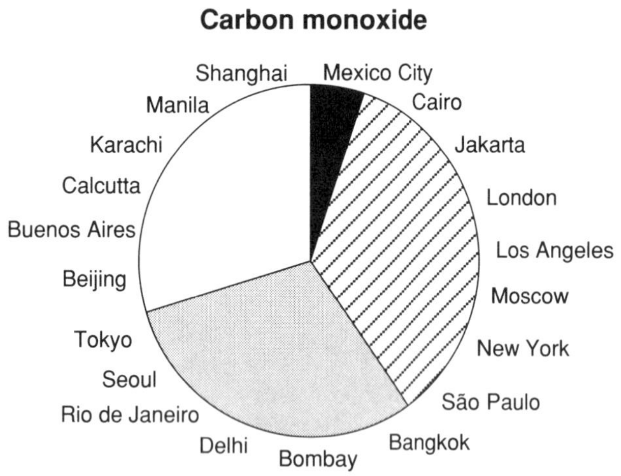

Figure 4.6 *Levels of carbon monoxide in the 20 megacities (see key in Figure 4.2)*

cities the concentrations are not as high, but the air quality standards are being exceeded quite frequently - at some locations in Los Angeles on 15 per cent of the days in the year. In Bombay the results to date indicate that, by and large, the WHO air quality guidelines are being met, except in areas of high traffic density.

Among the cities with insufficient data to make an assessment are Beijing, Buenos Aires, Calcutta, Manila and Shanghai. Where high traffic densities occur, as is the case in Mexico City, road-side and in-traffic concentrations could well be very high.

4.4.8 Nitrogen dioxide

Aside from a few of the megacities, NO_2 data are scanty, often available only from special studies and not from continuous routine measurements. None of the 20 megacities for which data were available shows NO_2 concentrations which are extremely high (Figure 4.7). In 10 of the cities, NO_2 concentrations are generally near the guideline values, although with periodic exceedences particularly as regards one-hour or 24-hour maximum concentrations. (The data from Mexico City show the hourly maxima range up to 700 $\mu g\ m^{-3}$. In Los Angeles hourly averages up to 500 $\mu g\ m^{-3}$ are measured. The WHO hourly guideline is 400 $\mu g\ m^{-3}$.)

In cities such as Moscow and São Paulo the measured concentrations are also relatively high. There are six cities for which there were either no data available for this study or for which the data were grossly insufficient; for these, no evaluation was possible. It could well be that in cities with high

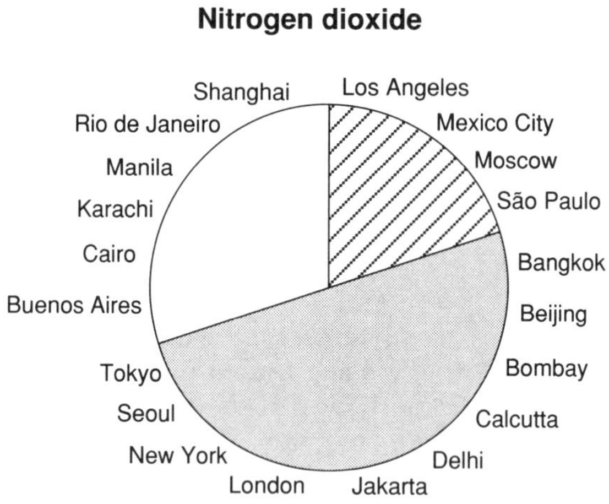

Figure 4.7 *Levels of nitrogen dioxide in 20 megacities (see key in Figure 4.2)*

traffic volumes, as in Buenos Aires and Manila, the levels of NO_2 are high at least in parts of these cities.

In cities where the use of natural gas or kerosene for domestic purposes is prevalent, the exposures of the inhabitants will be higher due to the oxides of nitrogen produced indoors in addition to the ambient levels.

4.4.9 Ozone

Ozone concentrations are high in Los Angeles, Mexico City, São Paulo and Tokyo (Figure 4.8). In

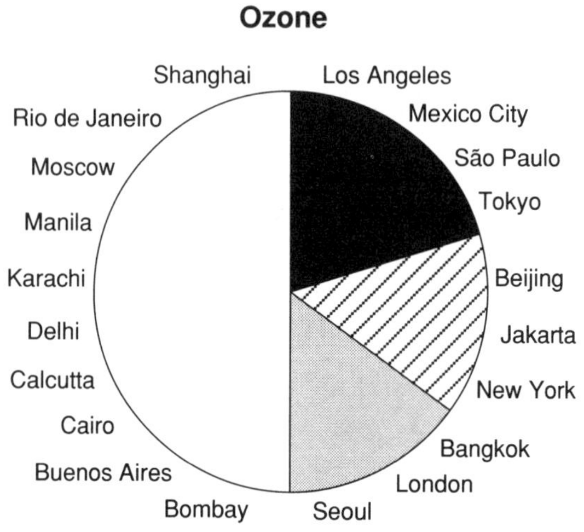

Figure 4.8 *Levels of ozone in the 20 megacities (see key in Figure 4.2)*

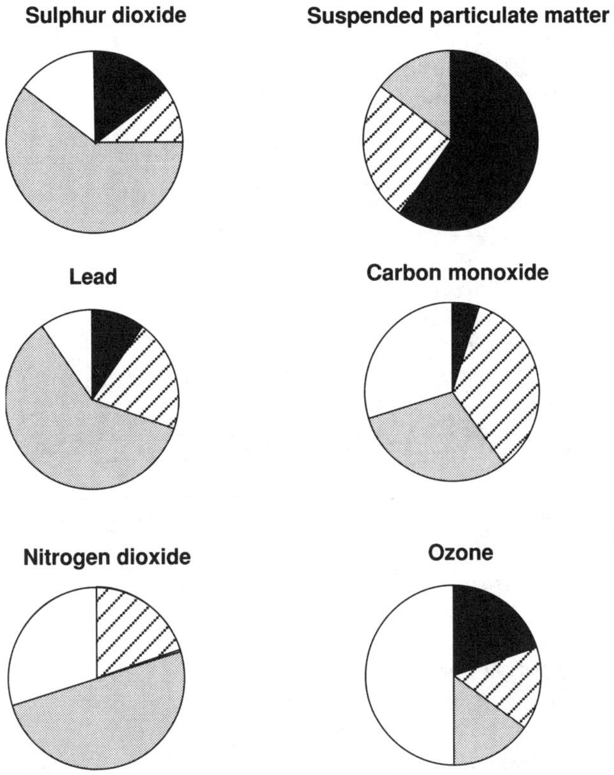

Figure 4.9 *Collective air quality status in 20 megacities (see key in Figure 4.2)*

1988, in Los Angeles, Mexico City and Tokyo (in certain areas) national air quality standards were exceeded 50 per cent, 70 per cent and 12 per cent of the days of the year respectively. In Mexico City hourly concentrations reached 600 µg m^{-3}, at times peaking to 900 µg m^{-3}, i.e., four times the maximum WHO guideline value. The health effects at these levels are associated with eye irritation, severe respiratory irritation and increased frequency of asthmatic attacks of susceptible individuals. Beijing, Jakarta, and New York also show concentrations above the guideline values, although less frequently. In London, Bangkok and Seoul the measurements of O_3 so far have indicated levels normally within the guidelines, with occasional excursions above them.

Of the cities for which data are not available, it could be expected that the levels of O_3 would be high in Bombay, Cairo, Calcutta, Delhi and Manila. Some of the necessary conditions for the formation of O_3, e.g., heavy traffic and a high degree of sunshine, occur in these cities.

4.4.10 Collective status

In degree of severity, Figure 4.9 shows that the high level of SPM is the major problem affecting the megacities as a group. It is the most severe problem in those megacities where high concentrations of SO_2 occur simultaneously. Next in severity are levels of O_3, Pb and CO. Many of these cities also have potentially serious NO_2 pollution; however, the levels recorded do not yet regularly exceed the WHO guidelines. It should be noted that, particularly for O_3, the data are lacking for about half the megacities.

4.5 Air Quality Trends

Air pollution has escalated greatly in most of the megacities over the past decade because of rapid population growth, industrialization and increasing energy use. The current predictions point to a substantial worsening of the situation in those cities where major air pollution control measures are not being implemented. These megacities could well see increases in their air pollution concentrations of levels as high as 75–100 per cent over the next decade. For example, in a city where the annual mean SPM concentration is currently 600 µg m^{-3} this could escalate to over 1,000 µg m^{-3} over the next decade if current trends persist. Nitrogen dioxide concentrations are also expected to increase substantially in all the cities due to the forecast increase in motor vehicle traffic. Consequently the effects on health will be magnified, e.g., from affecting only highly exposed sub-groups to affecting the general population.

It should be noted, however, that in some of the megacities studied the severe air pollution conditions observed today could have been much worse if certain control measures had not already been introduced. Examples are Beijing, Delhi, Seoul and Shanghai where, because of controls, the rise in air pollution levels has been slowed and, in some cases, stabilized before they could reach the high air pollution levels which, for example, were found in London 40 years ago. Over the past 30 years, London, Los Angeles, New York and Tokyo have reduced their air pollution dramatically.

From the information collected in this present report it can be seen that history repeats itself. A number of cities have approached the same degree of pollution that was present, for example, in London in the 1950s

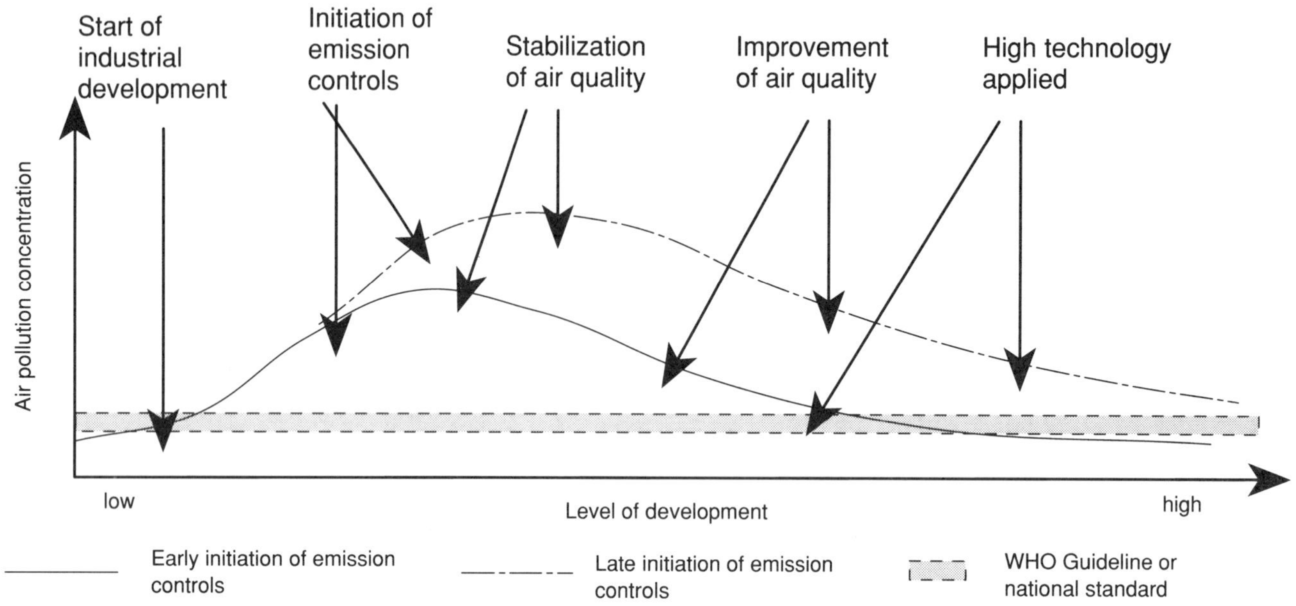

Figure 4.10 *Development of air pollution problems in cities according to development status*

before "clean air" legislation was introduced. This relationship between level and speed of development and environmental pollution is illustrated in Figure 4.10.

Before rapid industrial development takes place, air pollution is mainly from domestic sources and light industry; concentrations are generally low and increase slowly as population increases. As industrial development and energy use increase, air pollution levels begin to rise rapidly. Then urban air pollution becomes a serious public health concern and emission controls are introduced. Owing to the complexity of the situation, an immediate improvement in air quality cannot generally be achieved; at best the situation is stabilized, and serious air pollution persists for some time. Several of the megacities studied are now in the situation where controls must be implemented without delay. The introduction of emission controls has then been followed by a staged reduction of air pollution as controls take effect. The earlier that integrated air quality management plans are put into effect, the lower the maximum pollution levels that will occur. This is especially important for those cities of developing countries that are not yet of the size and complexity of present-day megacities.

4.6 References

Köppen, W. and Geiger, R. 1935 *Handbuch der Klimatologie,* Verlag von Gebrüder Borntraeger, Berlin.

Murley, L. (Ed.) 1991 *Clean Air Around the World,* Second Edition, International Union of Air Pollution Prevention Associations, Brighton.

NEERI 1991 *Air pollution aspects of three Indian megacities, Volume I Delhi,* National Environmental Engineering Research Institute, Nagpur.

Schaefer, M. 1992 *Combating Environmental Pollution – National Capabilities for Health Protection,* Report No. WHO/PEP/91.4, World Health Organization, Geneva.

UN 1989 *Prospects of World Urbanization 1988,* Population Studies No. 112, United Nations, New York.

Walsh, M. P. 1992 Personal communication, Alexandria, Virginia.

5

Summary Conclusions and Recommendations

5.1 Conclusions

- Air pollution is widespread in the 20 megacities – each has at least one major pollutant which exceeds WHO health guidelines, 14 have at least two and 7 have three or more major pollutants which exceed the guidelines.
- A high level of suspended particulate matter is the most prevalent form of pollution – in 17 megacities the health guidelines are exceeded and in 12 of these the values are double the guidelines. Sulphur dioxide and ozone pollution follow in terms of prevalence and severity.
- High concentrations of both sulphur dioxide and suspended particulate matter frequently occur in five of the megacities. This combination is particularly hazardous to health giving rise to increased mortality and morbidity.
- High-sulphur coal and oil are being phased out in industry and power generation in the megacities – the upward trend in sulphur dioxide pollution has been reversed in 10 megacities.
- Motor vehicle traffic is a major source of air pollution in all the megacities; in half of them it is the most important source. This leads to high concentrations of carbon monoxide, nitrogen dioxide and ozone.
- High domestic use of coal or biomass fuels is still a serious problem in 5 of the 20 megacities resulting in high human exposures to suspended particulates, sulphur oxides and carcinogenic polycyclic aromatic hydrocarbons.
- The current capabilities of the municipal or national authorities to monitor air quality in the megacities or to collect information on their sources and emissions are inadequate. Only 6 of the 20 megacities have satisfactory monitoring networks and data handling capabilities.
- There is no systematic collection of information on the health risks and effects of air pollution in most of the megacities.

5.2 Recommendations

- Air quality management should be developed and implemented as a matter of urgency in those megacities where strategic planning is weak or non-existent.
- Short-term, feasible approaches to reducing existing air pollution should be implemented as soon as possible. These include energy conservation, motor vehicle inspection and maintenance programmes, phasing out lead in petrol, promotion of mass transit and alternatives to open burning of refuse.
- In the longer term, increased emphasis should be placed on preventive measures in the development of management strategies to improve air quality, such as urban and transportation planning and the introduction of "clean" technologies.
- There is an immediate need to improve the monitoring and emissions inventory capabilities of the megacities. These are prerequisites for sound air pollution management strategies with the main aim of protecting public health.
- Simple and rapid techniques for air quality monitoring and assessment should be investigated and applied where appropriate.
- There are at present 69 cities of over three million people; of these cities, 45 are in developing countries and many are likely to become megacities. This report should be considered by the authorities in all cities to avoid repeating mistakes of the past.

PART II –
DETAILED MEGACITY REPORTS

6

Bangkok

6.1 General Information

Geography Bangkok is the capital and chief port of Thailand. It is located in the central region of Thailand on a flat coastal plain at the upper end of the Gulf of Thailand at Latitude 13°44′ N and Longitude 100°34′ E. The city was originally built on rich alluvial deposits on the deltaic plain of the Chao Phraya River 40 km from the Gulf of Thailand. Bangkok metropolis occupies an area of 1565 km^2.

Demography Bangkok's population is rapidly growing. A census in 1970 showed a population of 3.11 million, which had increased to 4.75 million by 1980 and to 6.1 million in 1985. This is approximately 10 per cent of Thailand's total population. The 1990 population was estimated to be around 7.16 million, and the UN projects that the population will reach 10.26 million by the year 2000 (UN, 1989). The population density of Bangkok is approximately 3,500 persons per km^2. The city continues to grow and is expanding into surrounding agricultural areas.

Climate Bangkok has a tropical savanna climate. The city is usually hot, with the mean temperature varying from about 25°C in December to 30°C in April. Annual rainfall is 1,500 mm, of which 80 per cent occurs in torrential downpours during the tropical monsoon rainy season.

The flat plains surrounding Bangkok allow free air movements and the near coastal location results in a high proportion of land-sea breezes. North-easterly (October–January) and south-westerly (February–September) winds predominate, depending on the season.

Emissions from the neighbouring industrial town of Samut Prakan usually by-pass Bangkok because it is situated off the NE/SW axis of prevailing winds. The city is also buffered from the heavy industrial zone by the Bangkrajao forest (green) belt.

Industry The Thai per capita GNP in 1988 was US $1,000, and it is estimated to be increasing at 8 per cent per annum. Bangkok has two thermal power plants in the metropolitan area, which provide 11 per cent of national capacity. Both power plants are fired with natural gas.

Numerous small and medium industrial facilities are located in metropolitan Bangkok, the principal manufacturing enterprises being textile and building materials, food processing, and electronic equipment assembly. There are many small-scale factories in the metropolitan area.

Since 1976, the government of Thailand has shifted industrial expansion into industrial parks on the outskirts of Bangkok. At present, the major industry is located in the town of Samut Prakan between Bangkok and the coast.

Transport Bangkok has a severe motor vehicle problem. Motor vehicle registration increased from around 600,000 in 1980 to 1,760,000 in 1988. The vehicle population is growing at a rate of 10 per cent per annum (Division of Land Transport, 1989).

It is estimated that the number of motorized trips will rise from 4.5 million trips per day in 1980 to 11.3 million trips per day by the year 2000. Analysis of the modal split of motorized passenger trips in 1980 reveals that 55 per cent of trips were by bus and 25 per cent by car. Taxis and motorcycles accounted for 10 per cent of trips each. It is reported that smoke from traffic is often so bad that reductions in visibility are a safety hazard.

Taxis and tricycles now mainly use liquid petroleum gas (LPG). There are many diesel vehicles in Bangkok because diesel fuel is about 25 per cent cheaper than petrol. The sulphur content of diesel is approximately 1 per cent. It is planned that by 1993 diesel sulphur content will be lowered to 0.5 per cent. Buses will have priority for the cleaner fuel.

6.2 Monitoring

Suspended particulate matter (SPM) has been monitored by the Ministry of Health for the GEMS/Air network since 1977 at three sites in Bangkok; sulphur dioxide (SO_2) monitoring started in

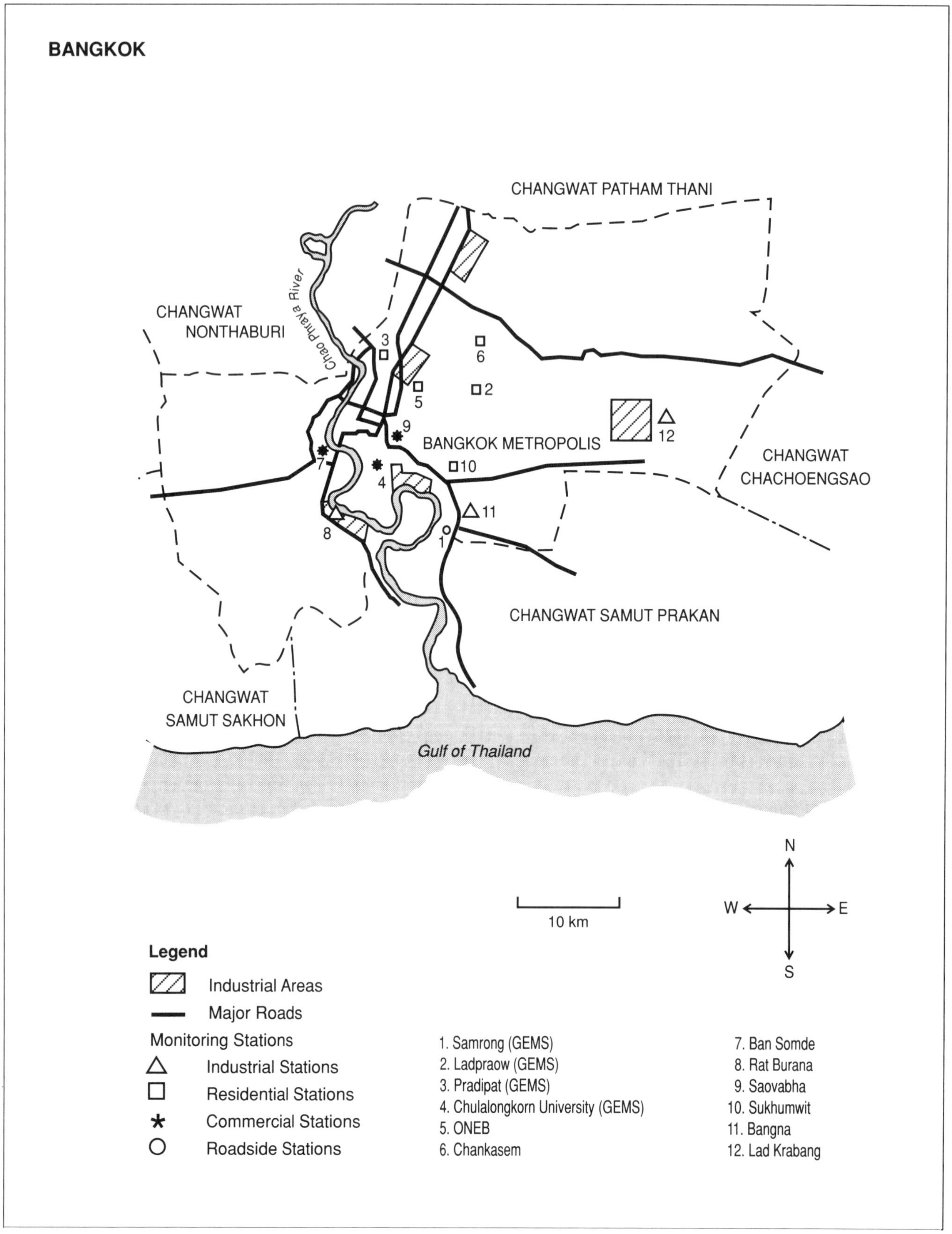

Figure 6.1 *Sketch map of Bangkok including GEMS and ONEB monitoring sites*

1980 at one site. In addition to the GEMS/Air sites, the Office of the National Environment Board (ONEB) has operated a monitoring network since 1983. In 1990, ONEB had seven permanent automated monitoring sites, and two mobile monitoring units. The mobile laboratories are used to monitor the direct effects of traffic pollution at nine to ten temporary sites on a regular basis. In addition, five permanent monitoring stations were established in the neighbouring industrial Samut Prakan Province in 1989. The locations of the GEMS and ONEB monitoring sites in metropolitan Bangkok are shown in Figure 6.1.

6.3 Air Quality Situation

Sulphur dioxide

Emissions Total emissions of SO_2 are inadequately known. Emission estimates have been prepared for Bangkok by Faiz et al. (1990), and for Thailand by the Thailand Development Research Institute (TDRI) as reported by USAID (1990). These estimates are sometimes contradictory owing to the different assumptions that are used in the absence of actual data. As there is no domestic heating and only a few heavy industrial sources, it is suggested that SO_2 emissions are moderate. Faiz et al. (1990) reports Bangkok SO_2 emissions in 1980 of about 120,000 tonnes per annum (Table 6.1). Meanwhile industrialization and increasing traffic have had considerable effects on emissions, as have fuel use changes and emission control efforts. Thus it is not possible to give a valid emission value for 1990.

The major sources of SO_2 are light and heavy industries using fuel oil and domestic lignite, the latter having a very high sulphur content of about 2.8 per cent. Diesel vehicles probably contribute less than 5 per cent of total SO_2 emissions (approximately 5,000 tonnes per annum), but their impact on roadside concentrations of SO_2 is certainly high.

It can be assumed that, with increasing industrialization and growing fuel consumption, SO_2 emissions will continue to grow in parallel.

Ambient Concentrations Concentrations of SO_2 in ambient air are very low. This is due to the meteorological conditions favouring the dispersion of industrial emissions outside metropolitan Bangkok. The WHO guideline maximum of 60 μg m^{-3} as an annual mean is not exceeded at any of the GEMS or ONEB monitoring stations. The WHO 98 percentile short-term guideline value of 150 μg m^{-3}, not to be exceeded on more than seven days per annum, is also met regularly.

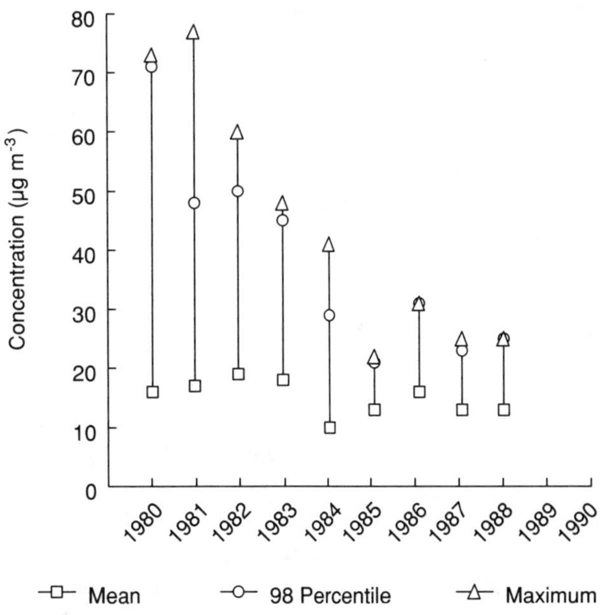

Figure 6.2 *Annual mean 98 percentile and annual maximum sulphur dioxide concentrations at the GEMS site*

Source: GEMS data

The data from the GEMS stations as given in Figure 6.2 show that the SO_2 annual maxima decreased from values of 50–80 μg m^{-3} in 1981–1983 to values of about 20 μg m^{-3} in 1986–1989. Similar concentrations were monitored at the ONEB stations. In 1989, the maximum 24-hour mean SO_2 concentration as measured by ONEB was 180 μg m^{-3}.

Sulphur dioxide levels have remained low since 1986. They will probably decrease further in 1993 when sulphur content in diesel fuel will be lowered to 0.5 per cent.

Table 6.1 *Estimated Total Emissions, 1980 and 1990*

Pollutant	Emissions ($t\ a^{-1}$)	
	1980[a]	1990[b]
SO_2	120,000	n.a.
SPM	40,000	80,000
NO_x	30,000	60,000
CO	120,000	280,000
Pb	20,000[c]	1,000

[a] Faiz et al., 1990
[b] Estimates from this report
[c] This figure is certainly too high
n.a. No data available

Sulphur dioxide pollution effects on the acidity of rain was reported by Khan (1980): mean pH values ranged from 6.32 to 5.57 with a minimum pH value of 4.45. These data indicate that rain acidity is moderate compared with the acidity of rain in more industrialized countries. This is probably due to a combination of the rather low SO_2 emissions and heavy tropical rainfall.

Suspended particulate matter

Emissions There are natural sources of particulate matter (such as wind-blown dusts) as well as anthropogenic sources (e.g., motor vehicle exhaust, industrial particulates). Most natural sources of SPM are not quantifiable; thus a comprehensive SPM emission inventory cannot be given. Detailed analyses of SPM samples indicated that in suburban residential areas only 30 per cent of SPM come from man-made activities, whereas at roadside locations 70–90 per cent of SPM are anthropogenic (ONEB, 1989).

Anthropogenic SPM emissions in Bangkok have been estimated to be about 40,000 tonnes per annum in 1980 (Faiz et al., 1990, Table 6.1). Since then, emissions have probably increased by a factor of two. Yet, given the actual ambient concentrations (see below), this figure is still likely to be a very low estimate.

The main sources of anthropogenic SPM are diesel-engine exhausts and emissions from light industry. Anthropogenic SPM emissions will grow in parallel with increasing industrialization and motor vehicle traffic, unless stringent pollution control measures are enforced.

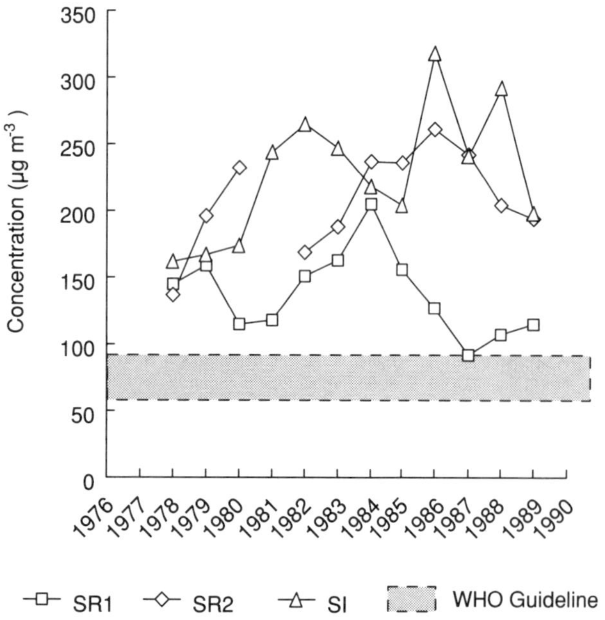

Figure 6.3 *Annual mean suspended particulate matter concentrations at GEMS sites*

Source: GEMS data

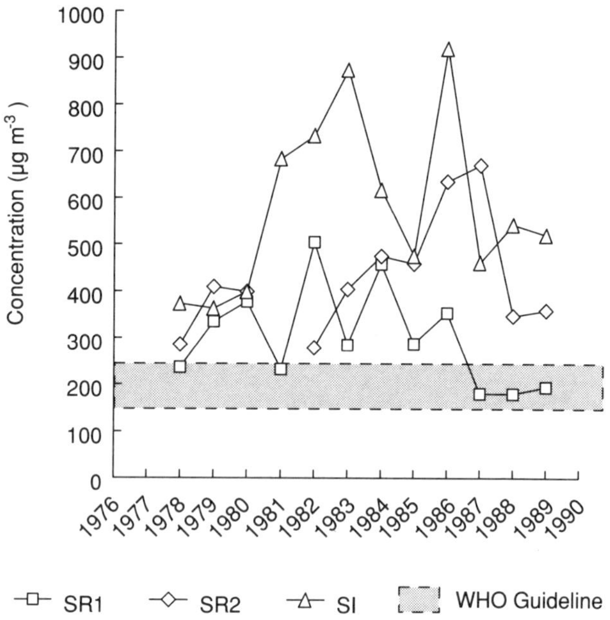

Figure 6.4 *Annual 98 percentile suspended particulate matter concentrations at GEMS sites*

Source: GEMS data

Ambient Concentrations Results from the GEMS and ONEB monitoring networks show that SPM is a serious air quality problem. Annual arithmetic means of SPM monitored at three GEMS sites since 1978 are shown in Figure 6.3. In the last ten years these values were in the range of 100–200 µg m^{-3} which is above the WHO annual mean guideline value of 90 µg m^{-3}. Comparison of GEMS data and ONEB data is complicated: ONEB annual means are expressed as geometric means in order to apply the Thai SPM air quality standard of 100 µg m^{-3} as the annual geometric mean. ONEB geometric annual means were reported to be about 100–140 µg m^{-3} from 1983–1989. As geometric means are about 10 to 20 per cent below arithmetic means computed for a given set of data, the ONEB data are consistent with the GEMS results.

As expected from the long-term means, the short-term SPM air pollution situation is also serious. GEMS data show that the WHO 98 percentile guideline of 230 µg m^{-3}, not to be exceeded on more than seven days per annum, is exceeded in two of the three monitoring sites (Figure 6.4). Data from ONEB and GEMS sites are similar. For example, during 1989 the maximum SPM value recorded by ONEB was 590 µg m^{-3} (at the Ban Somdet site), while the maximum value at a GEMS site (Occupational Health Centre) was 520 μg m^{-3}.

Besides regular ambient air quality monitoring, ONEB also monitors roadside air pollution in selected locations in order to assess traffic-related SPM exposures. Results show that in 1985–1987 annual geometric means ranged from 200–650 μg m^{-3}. The maximum 24-hour mean reported in 1988 was 1,400 μg m^{-3}, which is over four times higher than the Thai 24-hour standard of 330 μg m^{-3}.

In summary, particulate matter is a major air pollution problem in Bangkok. Even in suburban areas SPM concentrations exceed WHO guidelines. The situation is even more severe in traffic-exposed locations, where SPM concentrations are well above both the WHO and Thai national standards.

Particle size analyses of Bangkok SPM samples have shown that more than 60 per cent of particulates are smaller than 10 μm, particularly in the range of 0.6–1 and 5–7 μm. This indicates the seriousness of the Bangkok SPM problem as the particles fall into the inhalable size range (ONEB, 1989).

An "Action Plan to Solve Air and Noise Pollution Problems from Land and Water Transport" was introduced by ONEB in order to reduce black smoke emissions from motor vehicles by 1990.

Lead

Emissions In the absence of lead-emitting industrial sources (e.g., lead (Pb) smelters) almost all Pb emissions come from Pb additives in petrol.

Lead in commercial petrol was reduced from 0.84 g l^{-1} to 0.45 g l^{-1} in 1984 and to 0.4 g l^{-1} in 1989. From these numbers it can be estimated that present Pb emissions are in the range of 500–1,000 tonnes per annum. The Pb emissions of 20,000 tonnes per annum as reported by the Faiz et al. (1990) seem to be very high.

Emissions should drop significantly when the Pb content in petrol is reduced to 0.15 g l^{-1} in 1993.

Ambient Concentrations Data from the GEMS sites show that annual arithmetic mean airborne Pb concentrations are below the WHO long-term guideline of 1 μg m^{-3}. The data reported by ONEB indicate similar annual mean Pb concentrations ranging from 0.2– 0.7 μg m^{-3}. However, roadside monitoring results revealed much higher values: during 1985–1987 five of eight stations showed annual mean Pb concentrations above 1 μg m^{-3} with a maximum of 2.5 μg m^{-3} at one site (Figure 6.5). In 1988 an annual mean value of 4 μg m^{-3} was reported for a temporary roadside site.

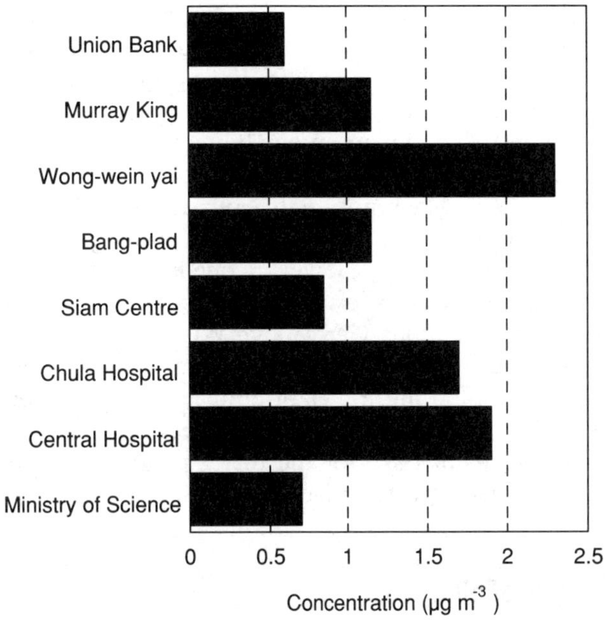

Figure 6.5 *Mean kerbside lead concentrations in Bangkok, 1985–1987*

After ONEB, 1989

Lead Exposure As in most urban areas, special attention has to be given to the potentially toxic levels of Pb. Blood Pb levels in children living near the Makasan station range from 15–28 μg dl^{-1}, indicating cause for concern. A study of blood Pb levels in bus drivers before and after the introduction of low Pb petrol in 1993 is being planned as a GEMS/HEAL study by the Ministry of Health, Division of Toxicology. The reduction of Pb levels in petrol should result in a significant reduction of Pb levels in blood.

Carbon monoxide

Emissions No comprehensive inventories of carbon monoxide (CO) emissions in Bangkok are available. Faiz et al. (1990) estimated CO emissions to be 120,000–160,000 tonnes per annum in 1980 and are projected to be 420,000 tonnes per annum in 2000. An interpolation of these figures would indicate 1990 emissions of 280,000 tonnes per annum.

In the absence of domestic heating nearly all CO emissions result from incomplete combustion of fuels in motor vehicles. Thus, CO emissions will increase with growing motor vehicle traffic unless effective emission controls are implemented. Such controls are part of the "Action Plan to Solve Air and Noise Pollution Problems from Land and Water Transport" which aims to control CO emissions from motor vehicles.

Ambient Concentrations The long-term level of CO in residential areas is generally around 1 mg m^{-3}. The hourly CO levels at the ONEB fixed monitoring sites are all below the WHO one-hour guideline of 30 mg m^{-3}. The maximum one-hour value monitored during 1986–1989 was 27 mg m^{-3} at the Chankasem site in 1989. Maximum eight-hour mean CO levels were also below the WHO eight-hour mean guideline value of 10 mg m^{-3}. However, one-hour and eight-hour mean CO levels reported at the roadside monitoring stations were above WHO guidelines at several stations. For example, a 49 mg m^{-3} one-hour mean and a 30 mg m^{-3} eight-hour mean value were recorded. This is important with respect to an assessment of health risks for the many people living or working near streets with heavy traffic.

Oxides of nitrogen

Emissions No comprehensive inventories for oxides of nitrogen (NO_x) emissions in Bangkok are available. Faiz et al. (1990) estimated NO_x emissions to be around 30,000 tonnes per annum in 1980, 30 per cent of which comes from motor vehicle traffic. This estimate seems to be rather low. Given the rate of industrialization, growing traffic and the introduction of natural gas-fired power plants, emissions probably doubled by 1990. Yet, in the absence of relevant data it is not possible to give a valid emission estimate for the 1990 situation.

Ambient Concentrations NO_x levels are routinely reported as nitrogen dioxide (NO_2) equivalents. The NO_2 levels reported at the GEMS sites are below the WHO 24-hour guideline of 150 μg m^{-3} as a 24-hour mean. ONEB began monitoring of NO_2 in 1987, and all values reported by three ONEB stations are below the Thai one-hour standard of 320 μg m^{-3}. The maximum one-hour mean reported in 1989 was 270 μg m^{-3} at the Saovabha site.

Ozone

Ambient Concentrations Ozone (O_3) is not a problem in the Bangkok metropolitan area because of the year-round monsoon winds which prevent the buildup of pollutants. Highest concentrations were found in the March–May season, when solar radiation is strongest. The maximum one-hour mean O_3 measured in 1989 at the ONEB sites was around 100 μg m^{-3} which is below the WHO guideline value of 200 μg m^{-3}. However, there may be areas of high O_3 concentration downwind of metropolitan Bangkok where O_3 is not monitored.

6.4 Conclusions

Air Pollution Situation Generally, the main air pollution problem in metropolitan Bangkok results from the very high SPM levels. In traffic-influenced areas, a large portion of the SPM consists of man-made respirable particles and could have potentially serious health effects on people living or working near streets with heavy traffic.

Most other air pollutants, like SO_2, CO, NO_2 and O_3, have low concentrations in most urban areas. However, the air pollution situation is more severe in roadside locations influenced by heavy traffic. Traffic-related pollutants, like SPM, Pb and CO, regularly exceed WHO guidelines at roadside monitoring sites.

Main Problems The lack of a valid emission inventory is a major problem hindering effective air pollution control planning. Only if the sources of pollutants are known sufficiently can cost-effective control measures be established and long-term air quality management plans implemented.

From the monitoring data it must be concluded that the main pollution problem in Bangkok stems from traffic, especially from the large number of cars with no emissions control devices. Light industry and domestic activities contribute relatively little to the pollutant load in the Bangkok atmosphere.

Control Measures It has been estimated that emissions from traffic will more than double between 1980 and 2000 because of the increase in the number of vehicles. Several attempts have been made to control traffic emissions. Through the "Action Plan to Solve Air and Noise Pollution Problems from Land and Water Transport" which was adopted in 1987, smoke and CO emissions from motor vehicles have been controlled since 1990. The plan provides manpower and resources for enforcement of applicable vehicle emission standards. In addition, it includes measures to reduce import taxes on pollution monitoring equipment and to allow certified private sector establishments to inspect vehicles.

As new vehicles are introduced they will replace some of the older vehicles which have "dirtier engines". However, the costs of modern emission control as currently enforced in the USA, Japan and Europe may be beyond the capacity of the Thai economy. Gradual introduction of more stringent standards and reliance on inspection and maintenance could be used to mitigate the expected emission increase. Furthermore, construction of a second expressway through Bangkok and of an elevated road to the airport which will be completed in 1995 will temporarily ease traffic congestion. An elevated municipal railway was planned for 1991 and should be completed by 1996. However, during this period of intense construction activity, traffic will be even more congested than it is today.

It is not possible to predict future pollutant concentrations from emission predictions because the traffic volume is almost at maximum capacity, and the central business area is saturated. Traffic volume will rise primarily in the surrounding areas where population will increase and development will take place. It is expected that SPM and CO concentrations will continue to rise and that exceedences of the CO eight-hour standards will begin to occur in the areas where the standard is at present met.

Industrial air pollution emissions are to be controlled through emission standards being drafted by a sub-committee of all concerned agencies and industries. ONEB requires submission of Environmental Impact Assessment Reports for acceptance prior to the issue of operating permits for new plants and older established plants will be required to submit plans to meet the new standards.

Table 6.2 *Number of reported cases of Pneumonia and Influenza in Bangkok*

	Number of cases	
Year	Influenza	Pneumonia
1981	2,971	n.a.
1982	1,835	n.a.
1983	2,286	350
1984	2,875	395
1985	3,001	595
1986	2,715	1,763
1987	2,733	4,288
1988	2,240	5,765
1989	1,707	5,646
1990	874	3,384

n.a. No data available
Sources: Ministry of Public Health, 1989, 1990

Health Effects At the present time no studies have been reported on the effect of air pollution on the respiratory health of Bangkok residents. The Ministry of Health collects statistics on incidence of respiratory diseases which could be influenced by air pollution exposures, but no attempt has been made to determine the relation of diseases and pollution levels in various areas. Numbers of reported cases shown in Table 6.2 indicate the order of magnitude of the number of people potentially affected by air pollution. A recent report (USAID, 1990) estimated that each year there are 10–50 million person-days of restricted activity for respiratory reasons in Bangkok which are not reported.

6.5 References

Division of Land Transport 1989 *Number of Vehicles Registered in Thailand,* Division of Land Transport, Thailand.

Faiz, A., Sinha, K., Walsh, M. and Varma, A. (Eds), 1990 *Automotive Air Pollution: Issues and Options for Developing Countries,* World Bank Policy and Research Working Paper WPS 492, World Bank, Washington DC.

Khan, M. S. 1980 The Relationship Between Acid Content of Particulates and Rainfall in Bangkok, *J. Environ. Sci. Health* **A15**, 561–572.

Ministry of Public Health 1989 *Reportable Cases and Deaths by Disease and Poisoning in Thailand; Preliminary Annual Summary,* Division of Epidemiology, Ministry of Public Health, Bangkok.

Ministry of Public Health 1990 Personal Communication, Division of Epidemiology, Ministry of Public Health, Bangkok.

ONEB 1989 *Air and Noise Pollution in Thailand 1989,* No. 07–03–33, Office of the National Environment Board, Bangkok.

UN 1989 *Prospects of World Urbanization 1988,* Population Studies No. 112, United Nations, New York.

USAID 1990 *Ranking Environmental Health Risks in Bangkok, Thailand,* US Agency for International Development, Washington DC.

7

Beijing

7.1 General Information

Geography Beijing is the capital city of China. It is located at the north-western border of the Greater North China Plain at 39°48′N latitude and 116°28′E longitude at an altitude of 44 m above mean sea level. The Mongolian Plateau is to the north of Beijing and the Yanshan Mountain range is to the north-east. The Beijing municipality includes 800 km^2 between two rivers, the Yongding and the Chaobai, both emptying into the Bohai Sea 160 km to the south-east (Figure 7.1).

The municipality consists of 10 municipal districts and nine counties comprising 16,800 km^2. Four of these districts (0.4 per cent of the metropolitan area) are within the central walled city. The inner city plan is based upon a north–south, east–west grid pattern. This area contains the government buildings, commercial centre and the old, densely populated residential districts. The three districts which make up the near suburbs and surround the walled city contain the majority of the factories and schools. The far suburbs make up 92 per cent of the metropolitan area and provide raw materials and food for the city.

Demography Beijing's population was 8.3 million in 1970 and 9.3 million in 1985. It was estimated that the population of the Beijing municipality in 1990 was 9.74 million, of which 6.1 million were spread over 2,740 km^2 and a central population of 2.4 million was concentrated in 87 km^2. This is equivalent to an urban population density of 27,600 people per km^2 in the central area. A small proportion live in the rural far suburbs. By the year 2000 Beijing is projected to have a total of 11.47 million inhabitants (UN, 1989). The city has 120 million m^2 of floor space. The municipality has 2.7 million homes, with an average of 3.6 people per household (Government of China, 1989). Thirty-two per cent of the population are between 20 and 35 years of age. The net internal annual growth rate of Beijing is 11.3 persons per 100,000, with a birth rate of 17.8 and a death rate of 6.5. Migration to/from the city from/to rural areas will be in addition to this rate.

Climate Beijing has a continental warm summer climate. The mean annual temperature in Beijing is 11.6°C and the mean total annual precipitation is 584 mm.

Winters are cold; the mean monthly temperature for January is – 4.6°C. However, Beijing experiences large daily temperature fluctuations, and minimum temperatures below –20°C are not uncommon. Precipitation is minimal in winter and the mean in December is about 1 mm.

Summers are hot and humid. The mean monthly maximum of 26.1°C is reached in July when a maximum hourly temperature of 40.6°C has been recorded. Maximum precipitation is 209 mm, which occurs in August.

The wind is predominantly from the north, especially from August to March. Southerly and south-westerly winds are most frequent from April to July.

Industry There are over 5,700 industrial enterprises operating in Beijing, including 24 electric power generating plants, 28 non-ferrous and 25 ferrous metal smelters, 18 coking plants, 194 chemical plants and 483 metal products factories (including electroplating) in the metropolitan area.

The municipal government has attempted to control industrial pollution by relocating certain industries, such as electroplating, away from the central area, and new enterprises are actively discouraged from being situated in the city. The major iron and steel complex of Shijing, 24 km to the west of the city, has both blast and electric furnaces and is a major source of industrial emissions. Mechanical and electrical factories are located to the south-east. Textiles are also a major industry in Beijing. Tourism and banking are important commercial activities.

Annual coal consumption in Beijing is 21 million tonnes. About 30 per cent of coal is used in industry, 27 per cent for coking, 23 per cent for residential heating, 14 per cent for electricity generation and 6 per cent for other uses. Of the 13.5 billion kWh consumed annually, 10.5 billion kWh are locally generated. Virtually all the city's electrical power generation is based on coal combustion (Krupnick and Sebastian, 1990; Wu et al., 1985). Beijing coal is of low sulphur

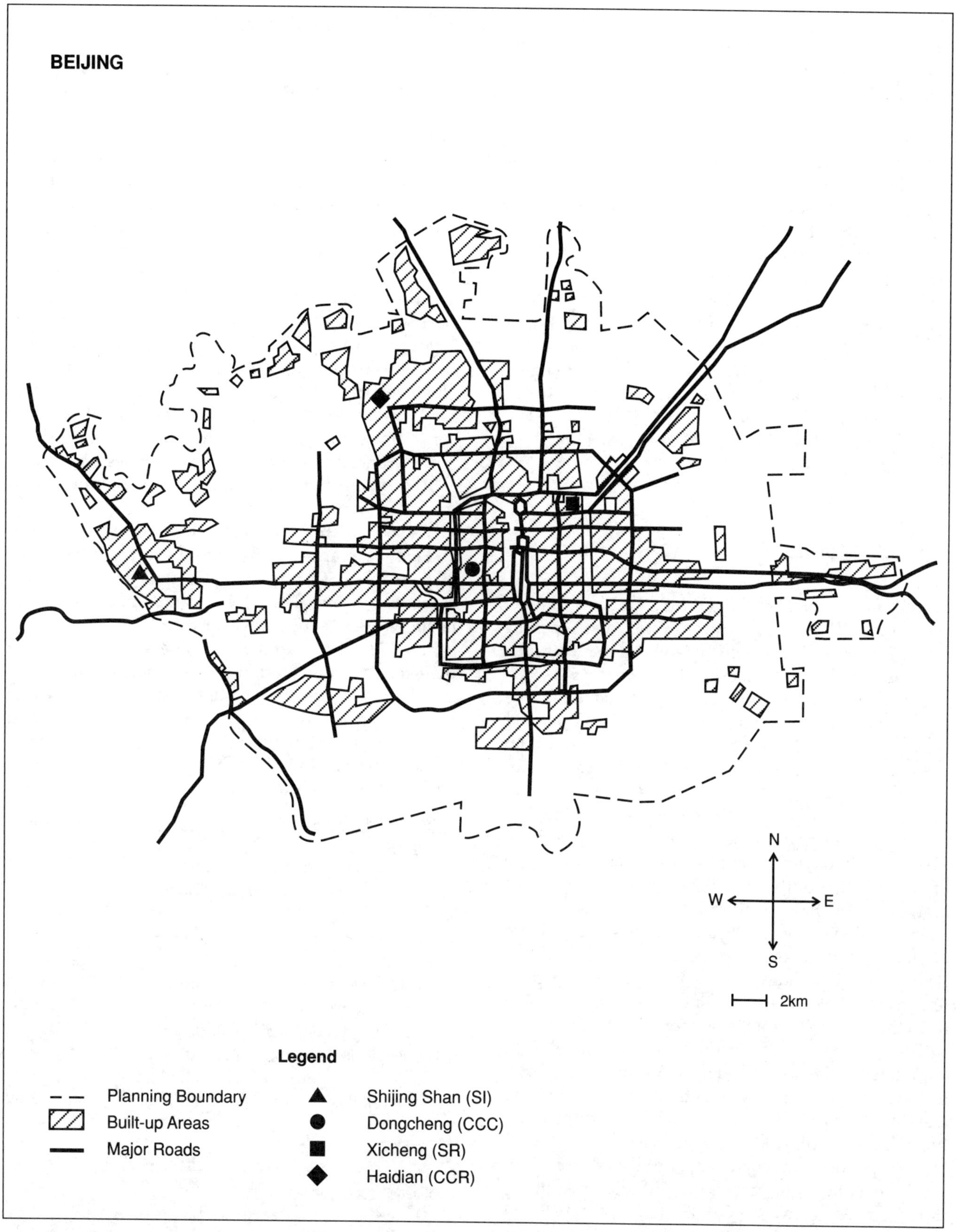

Figure 7.1 *Sketch map of Beijing showing the location of GEMS/Air stations*

content. Coal for domestic or residential combustion contains 0.3 per cent sulphur, and coal for small-scale industry and commercial establishments contains 0.5–1.0 per cent sulphur.

Transport Short travelling distances and flat terrain make bicycles the most popular transport. In 1988, four million bicycles were registered in Beijing, and public transport was provided by 4,100 buses and 12,500 taxis. In 1986 the total number of motor vehicles in Beijing was approximately 200,000 and in 1991 approximately 308,000. Although relatively small in number, many of these vehicles are highly polluting. The "Beijing" brand jeeps and the "Shanghai" brand cars, which together make up approximately 50 per cent of all light duty vehicles, have very low engine efficiencies and travel only about 6 km per litre of petrol. China's motor vehicle fleet (3,927,000 in 1988) grew at a rate of 13.9 per cent between 1984 and 1988 (Krupnick and Sebastian, 1990).

7.2 Monitoring

In the framework of the GEMS/Air programme, suspended particulate matter (SPM) and sulphur dioxide (SO_2) measurements have been made at four sites since 1981. Daily data for the period 1981–1989 are available (means, 98 percentile and maxima). The four Beijing GEMS/Air monitoring stations operated by the Ministry of Health are classified as suburban industrial (SI), suburban residential (SR), centre city commercial (CCC) and centre city residential (CCR). The SI site is within 3 km of a steel plant. GEMS/Air quality monitoring in Beijing is performed on a 15-day-on 15-day-off schedule, resulting in coverage of half the days of the year, while possible periods of high pollution may go unsampled.

In addition to the GEMS/Air network, the Chinese National Environmental Protection Agency (NEPA)/ Beijing Municipal Research Institute (BMRI) operates ten monitoring sites in Beijing. However, no data from this network are available for this report.

The GEMS data collection procedures in Beijing were reviewed during a WHO site visit (Mage, 1988) to the Institute of Environmental Health Monitoring (IEHM). At that time, a GEMS flow calibrator was not functioning leading to the conclusion that "recent total suspended particulates data collected from Beijing . . . may be of questionable validity". In addition, discrepancies between GEMS and NEPA/BMRI measurements were noted and suggestions for comparisons and standardizations were made.

7.3 Air Quality Situation

Sulphur dioxide

Emissions Beijing has made a comprehensive emission inventory and created a series of maps of emission densities taking into account planned industrial expansion, new industrial areas and population growth in new residential areas. However, the inventory was not available for this report.

An emission estimate for SO_2 was prepared for Beijing by the World Bank (Krupnick and Sebastian, 1990). According to these estimates, SO_2 emissions in Beijing were about 526,000 tonnes per annum in 1985. However, as may be seen in Table 7.1, these estimates for 1985 are not comprehensive and do not consider some emission sources, e.g., motor vehicles, heating boilers, and commercial stoves. Besides that, since 1985 growing industrialization has certainly caused increased emissions.

The dominant sources of SO_2 are heavy industries and power plants. All industrial sources account for 87 per cent of SO_2 emissions. Of these, more than half come just from 24 power plants plus 18 coking facilities. Household stoves account for yearly emissions of 73,000 tonnes of SO_2 (13 per cent of total), but their relative contribution is certainly much higher in the cold winters. Implementation of processes to reduce further the sulphur content of coal and promotion of briquetting for more efficient burning of domestic coal have been initiated.

Table 7.1 *Sulphur dioxide emission estimates for Beijing, base year 1985*

Source	Sulphur dioxide ($t\ a^{-1}$)	Per cent*
Industry	203,000	39
Heating boilers	n.a.	
Commercial stoves	n.a.	
Household stoves	73,000	14
Power plants	126,000	24
Coking	124,000	24
Motor vehicles	n.a.	
Total	526,000	

n.a. No data available
*Owing to rounding, percentage totals may not equal 100%
After Krupnick and Sebastian, 1990

Diesel vehicles, which are not included in the emission inventory, probably contribute very little to total SO_2 emissions, but their impact on roadside concentrations is likely to be high. It can be assumed that with increasing industrialization and growing fuel consumption SO_2 emissions will continue to grow in parallel.

Ambient Concentrations Sulphur dioxide has been monitored at Beijing GEMS/Air sites since 1981. As shown in Figure 7.2, annual mean SO_2 concentrations have remained relatively constant during the past four years. The city centre stations (CCC and CCR) had annual means over 100 μg m^{-3} until 1989, and thus exceeded the WHO annual mean guideline values of 40–60 μg m^{-3}. The suburban stations (SI and SR) meet the WHO annual guidelines. The WHO 98 percentile guideline (150 μg m^{-3}, not to be exceeded on more than 2 per cent of the days) is exceeded especially at the city centre stations which reach daily means above 500 μg m^{-3} on 2 per cent of the days. The SI station still exceeds 250 μg m^{-3} on 2 per cent of the days. Only the SR station meets the long-term as well as the short-term WHO guidelines.

Figure 7.3 gives an example of the typical seasonal variation of SO_2. In this SI station the days with the highest SO_2 pollution (up to 250 μg m^{-3}) are between November and March, which is the heating period, when large amounts of coal are used for domestic heating. Despite industrial sources nearby, summer SO_2 levels are very low (up to 30 μg m^{-3}). Maximum hourly concentrations of SO_2 occur between 0700–0800 hours and 1800–2000 hours and correspond with domestic cooking periods.

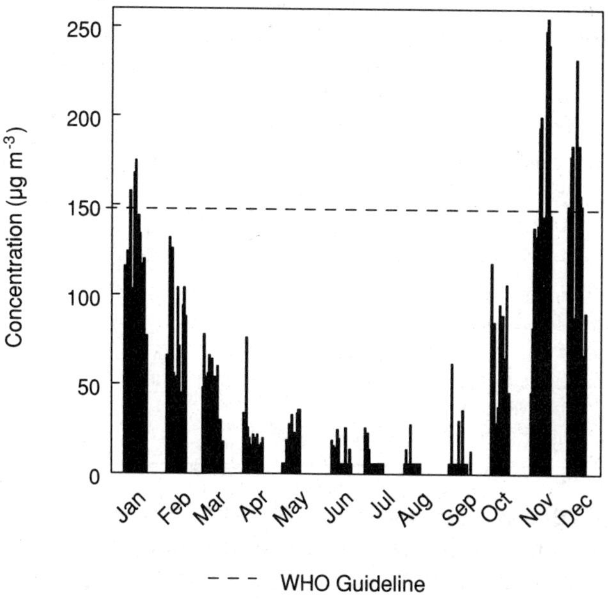

Figure 7.3 *Daily sulphur dioxide concentrations at Shijing Shan (SI), 1989*

Note: 16 days per month sampling frame. Zero levels indicate days when no measurements were taken.
Source: GEMS data

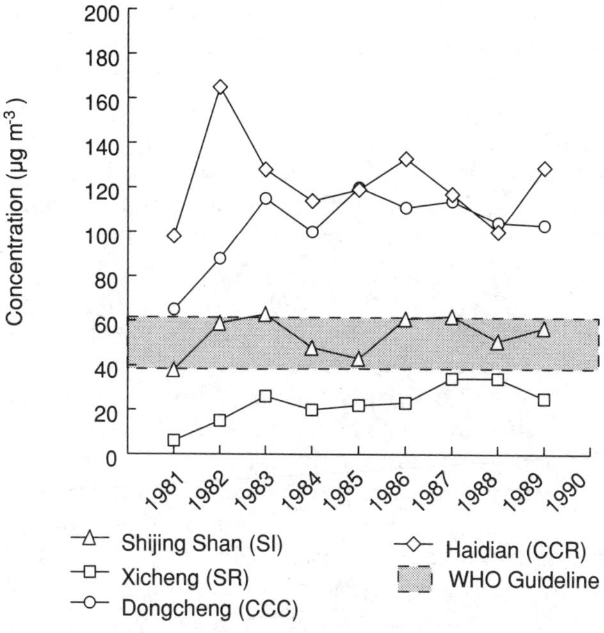

Figure 7.2 *Annual mean sulphur dioxide concentrations at GEMS/Air stations*

Source: GEMS data

Suspended particulate matter

Emissions Both natural emission sources of particulate matter (such as wind-blown dusts) and anthropogenic sources (e.g., motor vehicle exhaust, industrial particulates) are of importance in Beijing. Most natural sources of SPM are not quantifiable; thus a comprehensive SPM emission inventory cannot be given. Beijing is subject throughout the year to severe dust fallout from dust storms originating in the western plains. These dust storms can be especially severe in the late winter and early spring. During summer, 60 per cent of SPM is estimated to arise from natural sources, while during winter, 40 per cent of SPM is of natural origin (Zhao and Sun, 1986).

The detailed emission inventory established by the Beijing authorities was not available for this report. However, anthropogenic SPM emissions in Beijing have been estimated by the World Bank to be about 116,000 tonnes per annum in 1985 (Table 7.2; Krupnick and Sebastian, 1990). However, this inventory was not comprehensive and did not consider some emission sources, e.g., motor vehicles, power plants,

***Table 7.2** Suspended particulate matter (TSP) emission estimates for Beijing, base year 1985*

Source	TSP (t a^{-1})	Per cent
Industry	68,000	59
Heating boilers	34,000	29
Commercial stoves	7,600	7
Household stoves	6,000	5
Total	115,600	

Note: Does not include emissions from power plants, coking or motor vehicles
After Krupnick and Sebastian, 1990

or coking. Thus these figures must be considered to be a lower estimate of the actual situation. This inventory shows that the main sources of anthropogenic SPM are emissions from industrial stationary sources which account for 59 per cent of anthropogenic SPM emissions. Heating boilers (which include residential central and district heating) and household stoves, which contribute 34 per cent of total SPM emissions, are important sources of emissions especially during the heating period. Due to low emission heights and high emission densities in the heating period, the contribution of this source to the ambient air quality in winter seems to be high. All emissions sectors use local coal with a very high ash content which leads to inefficient burning and high SPM emissions.

Trace metals are emitted primarily from coal combustion, although refuse incineration, motor vehicles and crustal dust contribute a small percentage. Most fluoride, chloride, nitrate, sulphate and ammonium ions found in particulate samples presumably result from coal combustion.

Ambient Concentrations Suspended particulate matter has been monitored by the GEMS/Air network in Beijing since 1981. High-volume samplers capture total suspended particulates (TSP) including respirable particles (less than 10 μm in aerodynamic diameter), and non-respirable particles (greater than 10 μm).

Annual mean SPM (TSP) levels at the GEMS/Air monitoring sites are shown in Figure 7.4. The WHO annual mean guideline of 60–90 μg m^{-3} is exceeded at all four stations. Even if one considered half the TSP to be wind-blown natural dust, the TSP from coal combustion alone would be above the WHO guideline value. There appears to have been a decrease in TSP from 1981 to 1986 with a levelling off from 1986 to 1989. However, as the sampling schedule for the GEMS/Air data was 15-days-on and 15-days-off, the

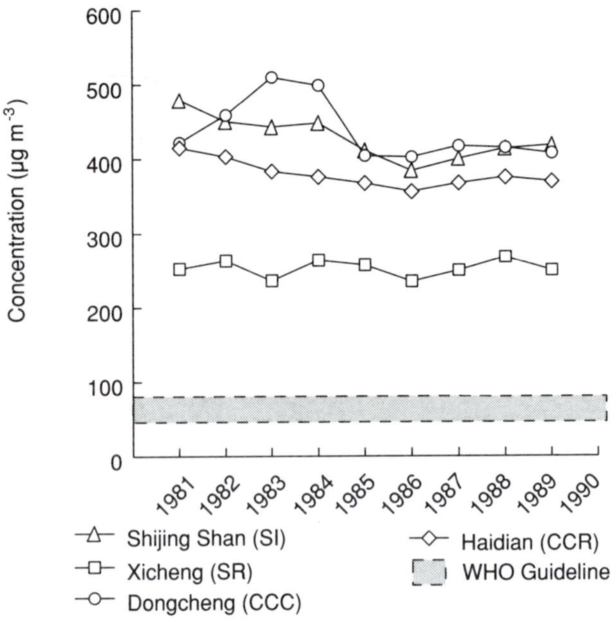

***Figure 7.4** Annual mean suspended particulate matter (TSP) concentrations*

Source: GEMS data

confidence intervals for these values are larger than if an every-other-day schedule had been used.

The SPM (TSP) problem of Beijing is also reflected in the daily TSP values (Figure 7.5). The WHO 98 percentile guideline of 230 μg m^{-3} (not to be exceeded on 2 per cent of the days) was exceeded on almost all days of the year 1989 in the SI site as shown in Figure 7.5. Highest daily TSP measurements were above 1,000 μg m^{-3}. The seasonal variations of TSP are not as evident as those of SO_2 because of the natural dust fraction. However, there are higher daily mean concentrations in the winter heating period. It is clear that the high SPM (TSP) pollution loads in winter represent a severe health risk for susceptible people, especially if associated with high SO_2 concentrations.

A study of particle chemistry during the period 1982–1983 (Wu et al., 1990) showed the major enrichment of the urban particulate from coal combustion, industry and a surprisingly high contribution from automotive sources if all lead (Pb) enrichment is from lead additives to automotive fuel.

A study of personal exposures with various SPM size-specific samplers was performed in Beijing during summer and winter of 1985 (WHO, 1985). In winter the time-weighted mean concentration of particles less than 10 μm was 360 μg m^{-3}, and that of particles less than 3.7 μm was 190 μg m^{-3}, but in summer the corresponding values were 180 μg m^{-3} and 68 μg m^{-3}. The higher ratio of large to small particles in summer is consistent with the view that a higher

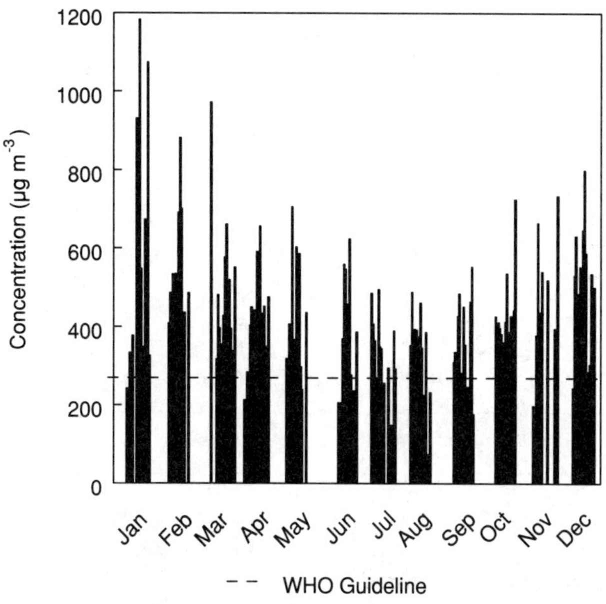

***Figure 7.5** Daily suspended particulate matter (TSP) concentrations at Shijing Shan (SI), 1989*

Note: 16 days per month sampling frame. Zero levels indicate days when no measurements were taken.
Source: GEMS data

percentage of large particles from wind-blown dust are present in summer months.

Airborne particulate samples are determined for the highly carcinogenic polycyclic compound Benzo-[a]-Pyrene (BaP) once per month by the Beijing Municipal Environmental Monitoring Centre (BEMC). The maximum of those 24-hour mean samples from 1984–1990 was 30.5 ng m^{-3} BaP (Liang, 1991). This is much lower than in winter 1980–1981, when BaP concentrations of 266 ng m^{-3} were found in urban areas, compared with 4.9 ng m^{-3} in clean rural and residential areas (Zhao and Sun, 1986).

Lead

Emissions No emission estimates were available for this report. There is a 0.8 g l^{-1} limit for Pb content in petrol, and 80 per cent of the petrol consumed is Pb free. Furthermore, traffic in Beijing is still limited, so emissions of Pb in Beijing are probably small and should be well below 100 tonnes per annum. However, Pb is also emitted from Pb smelters and Pb products industries. There is no detailed information on industrial production, but it is likely that at least one of the 28 non-ferrous metal smelters produces and emits Pb.

Ambient Concentrations Lead concentrations are not monitored in the GEMS/Air network. Lead analyses are performed by the BMRI, but these data were not available for this report. A study by Wu et al. (1990) reported Pb concentrations measured in 1982 and 1983 using a high volume sampler in Beijing at three types of stations. The annual mean Pb concentrations were 0.15 μg m^{-3} at a rural site, 0.35 μg m^{-3} at an industrial site and 0.5 μg m^{-3} at an urban site, with all values well below the WHO guideline value of 1 μg m^{-3} as an annual mean.

A personal monitoring study (Vahter and Slorach, 1990) included measurements of airborne Pb (and cadmium (Cd)) concentrations in Beijing. The mean concentration of Pb measured was 0.116 μg m^{-3}, with a standard deviation of 0.02 μg m^{-3}. The range of daily average Pb concentration values was 0.02– 0.32 μg m^{-3}. The average Cd exposure was 1.2 ng m^{-3} with a range of daily averages from 0.3 to 3.7 ng m^{-3} which is within the WHO guideline of 5 ng m^{-3} as an annual mean. The results of the study also showed an increase in blood Pb from 5.9 μg dl^{-1} (Vahter, 1982) to 7.3 μg dl^{-1}. This is an indication of increased Pb emissions in Beijing.

Carbon monoxide

Emissions An emission inventory for carbon monoxide (CO) in Beijing was not available for this report. The World Bank estimates that, owing to few mobile sources, CO should be a minor problem (Krupnick and Sebastian, 1990). However, it must be anticipated that the many small sources of coal combustion would emit considerable amounts of CO. The effect of small domestic combustion sources on the ambient air quality in winter could be quite high due to low emission heights and high emission densities in the heating period.

Ambient Concentrations Carbon monoxide is not measured by the GEMS/Air network. BMRI measures ambient CO routinely, but these data were not available for this report. The World Bank reports that a mean CO concentration of 3.4 mg m^{-3} was monitored in the fourth quarter of 1988, which is high compared with the WHO eight-hour guideline of 10 mg m^{-3}.

A study of CO personal exposures of non-smokers in Beijing (WHO, 1985) reported that seasonal mean outdoor CO exposures were of the order of 1.9 mg m^{-3} (1.6 ppm) in summer and 5 mg m^{-3} (4 ppm) in winter. Personal (indoor) exposures, however, averaged 2.8 mg m^{-3} (2.4 ppm) in summer and 17.9 mg m^{-3} (15.4 ppm) in winter. These data reflect the relative absence of traffic-generated CO in Beijing and the high indoor exposures due to domestic cooking with coal briquettes and other fuels.

Table 7.3 *Carbon monoxide personal monitoring record in winter 1985 (ppm)*

Time	18 Jan. Fri.	19 Jan. Sat.	20 Jan. Sun.	21 Jan. Mon.	22 Jan. Tue.	23 Jan. Wed.	24 Jan. Thu.
07:00	55.0	45.5	58.5	66.0	40.0	53.0	40.2
08:30	5.0	3.0	80.0	6.5	5.0	6.0	7.2
11:30	4.5	2.8	86.0	8.0	3.8	5.5	5.0
12:30	72.0	48.0	81.0	77.0	41.0	54.0	14.5
13:30	5.0	2.5	71.5	7.5	4.0	3.0	5.2
16:30	5.0	3.0	52.0	6.0	2.5	3.2	2.0
18:30	71.0	30.5	72.0	57.5	61.0	96.0	12.0
21:00	60.0	80.5	68.5	71.0	76.0	59.0	16.0

After WHO, 1985

Indoor CO levels as high as 104 mg m^{-3} (90 ppm) have been measured during cooking periods as shown in Table 7.3. This indicates high personal exposure for the most seriously exposed people. The importance of domestic coal combustion for indoor air quality is further underlined by the fact that, in multi-storey buildings without central heating, the mean carboxyhaemoglobin (COHb) concentrations in human blood went from 0.4 per cent in summer to 4 per cent in winter, a tenfold increase. As the WHO guideline value of 10 mg m^{-3} (9 ppm) as an eight-hour mean is intended to prevent the attainment of 2.5–3 per cent COHb in sensitive individuals, the CO exposure from indoor sources is certainly higher than the WHO guideline and must be considered to be a potential major health problem.

Oxides of nitrogen

Emissions An emission inventory for oxides of nitrogen (NO_x) in Beijing was not available for this report. The World Bank estimates that NO_x emissions in Beijing should not be of major concern (Krupnick and Sebastian, 1990). This is because there are relatively few mobile sources and NO_x emissions from the combustion of high-ash coal in small sources would generally be low.

Ambient Concentrations Nitric oxide (NO) and nitrogen dioxide (NO_2) are monitored by BMRI, but these data were not available for this report. A study by Wu et al. (1990) reported monthly NO_2 means from filter badge measurements during the period March-December 1983. The measured annual mean values were 54 µg m^{-3} at an urban site, 41 µg m^{-3} at an industrial site, and 14 µg m^{-3} at a rural location. The maximum monthly NO_2 concentration was 80 µg m^{-3} at the urban site in July 1983. However, it is not clear from the data whether the WHO 24-hour guideline of 150 µg m^{-3} for NO_2 was exceeded on any of the days monitored.

Tang et al. (1988) reported NO_2 data for Beijing from short-term studies of photochemical smog in July 1986 and 1987. The maximum hourly value of NO_2 in 1986 was 76 µg m^{-3}. Measurements in 1987 were below the WHO 24-hour guideline of 150 µg m^{-3}. Consequently, NO_2 may not currently be a major air pollution problem in Beijing. However, the number of motor vehicles is increasing and other unmonitored locations may soon be approaching WHO guideline values.

Ozone

Emissions Emission inventories for NO_x and volatile organic compounds (VOC), the main precursors of ozone (O_3), were not available for this report. As stated above, emissions of NO_2 are probably low. Volatile organic compound emissions result from a variety of sources, including solvent and petrol evaporation, incomplete combustion processes and natural sources. In the absence of statistical industrial data the amount of VOC emissions in Beijing cannot be estimated.

Ambient Concentrations Ozone is measured by the BMRI, but these data were not available for this report. Ozone data from a research study in July 1986 and June 1987 are shown in Table 7.4 (Tang et al., 1988). Maximum hourly means at the monitoring sites were 158-262 µg m^{-3} (79–131 ppb) in June 1986 and 138–320 µg m^{-3} (69–160 ppb) in July 1987. The WHO one-hour mean guidelines of 150–200 µg m^{-3} (75–100 ppb) were exceeded at most monitoring sites. These high ozone concentrations were connected with a very high VOC to NO_x ratio, which indicates that the ozone production may be NO_x-limited.

Table 7.4 *Maximum hourly mean ozone concentrations*

Site	Ozone (ppb)	
	July 1986	June 1987
1	79	69
2	105	160
3	131	100
4	83	109
5	100	128
6	n.a.	102

n.a. No data available   After Tang et al., 1988

Consequently, an increase in NO_x emissions (e.g., because of a growth in the number of motor vehicles without control measures) can be expected to increase ozone production. This possibility might represent a serious problem for Beijing because automotive emission controls for reducing NO_x emissions require catalytic converters which are expensive to purchase and maintain.

7.4 Conclusions

Air Pollution Situation The main ambient air pollution problem in Beijing results from the very high SPM and SO_2 levels. Short-term and long-term means regularly exceed the WHO guidelines in large parts of the urban area. In the heating season (November–March) high SPM and SO_2 pollutant concentrations occur regularly. During that time, respirable acid aerosols from domestic coal combustion form an additional risk to human health. However, because 40–60 per cent of SPM are wind-blown dusts, measures to control anthropogenic SPM emissions have only limited effects.

For the other pollutants, lack of available routine monitoring data makes an assessment difficult. Carbon monoxide pollution in ambient air seems to be high, mainly in winter and indoor CO pollution from domestic coal combustion is extremely high and is potentially a major health problem. In summer, O_3 might exceed WHO short-term guidelines.

Nitrogen dioxide concentrations are relatively low in most urban areas, as are levels of Pb (Matsushita and Tanabe, 1990). Both pollutants are below the WHO guidelines.

Main Problems From the monitoring data it must be concluded that the main air pollution problem in Beijing stems from coal combustion. Suspended particulates and SO_2 are emitted in large quantities as a result of burning coal to supply the city with its energy needs. Coal is the source of 70 per cent of the energy in Beijing. Despite the sulphur content of coal used in Beijing being low (0.3–1 per cent), the input of poor quality coal (i.e., unwashed and unscreened coal) and the use of inefficient combustion technologies in industrial boilers and urban heating systems result in high emissions.

A specific pollution problem results from the many small domestic combustion sources. Although the amount of emissions is moderate, these sources contribute a high proportion of ambient air pollution because stacks are low and pollutant dispersion is limited. For instance, household stoves contribute 14 per cent of total SO_2 emissions, but about 38 per cent of ambient SO_2 concentrations (Krupnick and Sebastian, 1990).

Motor vehicle traffic, which is the major air pollution problem in many of the world's cities, still makes a relatively small contribution to the pollutant load in the Beijing atmosphere.

Control Measures Air pollution problems in Beijing are typical of rapidly growing industrialized cities. High levels of air pollutants are emitted by poorly controlled sources. In 1979 an Environmental Protection Law was created to control emissions (Ross and Silk, 1987), but SPM and SO_2 concentrations still exceed WHO guidelines and Chinese national environmental standards (see Appendix I).

In 1988, the Beijing Municipal Government (BMG) passed regulations to implement the 1987 Air Pollution Prevention Act that requires industry to monitor its own emissions and report them to the BMG. The BMG has been authorized to set stricter emission standards than the national standards and it has done so. Emission sources are given deadlines to meet these standards and if they are unable to do so they must either shut down or move from the area. As of 1988 only a few emission sources had problems with compliance.

Emission control programmes recently put into effect are as follows:

- increase the supply of coal gas and natural gas for industrial use;
- convert urban residential fuel from coal to liquified petroleum gas (LPG) and natural gas;
- require residential sources still using coal to burn briquettes and shaped coal to reduce emissions;

- develop central heating plants to replace distributed smaller boilers by a single large installation with emission controls;

- modify existing boilers where the location of industry makes central heating impractical, to reduce emissions by automatic feeding and ash removal, and all boilers with a capacity of greater than one tonne must install scrubbers; and

- pave dirt roads and plant trees, flowers and grass to reduce the wind-blown dust component of TSP.

Efforts to control air pollution are limited by the huge capital investments necessary for control technologies. Furthermore, the limitations of assimilative capacity of other environmental media must be considered. For instance, reducing air pollution may increase solid wastes. However, the disposal of ash from coal combustion is already a severe problem and may result in surface- or ground-water pollution: in Beijing, shortages of surface drinking water are compensated by ground-water withdrawals. However, the quality of such ground water is already suspect.

In spite of the use of environmental impact assessments for new projects, it is anticipated that emissions will continue to increase and that there will be difficulty in meeting the air quality standards. The problem is a financial one since the costs of pollution control equipment are considerable. As in the other cities in China, the development of air pollution control technology which is both cost-effective and efficient is a major concern.

In summary, Beijing has an effective monitoring and control programme for air pollution. Its problems are typical of an urban high population density area and the proposed emission controls and planning mechanisms should lead to the required improvements in air quality. However, the cost of emission controls will be a difficult factor to overcome in implementing the strategy of emission reductions being planned.

Health Effects Studies on the effect of air pollution on the respiratory health of Beijing residents were not available for this report. From the scientific data it must be concluded that the continued exposure of populations to combined daily levels of SO_2 and SPM (TSP) of the order of 500 $\mu g\ m^{-3}$ and above (for example, on a day like 19 January 1988 with a 24-hour mean of 490 $\mu g\ m^{-3}$ SO_2 and 810 $\mu g\ m^{-3}$ TSP at the CCR station) is associated with increased mortality of susceptible individuals. This is a likely consequence of the severe chronic effects on the population of exposures to high annual averages of SPM (TSP) and SO_2 which lead to development of pulmonary insufficiency in sensitive and ageing people followed by the premature death of the most susceptible. In addition, indoor CO exposures may be cause for concern.

7.5 References

Government of China 1989 *China Statistical Yearbook 1988*, Government of China, Beijing.

Krupnick, A. and Sebastian, I. 1990 *Issues in Urban Air Pollution: Review of the Beijing Case*, Environment Working Paper No. 31, World Bank, Washington, DC.

Liang, X. 1991 Personal communication, Beijing Municipal Environmental Monitoring Centre, Beijing.

Mage, D. T. 1988 *Air Quality Management: Beijing, People's Republic of China*, Unpublished WHO Assignment Report, World Health Organization, Geneva.

Matsushita, H. and Tanabe, K. 1990 Status of Air Pollution in Beijing, *Journal of Environmental Sciences (China)* **2**, 27–39.

Ross, L. and Silk, M. A. 1987 *Environmental Law and Policy in the People's Republic of China*, Quorum Books, New York.

Tang, X., Li, I., Chen, D., Bai, Y., Li, X., Wu, X. and Chen, J. 1988 *The study of photochemical smog pollution in China*, Proceedings of the Third Joint Conference of Air Pollution Studies in Asian Areas, Japan Society of Air Pollution, Tokyo, 238–251.

UN 1989 *Prospects of World Urbanization 1988*, Population Studies No. 112, United Nations, New York.

Vahter, M. and Slorach, S. 1990 *Exposure Monitoring of Lead and Cadmium: an International Pilot Study within the WHO/UNEP Human Exposure Assessment Location (HEAL) Programme*, United Nations Environment Programme, Nairobi.

Vahter, M. 1982 *Assessment of human exposure to lead and cadmium through biological monitoring,* Report prepared for UNEP/WHO by National Swedish Institute of Environmental Medicine and Department of Environmental Hygiene, Karolinska Institute, Stockholm.

WHO 1985 *Human Exposure to Carbon Monoxide and Suspended Particulate Matter in Beijing, People's Republic of China,* GEMS: PEP/85.11, World Health Organization, Geneva.

Wu, J., Wang, A., Huang, Y., Ma, C., Iida, Y., Daishima, S., Furuya, K., Kikuchi, T. 1985 *China: The Energy Sector,* World Bank, Washington, DC.

Wu, J., Wang, A., Huang, Y., Ma, C., Lida, Iida, Y., Daishima, S., Furuya, K., Kikuchi, T., Matsushita, H., Tanabe, K. 1990 Status of air pollution in Beijing, *Journal of Environmental Sciences (China)* **2**, 27–39.

Zhao, D. and Sun, B. 1986 Atmospheric pollution from coal combustion in China, *Journal Air Pollution Control Association,* **36**, 371–374.

8

Bombay

8.1 General Information

Geography Bombay (Mumbai) is located (Latitude 18°54′N, Longitude 72°49′E) in Mahàràshtra State and is the principal Indian port on the Arabian Sea. The original city is confined by its island location; Bombay island is one of a group of islands off the Konkan coast which are connected to the mainland by a bridge to the north (Figure 8.1). Bombay's central business district (CBD), comprising most of the commercial and business centres and government offices, is located in the extreme southern part of the city. Bombay harbour is to the east of the city. The commercial Fort area lies to the south of the island while the commercial, residential and industrial areas are located north of the Fort area. The mean elevation of Bombay is 11 m above mean sea level. The metropolis covers an area of 603 km^2 of which Greater Bombay covers 438 km^2.

Demography The last census in 1981 recorded a population of 8,243,405, which had increased from 5.81 million in 1970. The population in 1990 was estimated to be 11.13 million and by the year 2000 the projected population will be 15.43 million (Figure 2.1) (UN, 1989). Bombay is extremely densely populated with an average of 15,000 person per km^2. It is estimated that the most densely populated areas contain over 100,000 persons per km^2. In 1981, it was estimated that 34.3 per cent of Bombay's population were slum dwellers. In the past, urban spread has been constrained by physical factors such as marshland and hills; however, shanty dwellings now predominate on the ill-drained marshy ground which lies between the two main railway lines.

Climate Bombay has a tropical savanna climate; mean humidity ranges between 57–87 per cent. The annual mean temperature is 25.3°C rising to a monthly maximum of 34.5°C in June and with a minimum of 14.3°C in January. Total annual mean precipitation is 2,078 mm with 34 per cent (709 mm) falling in the month of July. Due to the summer monsoons, maximum sunshine occurs in winter (291 hours in January) (WMO, 1971). In the winter the predominant wind direction is northerly (NW–NE). However, in the monsoon season westerly and southerly winds predominate. There is virtually always a sea breeze during the day with mean windspeeds between 5–8 km h^{-1}; calms occur at night between 2200 hours and 0600 hours. The mixing depth varies from 30 to 3,000 m (NEERI, 1991a).

Industry Bombay is the financial and commercial centre of India and also the most industrially developed Indian city. The city's port is the busiest in India handling approximately half of the country's foreign trade. There are between 6,000 and 7,000 industrial units in Bombay, mainly textile, chemical and engineering works. Bombay is the centre of the Indian cotton-textile industry. The largest concentration of industries is in the Trombay-Chembur area, along the eastern coast of the island, and also in the Lal Bag area. The Lal Bag area also contains the Tata thermal power plant (the only one in Bombay), the Rashtriya chemical and fertilizer plant, the Bombay Gas Company and numerous petrochemical plants, refineries, and textile mills. Industrial production is increasing at a rate of approximately 90 per cent per annum. Bombay is also home to a nuclear reactor and plutonium separator.

Due to acute overcrowding in the city, planners have prohibited the addition of commercial and industrial establishments on Bombay Island. Companies are forced to locate all new industry in a second city – New Bombay – which is situated on the mainland across the Thane Creek. Some existing established firms have been relocated to the new city.

Transport Being a major port, Bombay has traditionally been known as the Gateway of India. The burning by ships of furnace oil with a high sulphur content (3 per cent) is estimated to account for 34 tonnes of sulphur dioxide (SO_2) per day in Bombay.

During the morning and evening "rush hour" the majority of the working population have to travel in and out of the CBD from the residential areas to the north.

Due to the massive influx of commuters and the geographical constraints of the city, Bombay relies

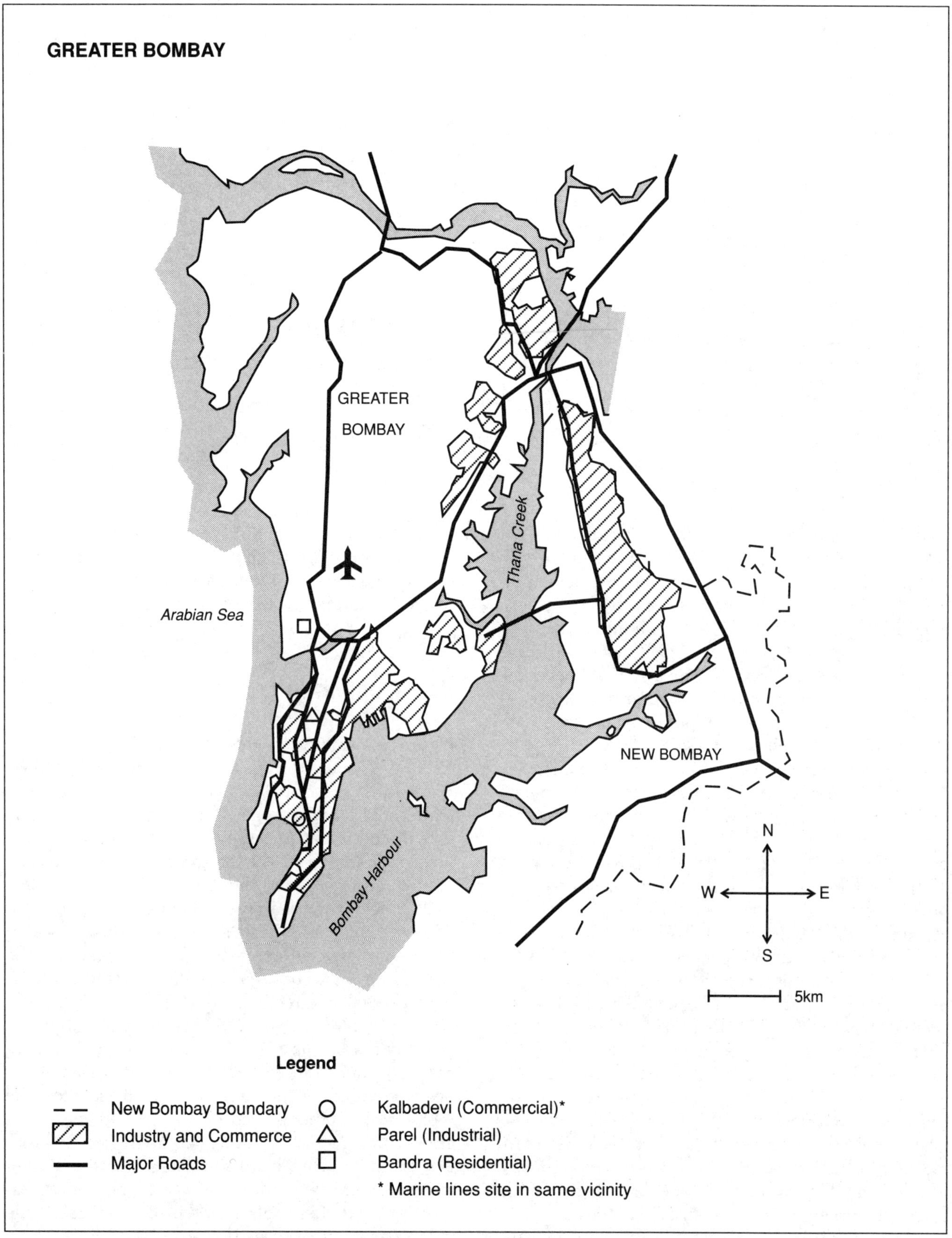

Figure 8.1 *Sketch map of Greater Bombay*

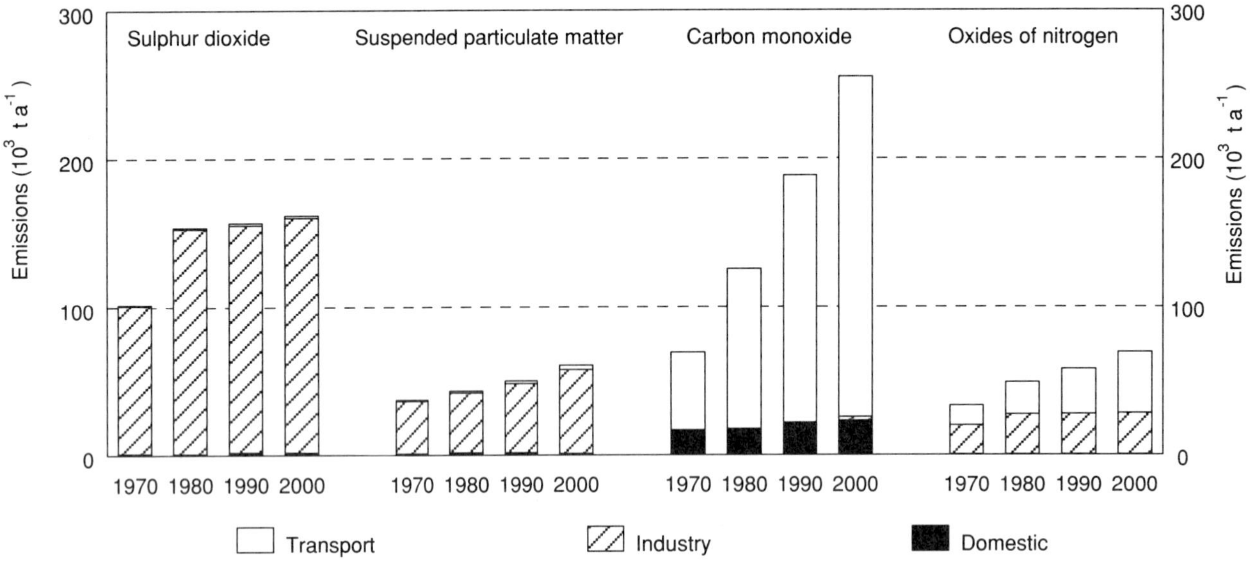

***Figure 8.2** Estimated and projected anthropogenic pollutant emissions in Bombay by source, 1970–2000*

Source: NEERI, 1991a

heavily upon public transport. Buses and railways account for two-thirds (34 per cent each) of all motorized trips; cars account for only 8 per cent of trips. Bombay has two suburban electric train systems which carry over 800,000 commuters daily. It has been estimated that over 25 per cent of all trips are made on foot. Despite the heavy reliance on public transport, motor vehicle numbers are increasing. The estimated number of motorized trips per day in 1980 was 5.25 million. This is projected to rise to 17.69 million by the year 2000 (Faiz et al., 1990). Bombay's vehicular population is growing at a rate of 10 per cent per annum which is reflected in automotive pollutant emissions. The total motor vehicle population in 1989 was 588,000 which was 5 per cent of the Indian total. Motor vehicle density is approximately 1,197 vehicles per km^2 or 364 vehicles per km road length (NEERI, 1991a).

Three-wheeled autorickshaws (a form of taxi) are a major mode of transport in Bombay. The number of registered autorickshaws in Greater Bombay has increased from 6 in 1971 to 21,577 in 1987. These vehicles are powered by a two-stroke engine which requires oil to be added to the fuel; on combustion this oil produces (blue) smoke and thus is a major source of pollution in terms of direct human exposure and is also a nuisance. This problem has been recognized by Bombay Municipal Corporation (BMC) which has resulted in this type of vehicle being banned from certain (central) areas of the city.

Motor vehicles are, and will become, an increasingly important source of air pollution in Bombay as old and often inefficient vehicles remain in circulation for a long time. A combination of traffic congestion, high-rise buildings, slow wind, high humidity and near isothermal conditions throughout the year, aggravate the effects of vehicular emissions.

8.2 Monitoring

Figure 8.1 indicates the location of the current GEMS/Air monitoring sites operated by NEERI. These three sites monitor SO_2, suspended particulate matter (SPM) and nitrogen dioxide (NO_2). Monitoring was discontinued at these sites in 1988 and recommenced in January 1990. All three stations have been moved in the past and therefore only long-term data from 1980 onwards are presented here. It should also be noted that the NEERI site classifications used in this report are different from those given by the GEMS/Air data base. The data presented in this report are taken directly from NEERI sources.

In 1978 BMC set up nine monitoring stations in Bombay measuring SO_2, SPM and NO_2. In 1984 there were 20 stations in operation. Very few details have been provided as to the location of these sites, the monitoring methods used or any results other than annual mean concentrations (for an unknown number of sites). This would suggest that the BMC's results should be made widely available for assessment. If valid, these data would prove invaluable due to their geographic spread.

Spot carbon monoxide (CO) and hydrocarbon (HC) monitoring have been undertaken at a number

Air Quality Monitoring in India

The data base on ambient air quality status in major Indian cities is still confined to the information generated by the National Environmental Engineering Research Institute (NEERI), Nagpur, through its National Air Quality Monitoring Network (NAQMN). The NAQMN started in 1978 and was discontinued in 1988 and 1989. The sites were subsequently reopened in 1990 and the relevant data are presented here.

Ten cities were identified to take part in the NAQMN, primarily because of the location of NEERI zonal laboratories at Ahmedabad, Bombay, Calcutta, Cochin, Delhi, Hyderabad, Jaipur, Kanpur, Madras and Nagpur. These cities also represent a country-wide, cross-section of different industrial, geographic and climatic conditions. Data from the three NEERI stations in each megacity (Bombay, Calcutta and New Delhi) were also reported to GEMS/Air. Some of the megacities data presented predate the NAQMN and were collected by NEERI specifically for the GEMS/Air programme.

Stations are situated at three sites in each city which are deemed to be representative of industrial, commercial and residential locations. In some cities it has been necessary to relocate monitoring stations since 1978 for various reasons. Where relocation has occurred, replacement sites have been chosen which are equally representative of the designated classification. Four air quality parameters were originally chosen for measurement: Suspended particulate matter (SPM) – daily, monthly and yearly averages; Sulphur dioxide (SO_2) – daily, monthly and yearly averages; Sulphation rate – monthly average; and Dust fall – monthly average. In addition nitrogen dioxide (NO_2), ammonia (NH_3) and hydrogen sulphide (H_2S) were also monitored where appropriate. Meteorological information was also collected for the cities for use in data interpretation. Only data for SO_2, NO_2 and SPM are presented in this assessment.

Sulphur dioxide, NO_2 and SPM samples were collected every tenth day thus maintaining a sampling frequency of three times a month. Sulphur dioxide samples were collected on an hourly or four-hourly batch basis for a period of 24 hours using the pararosaniline gas bubbler method. Nitrogen dioxide was sampled using the sodium arsenite gas bubbler method. Suspended particulate matter samples were also taken on a 24-hour basis in parallel with SO_2 and NO_2 monitoring. Air is passed through filter paper by a high volume sampler; the deposited particulates are determined gravimetrically. It has been necessary for NEERI to develop their own sampling equipment specifically designed for Indian conditions. Measurements of SPM have recently been switched to PM_{10} using NEERI designed and built samplers.

Unfortunately problems have been encountered with the data summaries. The data presented for the Indian megacities are taken from the Air Quality Status in Ten Cities reports. Data presented in other reports and submitted directly to GEMS for the same sites do not always correspond with the data presented here and it has not been possible to resolve all such anomalies. On the whole, it would appear that the majority of such errors are typographical mistakes. Such problems do, however, emphasize the need for quality assurance (QA) within the framework of the NAQMN. At present, no QA or quality control (QC) is carried out by NEERI or the Central Pollution Control Board (CPCB) and the initiation of some form of QA auditing system needs to be established.

The CPCB initiated its own National Ambient Air Quality Monitoring (NAAQM) Programme in 1985. The main function of the CPCB under the Air (Prevention and Control of Pollution) Act 1981 is to improve the quality of air and to prevent, control and abate air pollution throughout India. By 1990, 116 monitoring stations (out of 220 sanctioned), operated by the State Pollution Control Boards, were generating data. Some CPCB data are presented in this assessment; however, generally they are of little use in comparison with the data from GEMS/NEERI stations. A first data summary – National Ambient Air Quality Statistics of India 1987-1989 – has been produced by CPCB although, as yet, no assessments have been carried out.

Very little monitoring of additional pollutants, such as carbon monoxide or ozone, is undertaken in India's cities and at present there are no permanent urban monitoring stations for these pollutants. In addition, little has been done to characterize the particulate fraction or to determine atmospheric lead levels following reductions in the lead content of petrol, although work on this has recently begun. It is recommended that monitoring activities in the major cities of India should be expanded to cover more pollutants with a wider geographical coverage in order to quantify urban air pollution problems better. Localized air quality assessments should be carried out to identify specific problems and to provide detailed recommendations for pollution abatement.

Table 8.1 Industrial emissions inventory for Bombay, 1973

Industry	Sulphur dioxide		SPM		Oxides of nitrogen	
	($t\ a^{-1}$)	%	($t\ a^{-1}$)	%	($t\ a^{-1}$)	%
Chemical	48,180	45	2,920	19	7,045	29
Textile	27,010	24	1,351	9	4,453	18
Power Plants	22,630	21	10,366	67	10,950	46
Others	10,950	10	803	5	1,643	7
Total	108,770		15,440		24,091	

After NEERI, 1991a

of roadside sites in the past. However, at present there is no co-ordinated monitoring programme for CO, ozone (O_3) or HC.

8.3 Air Quality Situation

Sulphur dioxide

Emissions Figure 8.2 shows that industrial sources account for nearly all SO_2 emissions in Bombay. Following a 66 per cent increase in SO_2 emissions between 1970 and 1980 the rate of increase has slowed significantly over the past ten years. In 1990 estimated total SO_2 emissions were around 157,000 tonnes per annum. This levelling off in emissions is largely due to the introduction of natural gas as a major fuel source, from the newly opened gas fields located off the west coast. An inventory of SO_2 emissions from industrial sources conducted in 1973 is presented in Table 8.1. Although the industrial structure of Bombay has changed considerably since then, it can be assumed that the current distribution of emissions is similar. The chemical industry was found to be the principal source of SO_2 (NEERI, 1991a).

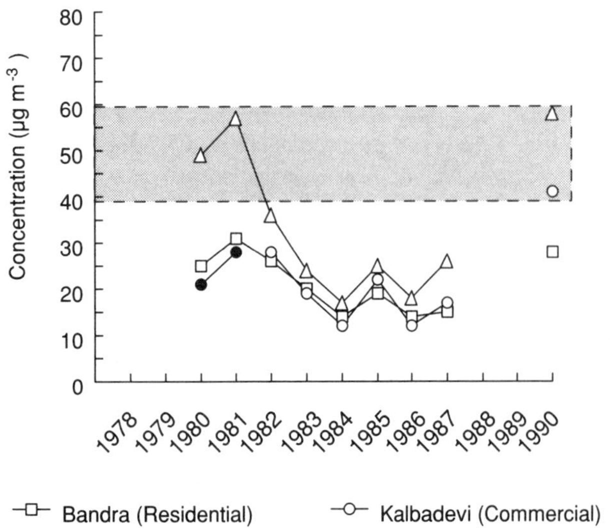

Figure 8.3 Annual mean sulphur dioxide concentrations

Sources: NEERI, 1983, 1988, 1990, 1991b

Ambient Concentrations Ambient SO_2 monitoring at the three GEMS/NEERI sites reveals that annual mean concentrations having dropped significantly between 1980 and 1987 to below WHO annual guide levels, had increased substantially in 1990 (Figure 8.3). However, annual mean SO_2 concentrations are still within the WHO guideline range (NEERI, 1980, 1983, 1988, 1990, 1991a, 1991b). Annual data from Bombay Municipal Corporation for 1978–1988/89 show a similar trend (NEERI, 1991a).

Monitoring conducted in the Air Pollution Survey of Greater Bombay 1970–73 indicates that SO_2 levels first started to decrease in the 1970s, probably due to planning measures such as the relocation of industry and increased stack heights. Annual mean SO_2 concentrations of 67 $\mu g\ m^{-3}$, 63 $\mu g\ m^{-3}$ and 53 $\mu g\ m^{-3}$ were recorded in 1971, 1972 and 1973 respectively (NEERI, 1991a).

Seasonal variation in SO_2 concentrations is minimal; peak concentrations occur in January – the coldest month of the year (NEERI, 1988).

Suspended particulate matter

Emissions Suspended particulate matter emissions have increased significantly in recent years and are projected to continue rising into the next century

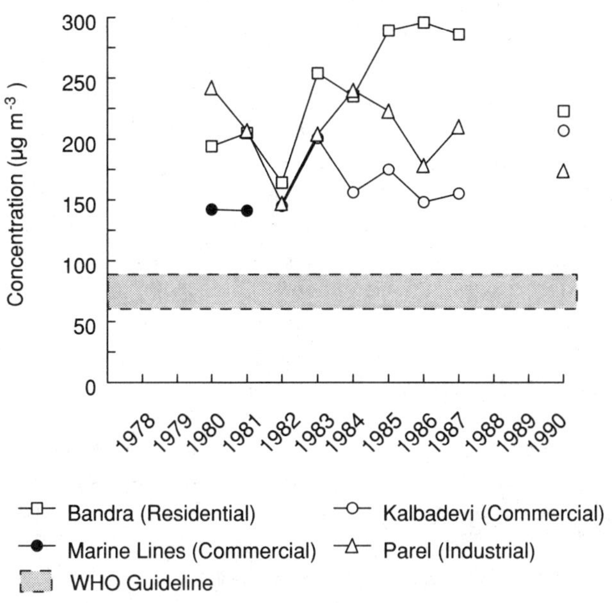

Figure 8.4 *Annual mean suspended particulate matter (TSP) concentrations*

Sources: NEERI, 1983, 1988, 1990, 1991b

(Figure 8.2). As with SO_2, industry is the dominant source, accounting for 93 per cent of all emissions. Table 8.1 indicates that in 1973, 67 per cent of all industrial SPM emissions were caused by power plants (NEERI, 1991a).

Domestic emissions have remained relatively constant in the past and are forecast to remain stable despite the projected increase in population. This is in part through the switch from biofuels, such as wood, charcoal and animal dung and also coal, to less "dirty" fuel such as liquid petroleum gas (LPG) and kerosene (NEERI, 1991a).

Suspended particulate matter emissions attributable to transport have increased greatly. The transport sector doubled its share in total estimated SPM emissions between 1970 (2 per cent) and 1990 (4 per cent). However, these estimates are extremely low considering Bombay's motor vehicle population. Recent estimates suggest that transportation, especially motor vehicles, accounts for approximately 35 per cent of particulate emissions in the Greater Bombay area. (Shah, 1992). This proportion is likely to increase further with increasing motor vehicle traffic. Diesel vehicles and very old vehicles are the main source of particulates in the transport sector. Cars and trucks are the major source of transport SPM (NEERI, 1991a). The low proportion of estimated domestic emissions may, in part, be due to difficulties in adequately quantifying domestic sources. Other inventories suggest that over 20 per cent of total SPM emissions originate from domestic sources (Shah, 1992).

Ambient Concentrations Ambient air quality monitoring at the three GEMS/NEERI sites reveals that both annual mean (Figure 8.4) and 98th percentile (Figure 8.5) TSP levels are higher at the residential monitoring station at Bandra than at the commercial and industrial sites, where one would expect motor vehicle and industrial emissions to dominate (NEERI, 1980, 1983, 1988, 1990, 1991a, 1991b). This may indicate an underestimation of the contribution of domestic sources in the emissions inventory. Alternatively, as suspended particulate matter concentrations at the residential site have increased in line with those at other sites, it is possible that this site may be heavily influenced by transport emissions as it is close to the busy Swami Vivekanand road. Transport SPM emissions may also have been underestimated in the NEERI inventory.

Smoke from diesel vehicles is of great concern as its constituents (e.g., polycyclic aromatic hydrocarbons) are carcinogenic. The popularity of two-stroke engines in autorickshaws and motorcycles is also a factor for consideration. Two-stroke fuel requires the addition of engine oil which, when burnt in the combustion process, produces the characteristic blue smoke.

Ambient SPM concentrations are likely to increase further in the coming decade unless control measures are adopted. The increase in industrial productivity is the greatest contributor to increasing SPM emissions.

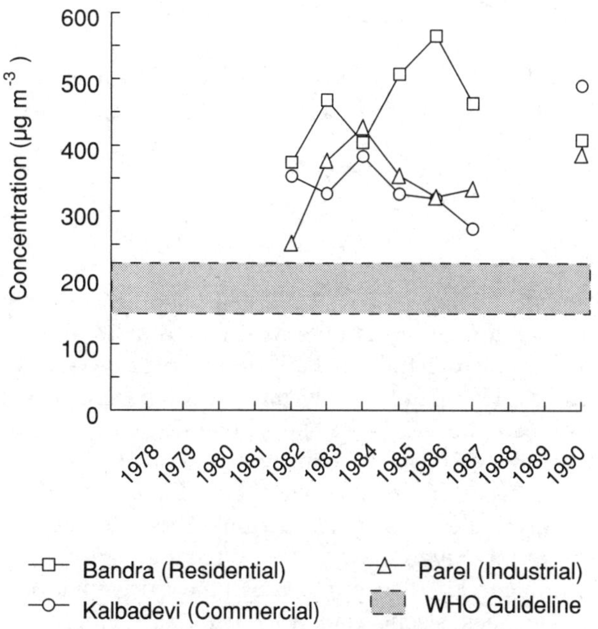

Figure 8.5 *Annual 98 percentile suspended particulate matter (TSP) concentrations*

Sources: NEERI 1983, 1988, 1990, 1991b

Table 8.2 *Monthly mean lead concentrations, 1971–73*

Monitoring Site	Monthly mean lead concentration ($\mu g\ m^{-3}$)
Bandra	2.2
Bhandup	1.8
Chembur	1.3
Dadar	1.7
Dongri	1.0
Ghatkopar	0.4
Goregaon	2.4
Kaliwada	0.4
Laibaug	1.1
Sion	2.0
Bombay Mean	1.4

After NEERI, 1991a

However, it should be noted that SPM concentrations in the early 1970s were higher than those during the 1980s; an annual mean of 380 $\mu g\ m^{-3}$ was calculated for 1973 (NEERI, 1991a).

Mean monthly SPM concentrations are considerably reduced during the monsoon (June to October). Monsoon levels are approximately half those of the winter (November to February).

Analysis of SPM in the Air Pollution Survey of Greater Bombay 1971-73 indicated a high proportion of organic matter; 10 sites throughout Bombay gave a mean of 46 per cent. It is this organic matter which is the greatest risk to health, thus suggesting that the natural dust component in Bombay, which is non-volatile, is not as great as in other Indian cities such as Delhi.

Lead

Ambient Concentrations Table 8.2 shows monthly mean lead (Pb) levels in SPM sampled at 10 sites in Bombay during the Air Pollution Survey of Greater Bombay 1971-73. Monthly mean values were found to exceed the WHO annual guideline (1 $\mu g\ m^{-3}$) at seven of the 10 sites; the mean across all 10 sites was 1.4 $\mu g\ m^{-3}$ with a maximum of 2.4 $\mu g\ m^{-3}$ at the Goregaon site. This wide distribution would suggest that petrol-driven motor vehicles are the main source of ambient Pb. The number of petrol-driven motor vehicles has risen from approximately 125,000 in 1971, to 468,000 in 1987 and to 588,000 in 1989. The Pb content of petrol in India in 1986 was 0.800 g l^{-1} for premium and 0.560 g l^{-1} for regular (Octel). The current mean Pb content of petrol from the Bombay refineries is 0.155 g l^{-1} (Indian Department of the Environment cited by NEERI, 1991a). Monitoring undertaken at the GEMS/NEERI sites in 1990 indicates that annual airborne Pb levels have fallen significantly since the 1970s to between 0.25 and 0.33 $\mu g\ m^{-3}$, well below the WHO guideline of 1 $\mu g\ m^{-3}$ (NEERI, 1991c). It is likely that kerbside levels will be much higher. Lead in street dust is also likely to be high and accumulation and resuspension will also result in increased personal exposure.

Carbon monoxide

Emissions The increase in motor vehicle population is reflected in CO emissions from this source. Estimated emissions have increased from 69,000 tonnes per annum in 1970 to 188,000 tonnes per annum in 1990-91 (Figure 8.2). This increase is likely to continue into the next century; NEERI estimate that CO emissions will be almost 255,500 tonnes per annum by 2000. Most of this increase is attributed to motor vehicle transport which is estimated to be responsible for 89 per cent of total CO emissions in 1990 (NEERI, 1991a).

It is estimated that domestic CO emissions account for 11 per cent of the total in 1990. Over the past 20 years the proportion of domestic CO has decreased as the transport sector has gained in importance. However, domestic emissions have increased overall and are likely to increase further in coming years. Changes in fuel use are likely to reduce domestic emissions (LPG produces CO). Increasing population will result in greater energy demand and hence increased emissions of CO. Recent research on CO emissions suggests that domestic sources and in particular biofuels, such as wood, charcoal and dung, make a larger contribution to anthropogenic and urban emissions than originally believed. Personal exposure, especially indoors (kitchen) is a very important factor for consideration when examining health effects. Forecasts based on increasing industrial activity indicate that 1 per cent of CO emissions will be from industrial sources by 2000 (NEERI, 1991a).

Ambient Concentrations Monitoring of CO is not undertaken by NEERI in Bombay at the present time. However, the Mahàràshtra State Pollution Control Board have undertaken a number of short-term roadside surveys between 1984 and 1987. Carbon monoxide levels were monitored at several roadside sites during periods of peak traffic flow (i.e., 0800-1200 hours and 1700-2100 hours). Eight-hourly mean concentrations ranged between 4-20.6 mg m^{-3}; a maximum hourly concentration of 50 mg m^{-3} was recorded at the Haji Bachoo Ali Hospital. Maximum

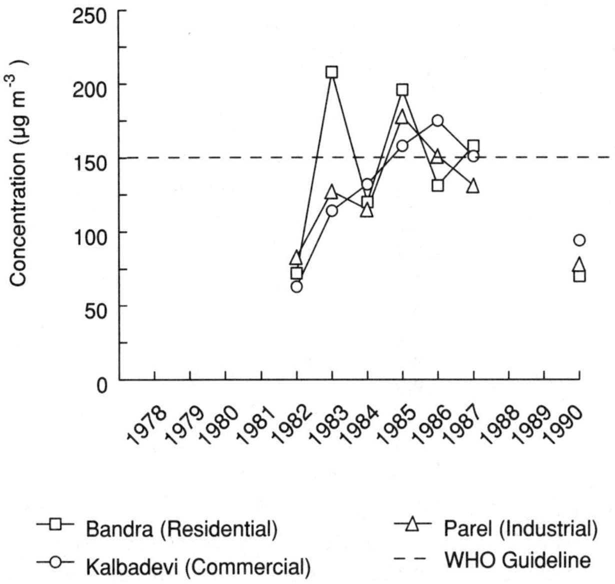

Figure 8.6 *Annual 98 percentile nitrogen dioxide concentrations*

Sources: NEERI 1983, 1988, 1990, 1991b

hourly concentrations are generally around 23–29 mg m^{-3} - below the WHO guideline value (30 mg m^{-3}). It is important to realize that these are results from roadside surveys and are not representative of ambient background CO levels which, due to Bombay's favourable meteorology, are likely to be quite low.

Oxides of nitrogen

Emissions Estimated and projected emissions of oxides of nitrogen (NO_x) (as NO_2) are presented in Figure 8.2. It would appear that although emissions are increasing, largely as a result of the contribution of increasing transport emissions, the rate of increase has slowed significantly in the 1980s and that since 1980 industrial emissions have to a large extent stabilized. NEERI projections indicate that despite increasing productivity industrial emissions will remain around 27,000 tonnes per annum by 2000. It is difficult to see how this stability has been achieved and how it will be maintained given the increase in industrial productivity. The switch to natural gas is likely to increase NO_x emissions significantly. In 1973 power plants were the main source of industrial NO_x followed by chemical works (Table 8.1) (NEERI, 1991a).

Transport is estimated to account for 52 per cent of NO_x emissions in 1990 and it is projected that transport emissions will have increased by approximately 14,600 tonnes per annum by 2000. Detailed vehicle emissions inventories produced by the Indian Department of the Environment (NEERI, 1991a) indicate that diesel vehicles (predominantly trucks) are the dominant source of motor vehicle-derived NO_x in Bombay.

Ambient Concentrations Nitrogen dioxide is monitored at all three GEMS/NEERI sites in Bombay. Figure 8.6 shows the trend in 98 percentile NO_2 concentrations between 1982 and 1990 at the GEMS/NEERI stations. Levels of NO_2 increased until 1985. In 1990 98 percentile values were similar to those in 1982 (70–85 µg m^{-3}). Concentrations at all three GEMS/NEERI sites remain relatively consistent suggesting that NO_2 concentrations are evenly distributed throughout the city (NEERI, 1980, 1983, 1988, 1990, 1991a, 1991b). This would point towards secondary NO_2 formation. Annual mean trends produced by BMC in Bombay and Chembur have been significantly lower than those observed at GEMS/NEERI sites since 1983. However, as no details of monitoring methodology and site location have been provided for the BMC sites, direct comparison of the data is not attempted (NEERI, 1991a).

Diminished insolation during the monsoon period reduces photochemical conversion of nitric oxide (NO) to NO_2 and therefore ambient levels are lower during this period. Maximum concentrations occur in winter, particularly in November and December.

Ozone

Ambient Concentrations Ozone is not routinely measured in Bombay. Increasing NO_x and hydrocarbon emissions and relatively high insolation imply that if O_3 is not already a problem in the Bombay region it is likely to become a problem during the 1990s. Continuous monitoring of urban O_3, however limited, should be regarded as a major priority by the authorities in order to identify if and when ambient urban O_3 becomes a threat to human health.

8.4 Conclusions

Air Pollution Situation Bombay has some specific air quality problems mainly attributed to increasing industrialization and motorization.

On a positive front, planning enforcement measures, such as the relocation of industries and increased stack heights, together with the introduction of natural gas have proved to be partially successful in slowing the decline in air quality and should be

Table 8.3 *Standardized prevalences of selected diseases in Bombay*

Disease	Pollutant load (Urban sulphur dioxide levels)			
	Low (<50 $\mu g\ m^{-3}$)	Moderate (51-100 $\mu g\ m^{-3}$)	High (>100 $\mu g\ m^{-3}$)	Rural (control)
Dyspnoea	3.2	6.0	7.3	5.5
Chronic cough	1.7	2.7	5.1	3.3
Intermittent cough	0.4	5.8	15.6	3.7
Frequent colds	12.1	20.8	18.0	11.0
Chronic bronchitis	2.3	4.5	4.5	5.0
Cardiac disorders	8.2	4.3	6.8	2.7

After Kamat and Doshi, 1987

encouraged further. Low-sulphur coal, a relatively small motor vehicle population and the scrubbing effect of the monsoons helps to reduce overall ambient concentrations in the city.

Main Problems Heavy industry on the island tends to create most of the pollution which is often blown by westerly winds over New Bombay on the mainland. Direct comparison of SO_2 levels in 1976 revealed that concentrations in New Bombay were twice those on the island.

The most important threat to air quality comes from the projected rise in urban population, which is forecast to increase by approximately 5 million between 1990 and 2000. This scale of increase will result in a huge increase in domestic emissions alone, simply for cooking and heating purposes. As population increase outstrips infrastructure development, such as electric supply from gas-powered power stations, people will be forced to use, or revert to, "dirty" fuels. The speed of development necessary simply to house and provide employment for this growing population means that proper planning measures cannot be adopted. Air quality considerations are very often a low priority.

Urban sprawl off the island will inevitably result in an increased number and length of motorized trips, thus leading to congestion and increased emissions.

Control Measures The most effective methods of preventing worsening air quality in Bombay relate to slowing the rate of urbanization, in order to allow proper development of infrastructure and urban planning. Any further relocation or decentralization of industry in the Greater Bombay region should recognize possible impacts upon surrounding and/or neighbouring areas. This also applies to the location of residential areas adjacent to large industrial sites.

Health Effects Between 1977 and 1983 the Environmental Pollution Research Centre, Department of Respiratory Medicine, at KEM Hospital conducted a comparative epidemiological study of 4,129 residents from the Khar (low), Chembur (moderate) and Lal Bag (high) districts of Bombay and a rural (control) area situated 40 km from the city. Standardized prevalences for a number of disorders were determined and classified according to SO_2 levels (Kamat and Doshi, 1987).

Table 8.3 shows that people living in areas with elevated SO_2 (i.e., > 50 $\mu g\ m^{-3}$) suffered an increased prevalence of dyspnoea (breathing difficulties), coughs and common cold. The "urban low" area had the lowest morbidity except for cardiovascular disease. The moderately polluted area showed a higher level of morbidity for common colds, intermittent cough and dyspnoea. Rural prevalences were not as low as might have been expected. However, the villages studied had no sanitation, no protected water supply, poor housing, widespread use of hazardous cooking fuel, poor nutrition and poorer medical care than in Bombay. All these factors will inevitably influence the health and lung function of the rural community. It should be noted that the rural sample showed better lung function at older ages. In all areas young children (under five) and the elderly showed the greatest respiratory morbidity.

The study also revealed a significant correlation between: NO_2 and frequent colds; NO_2 and SPM and coughs; and between SO_2, NO_2, SPM and dyspnoea and chronic cough. The frequency of common colds was observed to decline due to changes in industrial fuel use (switch to natural gas) and a strike.

Cross-sectional studies showed higher morbidity among slum residents, particularly in the urban "medium" pollution load area, probably due to confounding factors similar to those which influence morbidity in the rural population. Nutrition, occupation and smoking were shown to affect morbidity in addition to air pollution. Relatively constant weather conditions throughout the year mean that seasonal influences were found to be of little significance in terms of health effects.

Examination of standardized mortality data from between 1971 and 1979 also revealed that higher respiratory, cardiac and cancer mortality were associated with high air pollution.

A vehicle pollution component was added to the study in 1982. Carboxyhaemoglobin (COHb) levels in blood were related to traffic flow and clinical symptoms. Higher levels of COHb were found in areas with slow-moving traffic with heavy or moderate vehicle densities. People living in flats facing road traffic also showed higher COHb levels. The only significant correlation between elevated COHb and symptom morbidity was an increase in the incidence of chest pain and irritability.

8.5 References

Faiz, A., Sinha, K., Walsh, M. and Varma, A. 1990 *Automotive Air Pollution: Issues and Options for Developing Countries,* World Bank Policy and Research Working Paper WPS 492, The World Bank, Washington DC.

Kamat, S. R. and Doshi, V. B. 1987 Sequential health effects study in relation to air pollution in Bombay, India, *European Journal of Epidemiology* **3**, 265–277.

NEERI 1980 *Air quality in selected cities in India 1978–1979,* National Environmental Engineering Research Institute, Nagpur.

NEERI 1983 *Air quality in selected cities in India 1980–1981,* National Environmental Engineering Research Institute, Nagpur.

NEERI 1988 *Air quality status in ten cities: India 1982–1985,* National Environmental Engineering Research Institute, Nagpur, India.

NEERI 1990 *Air quality status in ten cities: India 1986–1987,* National Environmental Engineering Research Institute, Nagpur. (Unpublished report).

NEERI 1991a *Air pollution aspects of three Indian megacities, Volume II: Bombay,* National Environmental Engineering Research Institute, Nagpur.

NEERI 1991b *Air quality status 1990,* National Environmental Engineering Research Institute, Nagpur.

NEERI 1991c *Air quality status: Toxic metals, polycyclic hydrocarbons, anionic composition and rain water characteristics (Delhi, Bombay and Calcutta),* National Environmental Engineering Research Institute, Nagpur.

Shah, J. 1992 Personal communication, The World Bank, Washington DC.

UN 1989 *Prospects of World Urbanization, 1988,* Population Studies, No. 112, United Nations, New York.

WMO 1971 *Climatological Normals (CLINO) for Climate and Climate Ship Stations for the Period 1931–1960,* No. 117, World Meteorological Organization, Geneva.

9

Buenos Aires

9.1 General Information

Geography The Metropolitan Area of Buenos Aires (MABA) comprises 19 municipalities and the city of Buenos Aires, the federal capital of Argentina (Figure 9.1). The MABA (Latitude 34°35′S, Longitude 58°29′W) is situated on the east coast of Argentina on the southern bank of the River Plate (Rio de la Plata) estuary, 25 m above mean sea level. The MABA urban agglomeration covers some 7,000 km^2 (of which the city of Buenos Aires covers 200 km^2) consisting of a number of urban centres interspersed with small rural areas.

Demography Argentina is one of the least densely populated countries in the world. More than 70 per cent of the population live in urban areas; in 1990 an estimated 11.58 million Argentinians, or one-third of the total population, were concentrated in the metropolitan area of Buenos Aires (Table 2.1). This total is projected to increase to 13.05 million by the year 2000 (UN, 1989). The city's population has remained relatively constant over the past 40 years, while that of the metropolitan area more than doubled, indicating the degree of urban spread into the surrounding countryside.

Climate Buenos Aires has a humid sub-tropical climate. The yearly mean temperature at the central observatory of Buenos Aires fluctuates around 17°C with monthly means ranging from 10–11°C in winter (June–August) and 24–25°C in summer (December–February). On a yearly basis some 2,600 hours of sunshine are recorded, with a low of 130 hours per month in June and a high of 290 hours per month in January. The annual mean precipitation is 1,000 mm, with monthly extremes of 120 mm in March and 60 mm in July. The prevailing atmospheric conditions, and the flatness of eastern Argentina, ensure a good and quasi-continuous mixing in the tropospheric layers, resulting in the rapid dispersion of airborne contaminants. These conditions have been confirmed by means of the precipitation sampling carried out over a number of years in the highly populated areas and their suburbs. These samples show only a few cases of acid rain, with low acidification rates. During the 1983–1988 observation period, only 16 per cent of samples had values as low as pH4 (Canziani et al., 1989).

Industry There are an estimated 50,000 enterprises ranging from very small family businesses to big industrial complexes in the MABA. There are three thermal power stations in the federal capital with a total capacity of 2.3 GW. In 1988 they generated 11.7 TW and consumed 1.6 million tonnes of fuel oil and 1.7 Gm^3 of natural gas.

Transport More than one million people travel daily by various means, predominantly motor vehicles. The number of motor vehicles including private cars, minibuses, trucks and rental cars is estimated to approach one million in Buenos Aires.

9.2 Monitoring

Air quality monitoring data for Buenos Aires and its metropolitan area are scarce. The data presented here are mainly from the 1982 Redpanaire review (OPS, 1982) – which is without any specific siting information – and for suspended particulate matter (SPM) from two GEMS sites in 1985 and 1986.

Between 1974 and 1977 four successive (and somewhat different) random campaigns were organized in the central part of Buenos Aires to study specifically the problem of air pollution from motor vehicle traffic. These campaigns were a collaborative project of the National Sanitation Directorate of the Ministry of Social Welfare and the Association of Car Manufacturers ADEFA (Mazzola et al., 1991). In each of these one-year monitoring campaigns some 81 half-hour samples were collected in a random way in each of the 10 fixed monitoring locations scattered throughout Buenos Aries. Each location was supposed to be representative of the 10 km area under investigation. The 10 sampling points were entirely or

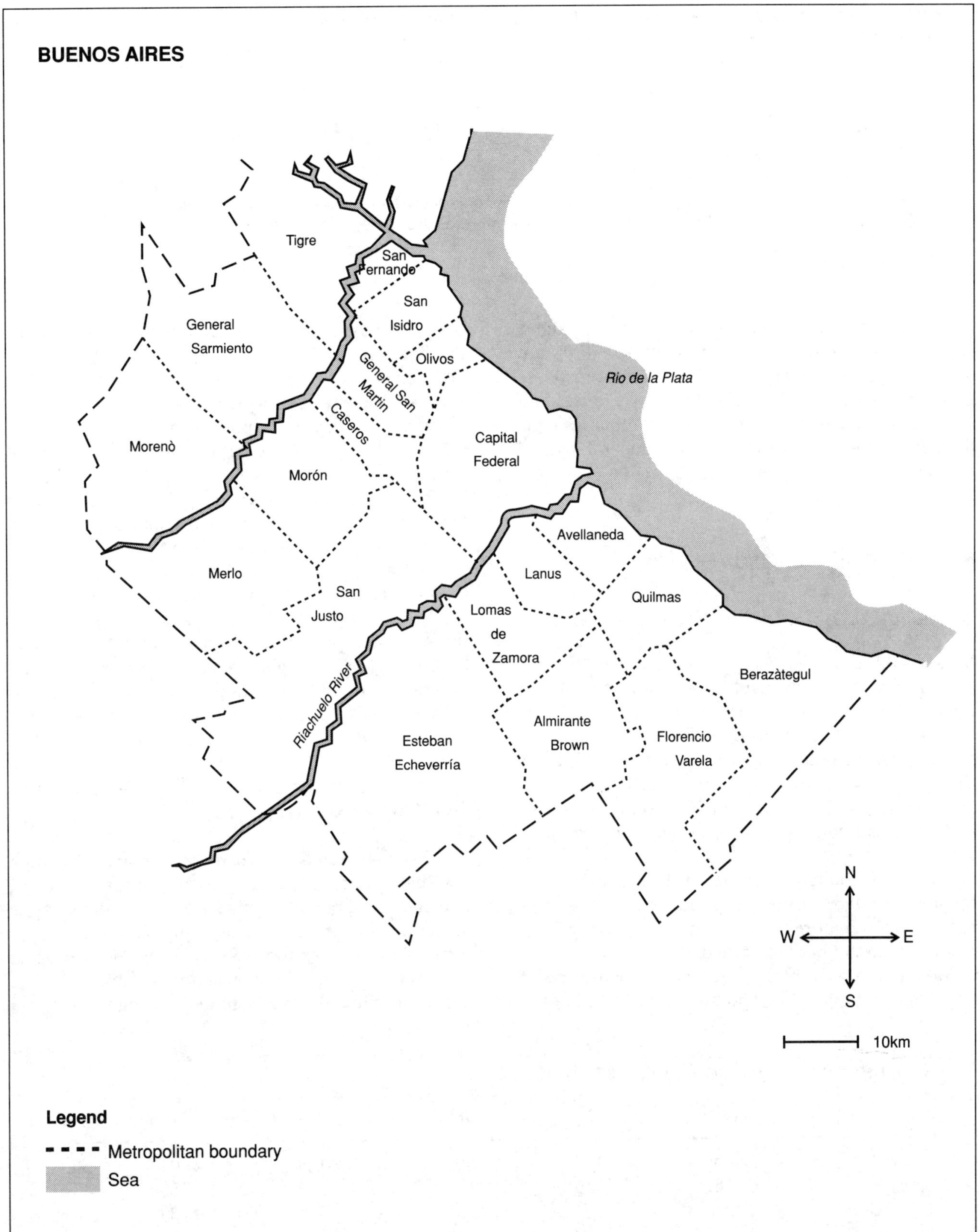

Figure 9.1 *Sketch map of the metropolitan area of Buenos Aires*

Table 9.1 *Random campaigns (Stratmann approach) from 1974 to 1977 in the federal capital*

	SO_2 ($\mu g\ m^{-3}$)	CO ($mg\ m^{-3}$)	NO_x ($\mu g\ m^{-3}$)	HC (ppm)	Total Oxidants (ppb)
Mean * levels	2–17	4–24	40–480	4–5	7–10
Maxima ** (30 min)	24–100	24–74	200–650	7–14	17–55

* Range for the four successive campaigns for CO, SO_2 and NO_x. Hydrocarbons and total oxidants were only measured in the first campaign
** Known only for the first 1974/1975 campaign, based on 10 sampling locations highly exposed to intensive traffic
After Mazzola et al., 1991

partly different in each of the successive campaigns, but over the four campaigns only 20 different sampling locations were used. Half of the sites were highly influenced by intensive traffic, the other half hardly influenced at all. Table 9.1 summarizes the data from these campaigns in the federal capital during the 1974–1977 period.

9.3 Air Quality Situation

Sulphur dioxide

Emissions A comprehensive emissions inventory for the MABA is not available. Only for the city of Buenos Aires (i.e., the federal capital, Figure 9.2) has an attempt been made to quantify the emissions of some common pollutants (Mazzola et al., 1991). A first-order approximation, based on the yearly amounts of the various liquid fuels sold in the city and the mean emission factors as specified in Table 9.2, was used.

The main results of this first-order approximation for the 1970–1989 period are summarized on a five-yearly basis in Figure 9.2. For sulphur dioxide (SO_2) a downward trend has been noted.

Taking into account that there are numerous other important sources of these pollutants, the estimates in Table 9.2 have to be considered as absolute minima. Furthermore, they are only representative for the city of Buenos Aires, representing only 200 km^2 of the 7,000 km^2 covered by the MABA.

As to the origin of the pollutants emitted, Figure 9.2 shows that motor vehicle traffic in the federal capital of Buenos Aires accounts for 40 per cent of SO_2 emissions but almost all the carbon monoxide (CO) emissions (Table 9.3). For SPM this drops to some 70 per cent and for oxides of nitrogen (NO_x) to 50 per cent.

The problems associated with these emission estimates are illustrated by the following information (Mazzola et al., 1991). In the federal capital there are three thermal power stations with a total capacity of 2.3 GW. In 1988 they generated 11.7 TW and burnt 1.6 million tonnes of fuel oil and 1.7 Gm^3 of natural gas. This resulted in an emission of 87,700 tonnes of SO_2. This is obviously significantly more than the estimates given in Figure 9.2 and Table 9.3 for the total emissions estimated on the basis of all liquid fuel "sold" in the federal capital.

Ambient Concentrations No recent ambient SO_2 data for Buenos Aires were made available for this report and it is not known whether any SO_2 monitoring is currently in progress. The only long-term data available were from the 1968–1973 period when apart from some extreme daily means, concentrations were below or within the WHO guideline range.

Table 9.2 *Emission factors (kg per tonne) for liquid fuels in the city of Buenos Aires*

Source	Fuel	SO_2	SPM	CO	NO_x	HC
Power plants	Fuel oil	19.90	1.04	0.65	13.20	0.13
Industrial	Gas oil	20.10	2.13	0.59	7.50	0.41
Motor vehicles	Petrol	0.54	2.00	377.00	10.30	14.50
Motor vehicles	Diesel	19.00	2.40	44.00	11.00	2.50

After Mazzola et al., 1991

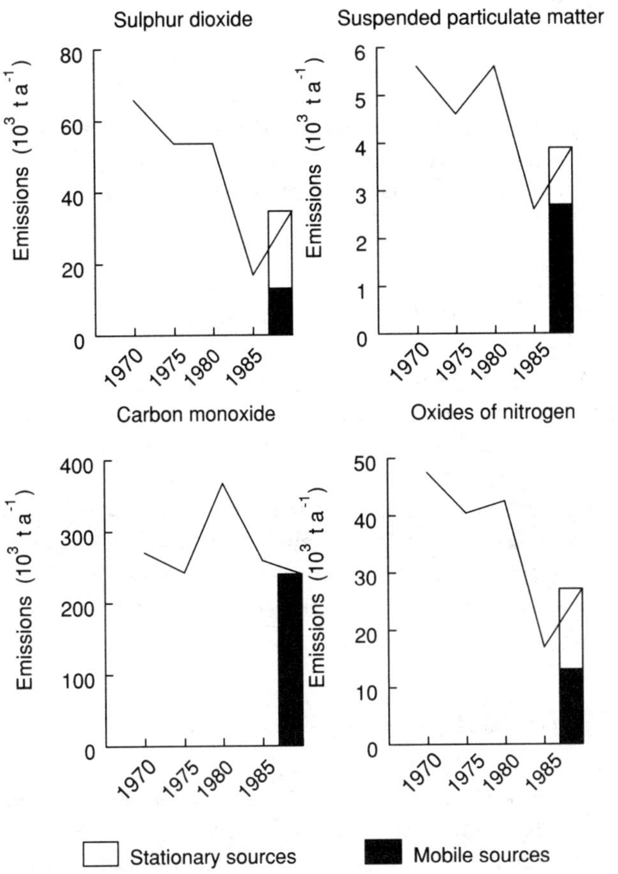

***Figure 9.2** Emission estimates (10^3 t a^{-1}) based on all liquid fuels sold in the city of Buenos Aires*

Source: Mazzola et al., 1991

Suspended particulate matter

Emissions Based on liquid fuel sales, SPM emissions were estimated to total approximately 4,000 tonnes per annum in 1989. Motor vehicles are estimated to account for 70 per cent of total SPM emissions. Emissions from the three thermal power stations are thought to add a further 5,700 tonnes per annum which would make them the largest source in the city (Mazzola et al., 1991).

Ambient Concentrations Figure 9.3 shows that mean and maximum daily SPM concentrations are clearly exceeding WHO guidelines at the GEMS2 site in 1985–1988 and at a number of stations in the 1968–1973 period (Rispoli, 1976). No detailed assessment is possible because of the lack of supporting information.

Lead

Emissions The lead (Pb) content of petrol in Buenos Aires is relatively high, ranging from 0.6–1.0 g l^{-1} which gives cause for concern since there is an increase in the level of motorization in cities throughout South America. Unfortunately, no gross estimate of emissions can be made as details of the motor vehicle fleet were not made available.

Ambient Concentrations Caridi et al. (1989) presented some typical Pb concentrations in air in central Buenos Aires and its suburban areas at different times of day. They range from 0.3 μg m^{-3} for clean suburban air to 1 μg m^{-3} for suburban areas with medium traffic and 3.9 μg m^{-3} for central areas with high density traffic. This latter value would seem to be quite high, although the averaging time is not provided. (It is believed to be a spot sample.)

Carbon monoxide

Emissions Estimated emissions attributable to the burning of liquid fuels accounted for 240,000 tonnes of CO in 1989, almost entirely derived from motor vehicle traffic. Carbon monoxide emissions have fallen steadily since 1970 when they were estimated to be 270,000 tonnes.

Table 9.3 *Emission estimates for mobile and stationary sources in the city of Buenos Aires, 1989 (10^3 t a^{-1})*

Source	SO_2	SPM	CO	NO_x	HC
Mobile	13.1	2.7	239.0	13.1	9.7
Stationary	21.5	1.2	1.0	14.0	0.2
Total	34.6	3.9	240.0	27.1	9.9

After Mazzola et al., 1991

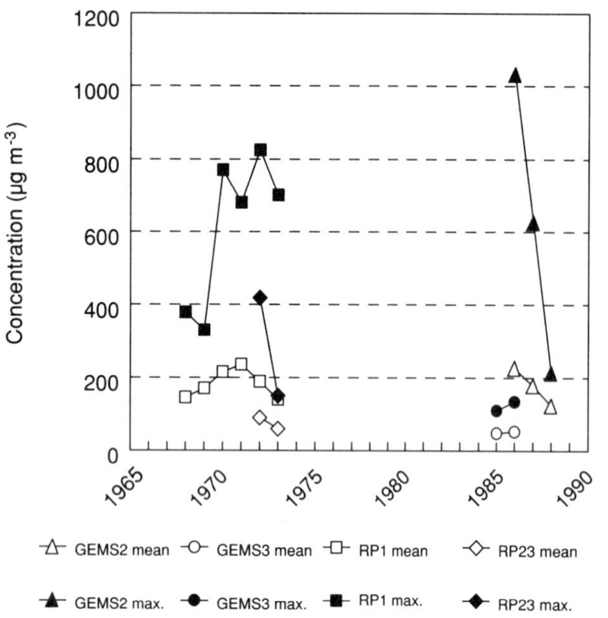

Figure 9.3 *Annual average and annual maximum suspended particulate matter (TSP) concentrations*

Note: RPl and RP23 are Redpanaire stations
Sources: GEMS data; OPS, 1982

Ambient Concentrations No recent ambient CO data were made available for this report and it is not known whether any CO monitoring is currently undertaken. The only data presented are from the random campaigns discussed in Section 9.2.

Oxides of Nitrogen

Emissions In 1989 approximately 27,000 tonnes per annum of NO_x were emitted as a result of liquid fuel combustion in Buenos Aires. This compares with 47,500 tonnes per annum in 1970 and an estimated 42,500 tonnes per annum in 1980. Emissions fell to 17,000 tonnes per annum in 1985, but have presumably increased as a result of increasing motor vehicle traffic in recent years. Motor vehicles account for approximately 48 per cent of liquid fuel attributable emissions. Buenos Aires' power stations are believed to produce an additional 38,000 tonnes of NO_x per annum.

Ambient Concentrations No recent ambient nitrogen dioxide (NO_2) data were made available for this report and it is not known whether any NO_2 monitoring is currently undertaken. The only data presented are from the random campaigns discussed in Section 9.2.

Ozone

Ambient Concentrations No recent ambient ozone (O_3) data were made available and it is not known whether any O_3 monitoring is currently undertaken. The only data available are for total oxidants, 1974-77, from random campaigns discussed in Section 9.2.

9.4 Conclusions

Taking into account the size and growth rate of the MABA and its importance for the country's socio-economic development, the limited and incomplete air pollution information available is extremely poor. It is obvious that appropriate monitoring has to be initiated as soon as possible. Efficient and effective air quality management cannot be implemented without proper knowledge of the present situation and trends.

9.5 References

Canziani, D. F., Cordero, J. A. and Milic, E. 1989 Struggle against environmental pollution in Argentina: Updating of information. In: *Man and his Ecosystems, Proceedings of the 8th Clean Air Congress 1989* **5** 213–221, Elsevier Science Publishers B.V., Amsterdam, .

Caridi, A., Kreiner, A. J., Davidson, J., Davidson, M., Debray, M., Hojman, D. and Santos, D. 1989 Determination of atmospheric lead pollution of automotive origin, *Atmospheric Environment* **23**, 2855–2856.

Mazzola, M. D., Gavarotto, M. C. and Petcheneshsky, T. 1991 *Calidad del Aire en el Area Metropolitana de la ciudad de Buenos Aires (periodo 1970-1989)*, Ministerio de Salud Publica y Medio Ambiente, Buenos Aires.

OPS 1982 *Red Pan Americana de Muestro de la Contaminacion del Aire, Informe Final 1967–1980*, Publication CEPIS 23, Organizacion Pan Americana de la Salud, Lima.

Rispoli, J. A. 1976 The struggle against air pollution in the Argentine Republic. In: *Proceedings of the 4th International Clean Air Congress*, Tokyo, 910–912.

UN 1989 *Prospects of World Urbanization 1988*, Population Studies No. 112, United Nations, New York.

10

Cairo

10.1 General Information

Geography Cairo (Al-Qāhira) is fan shaped and bounded by the Nile delta to the north. The city of Cairo is situated principally on the eastern bank of the River Nile a few kilometres south of the Rosetta and Damietta branches at an elevation of 74 m above mean sea level (Latitude 30°08′N, Longitude 31°34′E). The current city is approximately 1,000 years old. The central urban area of Cairo is primarily a commercial area comprising thousands of small workshops, industries and bakeries. Greater Cairo consists of three governorates; Cairo, Giza and Kalubia (Figure 10.1). Giza (or Al-Jiza) is a large residential sector situated to the west of the city.

Demography Cairo is the capital of Egypt and is the most populous city in Africa. An extensive flood control programme and improved transport facilities introduced over the past 30 years have accelerated urbanization of the Greater Cairo area. In 1950 Cairo had a population of 2.4 million. The UN estimate of the 1990 population was 9.08 million million. By the year 2000 Cairo is expected to have a population of 11.77 million. Egypt's estimated annual urban growth rate was 3.4 per cent in 1990 compared with 2.5 per cent in 1970 (UN, 1989).

Climate Cairo has a desert climate characterized by very dry heat. Monthly mean temperatures range from 14°C in January to 29°C in July. The maximum daily temperature in July can reach 43°C. Annual mean rainfall is only 22 mm, the monthly maximum (7 mm) occurring in December. Average relative humidity is 56 per cent. The average total annual duration of sunshine is 3,504 hours (WMO, 1971). Nights are generally cool and during the winter quite damp; radiational cooling often leads to shallow, stable inversions. A Nile breeze is also a characteristic of the night-time climate. Pollution is generally exacerbated by low wind speeds, lack of rain, tall buildings, narrow streets and traffic congestion in Cairo. More unstable wind conditions later in the summer tend to reduce pollutant concentrations.

Industry Cairo accounts for over half of Egypt's energy consumption yet only 15 per cent of the country's population. Table 10.1 shows the increase in fuel use at Cairo's 11 power stations from 1980 to 1989. Almost 50 per cent of all fuels are consumed in the Shoubra El-Khayma power station. Heavy fuel oil and natural gas are the major fuels used for power generation in Egypt, accounting for 52 and 46 per cent respectively. From fuel consumption figures given in Table 10.1, it can be concluded that power stations use about 1,800,000 tonnes of heavy fuel oil per annum (1988/9).

Cairo's main industries are textiles, iron and steel, motor vehicle manufacturing, cement, chemicals, refrigerator manufacturing, and food processing. Approximately 60 per cent of industrial enterprises in Egypt are controlled by the public sector. A major industry in the area surrounding the city having a major influence on Cairo's air quality is limestone quarrying and the adjacent cement factories. Three major cement factories are situated at Helwān 24 km to the south-east of Cairo. It is estimated that 5 per cent of all cement produced is lost as particulate emissions. Helwān is the biggest industrial area in Egypt employing over 100,000 people. In addition to cement factories, it is also the site of iron and steel works, a lead and zinc smelter, coke and chemical works and fertilizer factories. Manufacturing industries in the Helwān area produce cars, textiles, ceramics and bricks. Power for these industries is provided by electricity generating stations located in the area.

A second industrial complex, Shoubra El-Khayma, lies 30 km to the north-west of Cairo. It consists of over 450 industrial units of various sizes and employs approximately 85,000 people, most of whom are also residents in the area. Industries here include metallurgical work, ceramics, glass, bricks, textiles, and plastics. There is also a very large thermal power station which provides electricity for these industries (Table 10.1).

Transport The main mode of public transport in Cairo is the bus. The modal share of motorized trips in Cairo in 1980 was 15 per cent car, 15 per cent taxi, and 70 per cent bus. Yet, since 1980 the number of

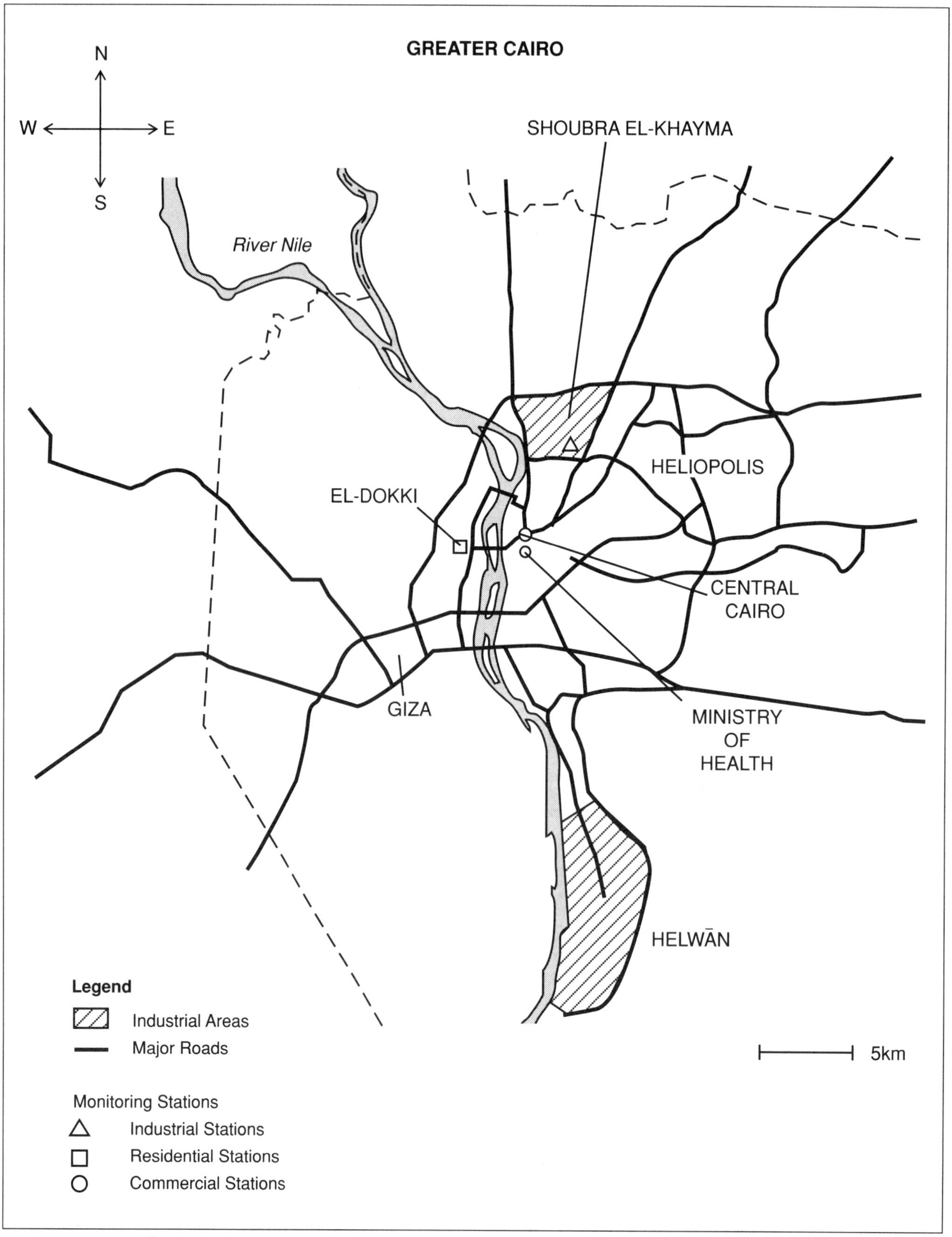

Figure 10.1 *Sketch map of Greater Cairo*

Table 10.1 *Fuel use in electric power stations in Greater Cairo. (Units are 10^3 tonnes heavy fuel oil equivalents per annum)*

Station name	1979/80	1982/83	1985/86	1988/89
West Cairo	378	557	526	446
South Cairo	414	533	515	506
North Cairo 1	179	229	150	207
North Cairo 2	16	35	13	–
East Cairo	23	102	42	11
Tebeen 1	77	124	98	51
Heliopolis	–	30	6	0.2
Helwān	25	281	253	214
Shoubra El-Khayma	–	–	1,069	1,809
Tebeen 2	22	116	91	16
Wadi Hoff	–	–	15	138
Total	1,134	2,007	2,778	3,398.2

Note: 52 per cent of total is heavy fuel oil and 46 per cent is natural gas.
– Not operational
Source: Nasralla, 1990

motor vehicles in Cairo has increased from about 400,000 to more than 900,000 in 1990. The number of motorized trips per day is projected to rise from 3.9 million in 1980 to 12.0 million by the year 2000 (Faiz et al., 1990). At present the city has only two main metro/train links. The metro links El-Marg in the north of the city with Helwān and a suburban train links the centre with Al Ma'adi in the south.

10.2 Monitoring

The Ministry of Health (MoH) is responsible for air pollution monitoring in Cairo. Ambient and occupational air quality standards were promulgated in 1971. Sulphur dioxide (SO_2) and smoke monitoring commenced in 1973; in addition, total suspended particulate matter (SPM/TSP) monitoring was initiated on a regular basis at various sites in 1985. The MoH's Imbaba Centre provides the facilities for the analysis of SO_2 and TSP for Cairo and Greater Cairo. Smoke measurements are carried out at the MoH Central Office. Only a limited number of staff are available for specialized air pollution monitoring.

There are currently three GEMS stations operated by the MoH (Figure 10.1). Site 1 is located in a city centre commercial (CCC) area at the Industrial Health Department, Maglis El-Omma ('Ministry of Health'), Site 2 in a suburban industrial (SI) area at Sahel Hospital, Shoubra ('Shoubra'), and Site 3 in a suburban residential (SR) area at El-Adamchia Square, NASR City Health Bureau ('Nasr City'). Funding for the monitoring programme is provided by the Environmental Affairs Agency. An assessment of air quality monitoring in Egypt by WHO (Commins, 1987) showed problems with respect to SO_2 measurements. It was found that the measurements were affected by partial neutralization of the absorbed SO_2 in the bubblers by ammonia. Until now, the influence of ammonia has not been quantified.

In addition to the GEMS/Air monitoring programme, several research studies on air pollution are carried out at the National Research Centre and at Cairo University.

10.3 Air Quality Situation

Sulphur dioxide

Emissions There is no SO_2 emission inventory available for Cairo. While natural gas-fired power plants emit very little SO_2 it can be estimated from the fuel consumption data for power plants given in Table 10.1 (52 per cent of 3.4 million tonnes of heavy fuel oil equivalent), that SO_2 emissions might be in the range of 50,000–80,000 tonnes per annum. Another 4,000 tonnes per annum of SO_2 are emitted by cars and buses, which is more than a sevenfold increase compared with 1970 emissions. As the sulphur content of diesel is much higher than in petrol, nearly 80 per cent of all motor vehicle SO_2 emissions come from diesel-powered buses, the remainder from diesel cars (Table 10.2).

Table 10.2 *Estimated pollutant emission by cars and buses in Cairo, 1980–2000*

	1980 ($t\ a^{-1}$)	1990 ($t\ a^{-1}$)	2000 ($t\ a^{-1}$)	Cars (%)	Buses (%)
CO	72,000	160,000	223,000	99	1
NO_x	3,300	7,300	10,600	66	34
SO_2	1,900	4,100	6,100	22	78
Smoke	600	1,300	1,800	88	12

After Faiz et al., 1990

There are, of course, numerous other sources for SO_2 emissions. Domestic emissions might be relatively small, but industrial sources (including iron and steel production, motor vehicle manufacturing and chemical industries) certainly emit considerable additional amounts of SO_2. It is recommended that a full and detailed SO_2 inventory for Cairo be carried out as soon as possible.

Ambient Concentrations The Egyptian MoH has been monitoring SO_2 in Cairo since 1973. Nine sites (including the three GEMS/Air sites) are situated in Cairo, four sites in the residential suburb of Giza and one site in the industrial area of Shoubra El-Khayma.

Until recently, acidimetric titration (hydrogen peroxide) was the main monitoring method employed in Greater Cairo. However, it was shown that due to high atmospheric ammonia concentrations and high temperatures this method was unreliable (Nasralla et al., 1984a; Commins, 1987). The more reliable colorimetric (pararosaniline) methods are now favoured. As the hydrogen peroxide titration method is still used at the three GEMS/Air sites, the GEMS data are not presented here due to doubts over their reliability. Furthermore, there are large gaps in the available GEMS data between 1978 and 1989.

Monthly mean levels were reported to be regularly in the range of 100–300 $\mu g\ m^{-3}$ at a number of sites (Commins, 1987). Similarly, other monitoring data shown in Table 10.3 (Nasralla et al., 1984a) showed that seasonal means of SO_2 concentrations were 140–430 $\mu g\ m^{-3}$ in city centre locations and 60–140 $\mu g\ m^{-3}$ in residential sites in the years 1979 and 1983/84 respectively. Annual means were 260–290 $\mu g\ m^{-3}$ in the city centre and 90–110 $\mu g\ m^{-3}$ in the residential areas. Generally, higher concentrations were found in the summer months and the lowest concentrations in winter. Maximum monthly mean SO_2 concentrations in the city centre were above 500 $\mu g\ m^{-3}$ in June 1984 (Nasralla et al., 1984a). These data show that the WHO annual mean guidelines of 40–60 $\mu g\ m^{-3}$ are exceeded by a factor of five in the city centre and by a factor of two in some residential areas.

The maximum 24-hour mean SO_2 concentrations in Cairo were reported to be over 1,000 $\mu g\ m^{-3}$ (Commins, 1987). Such high values would pose a considerable risk to human health.

Table 10.3 *Seasonal variation of SO_2 average concentrations in Cairo (units $\mu g\ m^{-3}$)*

	1979		1983/1984	
	City Centre	Residential	City Centre	Residential
Winter	140	90	170	90
Spring	230	140	230	90
Summer	430	140	370	140
Autumn	340	110	260	60
Mean	290	110	260	90
Summer/Winter ratio	3	1.7	2.2	1.7

Note: Original data in ppm. Conversion factor: 1 ppm = 2,860 $\mu g\ m^{-3}$. Converted data were rounded to the nearest 10 $\mu g\ m^{-3}$
After Nasralla, 1990

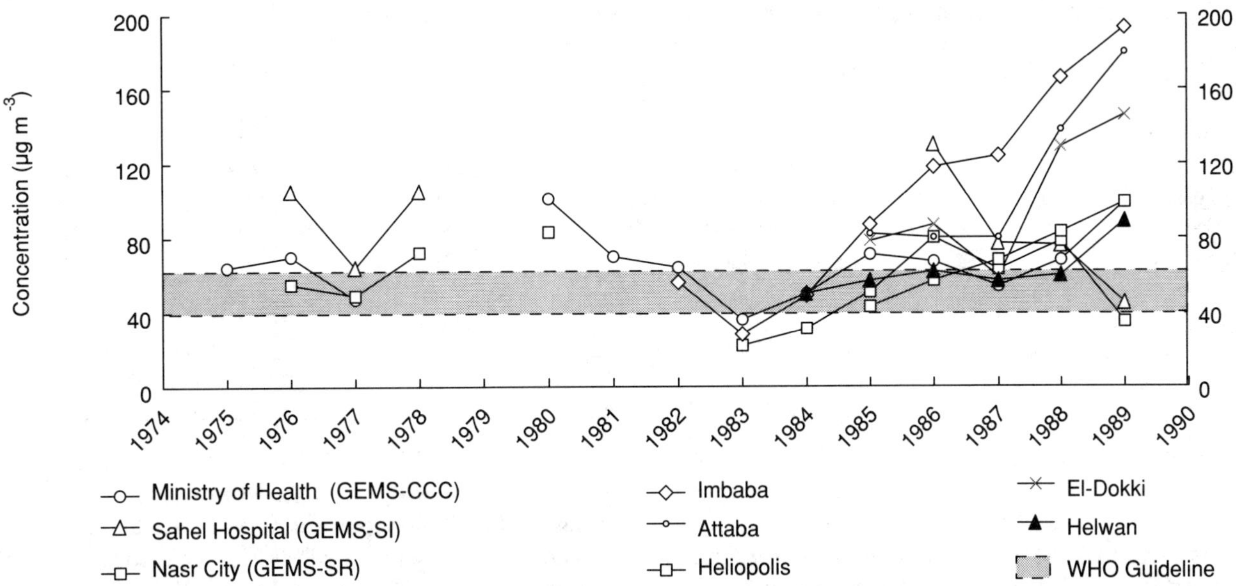

Figure 10.2 *Annual mean smoke concentrations*

Sources: GEMS data; Nasralla, 1990

Suspended particulate matter

Emissions There is no SPM emission inventory available for Cairo. There are natural emission sources of particulate matter (such as wind-blown dusts) as well as anthropogenic sources (e.g., motor vehicle exhaust, industrial particulates). A large contributor to natural SPM levels are the north-easterly winds in spring and the fresh-to-strong hot "Khamasin" southerly wind which are usually loaded with high levels of natural sand and dust. However, there are no estimates of the contribution of natural SPM to total ambient SPM levels.

There are many anthropogenic SPM emission sources, especially from incomplete combustion processes, industry (iron and steel, cement, and so on), and traffic. Smoke emissions from cars and buses have been estimated to be 1,200 tonnes per annum in 1990. This is a more than sevenfold increase since 1970. About 88 per cent of these emissions come from cars. Traffic SPM sources are believed to be relatively small compared with other SPM sources, but their impact on SPM pollution at roadside locations is probably severe.

Ambient Concentrations Suspended particulate matter is measured by the MoH network as both black smoke (BS) and TSP.

The annual mean smoke concentrations at eight sites in Greater Cairo (including the three GEMS sites) are given in Figure 10.2. It is important to note that with the exception of the three GEMS stations, only annual mean data are available, no statistics or number of observations are provided, and therefore the data validity is questionable. The monitoring data suggest that, on the whole, annual mean smoke concentrations have increased during the 1980s, with the notable exception of the GEMS "Sahel Hospital" site in a suburban industrial area. In 1989, annual mean smoke concentrations were 40–190 μg m^{-3}, with six out of the eight monitoring sites having smoke concentrations above the WHO annual guideline of 40–60 μg m^{-3}. In three of the stations (Imbāba, Attaba and El-Dokki), annual mean smoke levels in 1988 and 1989 even exceeded the WHO daily guideline of 125 μg m^{-3}. Only two of the eight stations were within or below the WHO annual guideline.

Annual mean TSP data for 1987 to 1989 are presented in Table 10.4. The measured annual TSP levels of about 500–1100 μg m^{-3} are far in excess of the WHO annual guideline of 60–90 μg m^{-3}. The large disparity between the black smoke and TSP data is probably caused by the large amount of suspended dust from natural sources, cement industries, and construction. Figure 10.3 shows monthly means of parallel smoke and TSP measurements at a residential (El-Dokki) and an industrial site (Shoubra El-Khayma) from December 1982 until November 1984 (Hindy et al., 1990). While monthly mean smoke concentrations are around 100–150 μg m^{-3}, the respective TSP levels are 400–900 μg m^{-3}. Generally, there are slightly higher SPM levels in winter. Monthly smoke levels were 10 per cent higher at the industrial site, while TSP levels were about 40 per cent higher at the same site.

Table 10.4 Annual mean total suspended particulate matter (TSP) concentrations (μg m^{-3}) in Cairo, 1987–1989

Station	Location	1987	1988	1989
Ministry of Health	City Centre	646	649	632
Attaba	City Centre (high traffic)	641	704	699
Heliopolis	Residential (high traffic)	502	602	548
El-Dokki	City Centre Commercial	935	591	602
Helwān	Residential Industrial	1,161	838	1,100
Mean		777	677	716

Source: Nasralla, 1990

Lead

Emissions There is no emission inventory for lead (Pb) available in Cairo. Usually, the major source for Pb emissions are Pb compounds added to petrol as antiknock agents.

The maximum Pb content of petrol in Cairo is 0.8 g l^{-1} which is relatively high. It may be estimated that a motor vehicle population of 900,000 could give rise to Pb emissions in the order of 1,000–2,000 tonnes per annum.

Besides vehicle exhaust, industrial emission sources (e.g., Pb smelters) emit only small amounts of Pb; nevertheless their local impact on air quality might be severe.

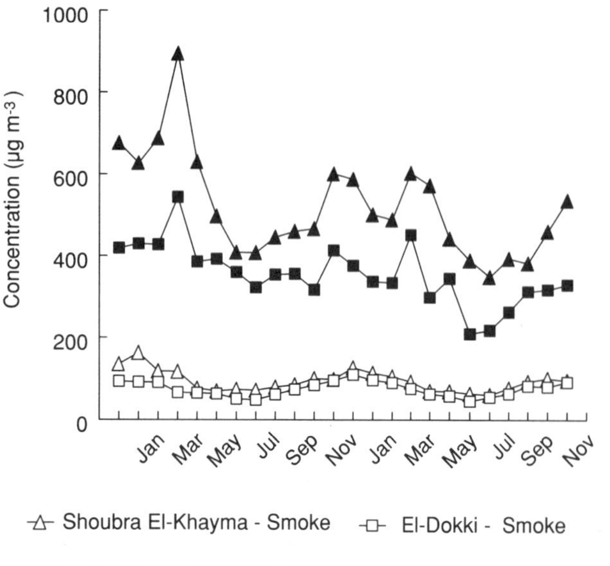

Figure 10.3 *Monthly mean total suspended particulates (TSP) and smoke concentrations, 1982–1983*

After Hindy et al., 1990

Ambient Concentrations A study of ambient atmospheric Pb concentrations at five sites in Cairo was conducted by the Air Pollution Laboratory of the National Research Centre (Ali et al., 1986). As can be seen in Table 10.5, Pb levels exceed the WHO annual mean guideline of 1 μg m^{-3} in all but one north-eastern residential site. Lead concentrations were highest in the city centre. During this monitoring campaign, Pb concentrations were greatest in the summer with the maximum of 6.4 μg m^{-3} occurring in June 1984.

A survey of blood Pb levels among Cairo males and traffic policemen revealed that the mean Pb concentration in blood was much higher in the urban population than in the rural population (Nasralla et al., 1984b). The mean Pb levels in blood of the urban population were about 31 μg dl^{-1}, and of heavily exposed policemen about 63 μg dl^{-1}, which is above the WHO 98 percentile blood Pb guideline of 20 μg dl^{-1}.

Carbon monoxide

Emissions An all-inclusive carbon monoxide (CO) emission inventory has not yet been established for Cairo. Generally, in urban areas exhaust of petrol motor vehicles is the largest CO emission source. In Cairo, CO emissions from cars and buses have been estimated to be 160,000 tonnes per annum in 1990, more than a sevenfold increase from 1970. About 99 per cent of traffic CO emissions are from cars, as buses (and other diesel-powered vehicles) have very low CO emission rates. Due to growth in traffic volume, a 40 per cent increase in CO emissions is expected by the year 2000.

Other CO emissions sources are probably relatively small compared with motor vehicles. A 1985 survey showed that the 20 major point sources in Greater Cairo had CO emissions of about 1,000 tonnes per annum with a single iron and steel plant

Table 10.5 *Lead concentrations (μg m⁻³)in ambient air in Cairo, 1983 –1984*

	City Centre	Residential North	Suburban North-east	Commercial South-east	Residential/Industrial North-west
Autumn	2.6	1.2	0.5	1.9	n.a.
Winter	2.2	1.4	0.6	2.1	1.2
Spring	2.3	1.3	0.5	2.0	1.2
Summer	4.9	1.8	0.7	2.6	1.4
Mean	3.0	1.4	0.6	2.2	1.3

n.a. No data available

After Ali et al., 1986

accounting for more than 60 per cent of those emissions (Nasralla, 1990). Power stations generally have very efficient combustion systems and thus emit relatively low amounts of CO.

Ambient Concentrations No long-term routine CO monitoring network exists in Cairo. However, some specific research studies on CO pollution from motor vehicles have been carried out over the past 20 years. For instance, measurements of CO during one day in 1984 at roof level at a city centre site with heavy traffic showed half-hourly CO concentrations of 7–19 ppm (about 8–22 mg m^{-3}), and the WHO eight-hour guideline of 10 mg m^{-3} was exceeded during most hours of the afternoon and the evening (Nasralla, 1990). Since 1984, traffic CO emissions have grown by about 15 per cent, which had an influence on the ambient CO levels. The projected further increase in motor vehicle traffic without any emission control devices will lead to even higher levels of ambient CO in the future.

Oxides of nitrogen

Emissions At present, no comprehensive emission inventory for oxides of nitrogen (NO_x) has been established in Cairo.

Oxides of nitrogen emissions from motor vehicles were estimated to be about 7,300 tonnes per annum in 1990, with two-thirds of it coming from cars and one-third from buses. This relatively low number is due to a relatively large number of cars with low-temperature, inefficient combustion engines.

A 1985 survey showed that the 20 major point sources (power stations and industry) in Greater Cairo had NO_x emissions of about 24,700 tonnes per annum. Since then, one very large power plant (Shoubra El-Khayma) was opened, which has led to increased NO_x emissions.

Ambient Concentrations There is no long-term routine nitric oxide (NO) or nitrogen dioxide (NO_2) monitoring network in Cairo. There has been only one major research study on NO_x and photochemical oxidants (NO and NO_2) which took place in 1979 (Nasralla and Shakour, 1981). During 1979, monthly mean NO_x concentrations in the city centre location were 0.2 ppm (380 μg m^{-3}) in January–March, 0.4–0.75 ppm (760–1,400 μg m^{-3}) in April–July, and 0.3–0.4 ppm (570–760 μg m^{-3}) in August–December, as shown in Figure 10.4. The marked maximum NO_x concentrations during May and June are connected with increased traffic during that time and probably enhanced by temperature inversions during those months. These values are far above the WHO 24-hour guideline of 150 μg m^{-3} for NO_2.

Similarly high values were reported from measurements of traffic-related NO_x during one day in 1984 at roof level at a city centre site with heavy traffic (Nasralla, 1990). Results showed half-hourly NO_x concentrations of 0.1–0.3 ppm (about 190–570 μg m^{-3}).

Ozone

Emissions Ozone (O_3) is a secondary pollutant which is a product of complex atmospheric reactions of NO_2 and reactive volatile organic compounds (VOC) under the influence of sunlight.

There is only a partial emission estimate for NO_2 (see above), and no emission inventory for VOC. As stated above, concentrations of NO_x are very high, which means that NO_x emissions should be high, too. Volatile organic compound emissions result from a variety of sources including solvent and petrol evaporation, incomplete combustion processes, and natural sources (e.g., vegetation). In the absence of statistical industrial data the amount of VOC emissions in Cairo cannot be estimated.

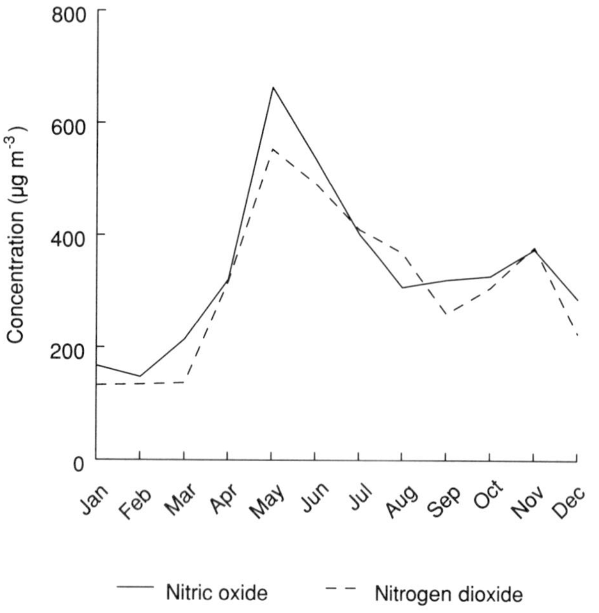

Figure 10.4 *Monthly variations of oxides of nitrogen concentrations, 1979*

After Nasralla and Shakour, 1981

Ambient Concentrations Photochemical oxidants are not monitored on a routine basis in Cairo. Some data on O_3 concentrations are available from research studies.

During a 1979 study (Nasralla and Shakour, 1981), O_3 was monitored on six to nine days per month. There was not a single month in which the WHO one-hour guideline of 150-200 μg m^{-3} was not exceeded. From April to September 1979 the hourly WHO guideline was exceeded during two to eight hours on every day on which O_3 was monitored. Peak hourly concentrations of more than 1,000 μg m^{-3} were recorded (Table 10.6). However, during this study a method was used (neutral KI method), which has a rather poor reliability compared with modern UV absorption techniques.

A more recent study on O_3 levels was performed between May and August 1989 at a residential site in Cairo (Nasralla, 1990). At this site, rather low O_3 concentrations were recorded. Maximum half-hourly concentrations of 110–140 μg m^{-3} were monitored which are below the WHO hourly guideline (150–200 μg m^{-3}). A clearer view of the O_3 pollution situation in Cairo cannot be given unless a more thorough monitoring of O_3 is initiated.

10.4 Conclusions

Air Pollution Situation Generally, the ambient air pollution situation in Cairo can be characterized by the very high levels of all major air pollutants (SO_2, SPM, Pb, CO, NO_x and O_3). Although only limited monitoring data are available, and the reliability of some of the available data is unclear, the data indicate that short-term as well as long-term mean pollutant concentrations regularly exceed the WHO guidelines, especially in the city centre.

Table 10.6 *Ozone monitoring in Cairo, 1979*

Month	No. of measured days	Days with 1-hr ozone levels above (μg m^{-3}) 100	200	300	400	600	800	1000	Duration over 200 μg m^{-3} (hrs)
January	7	6	3	2	1				2-3
February	6	6	3	1					1-3
March	8	7	5	4	2				2-4
April	8	8	8	4	3	1			2-5
May	8	8	8	7	7	5	2	1	3-8
June	8	8	7	6	5	3	1	1	3-7
July	9	8	8	8	7	3	1		2-7
August	7	7	7	5	4	1			2-6
September	8	8	8	5	2				2-5
October	8	7	4	1					2-5
November	8	6	4	2					1-3
December	8	1	4	2					1-3

After Nasralla and Shakour, 1981

Since Egypt has a desert climate, a considerable amount of SPM results from natural dusts. In addition, because of low rainfall, there is frequent re-entrainment of street dust and other particulates to the air, especially on windy days.

Main Problems From the monitoring data it must be concluded that the main pollution problem in Cairo results from motor vehicle traffic and from the high density of industrial activities in the greater Cairo area. Cement industries are of particular concern for the emission of vast quantities of particulate matter.

As Cairo is faced with increasing industrial development and vehicular traffic coupled with relatively little overall control of emissions, the air pollution situation will inevitably deteriorate.

Control Measures Air pollution problems in Cairo are typical of rapidly growing industrialized cities. High levels of air pollutants are emitted by poorly controlled sources which have not been adequately quantified. The lack of emission inventories make the formulation of efficient control strategies difficult.

The large number of motor vehicles in Greater Cairo contribute very significant quantities of both particulate and gaseous pollution. Commins (1987) reports that there will be enforcement of regulations regarding some gaseous emissions from motor vehicles in the future. However, the emission of lead will remain largely uncontrolled, which is of great concern due to the high lead content of petrol in Egypt.

Industrial emissions from large stationary sources have already been addressed by pollution control, but the large number of small industrial facilities makes control policies difficult to implement. There is also often a widespread open incineration of rubbish and this needs to be controlled.

In 1987 WHO conducted a detailed review of air pollution in Egypt (Commins, 1987). Some of the WHO recommendations for an improved air quality management system in Cairo were:

- Air pollution monitoring, including sampling and analytical quality control should be strengthened in order to ensure that continuous and reliable data are produced in the future. Measurement of NO_2, O_3, Pb and CO should be introduced on a permanent basis;
- Emissions inventories should be carried out for all the major gaseous pollutants in the Greater Cairo area;
- A 'Pollution Control Inspectorate' should be developed to supervise industrial plant location and to check compliance with emissions standards;
- Lead additives in petrol should be reduced as a major priority;
- A motor vehicle exhaust inspection programme should be introduced particularly for older vehicles;
- Compressed natural gas should be considered as a fuel for captive fleets;
- Traffic congestion should be reduced. An improved traffic flow can be achieved through the introduction of traffic restraint, one-way systems, parking control and specific vehicle bans;
- Dust control technology should be introduced at major industrial works;
- Refuse collection and disposal should be improved, and legislation should be introduced to prevent open burning of rubbish in urban areas;
- Strict industrial zoning should be applied for new industries. If possible, polluting industries should be relocated away from residential areas;
- Further use of natural gas, especially by industry and by electric power stations, should be encouraged.

10.5 References

Ali, E. A., Nasralla, M. M. and Shakour, A. A. 1986 Spatial and Seasonal Variation of Lead in Cairo Atmosphere, *Environmental Pollution* **11B**, 205–210.

Commins, B. 1987 *Air Pollution in Egypt*, (EGY/CEH/001) WHO Regional Office for the Eastern Mediterranean, Alexandria.

Faiz, A., Sinha, K., Walsh, M. and Varma, A. 1990 *Automotive Air Pollution: Issues and Options for Developing Countries*, World Bank Policy and Research Writing Paper WPS 492, The World Bank, Washington DC.

Hindy, K. T., Farag, S. A. and El-Taieb, N. M. 1990 Monthly and Seasonal Trends of Total Suspended Particulate Matter and Smoke Concentration in Industrial and Residential Areas of Cairo, *Atmospheric Environment* **24B**, 343–353.

Nasralla, M. M. and Shakour, A. A. 1981 Nitrogen Oxides and Photochemical Oxidants in Cairo City Atmosphere, *Environment International* **5**, 55–60.

Nasralla, M. M., Shakour, A. A. and Ali, E. A. 1984a *Using the British standard method to monitor SO_2 in Egypt air*, Second Egyptian Congress of Chemical Engineers.

Nasralla, M. M., Shakour, A. A. and Said, E. A. 1984b Effect of lead exposure on traffic policemen, *Egyptian Journal of Industrial Medicine* **8**, 87–104.

Nasralla, M. M. 1990 *Air Quality in Cairo Metropolitan Area*, Report for WHO, Air Pollution Department, National Research Centre, Cairo.

UN 1989 *Prospects of World Urbanization 1988*, Population Studies No. 112, United Nations, New York.

WMO 1971 *Climatological Normals (CLINO) for Climate and Climate Ship Stations for the Period 1931–1960*, World Meteorological Organization, No. 117, Geneva.

11

Calcutta

11.1 General Information

Geography Calcutta is the premier city of West Bengal and is situated in the Ganges delta (Latitude 22°32′N, Longitude 88°20′E) on the banks of the Hooghly River which divides the region in two (Figure 11.1). The conurbation on the eastern bank is known as Calcutta, the western part as Howrah. Calcutta was originally a port, but is now some 130 km from the Bay of Bengal due to alluvial deposition. The mean elevation of the city is 6 m above mean sea level rising to a maximum of 9 m. Calcutta Metropolitan District consists of two major municipal corporations, 33 municipalities and 37 non-municipal areas and covers an area of 1,295 km^2.

Demography Calcutta has the largest urban population in India. In 1990 the estimated population of the urban agglomeration was 11.83 million. UN estimates indicate that the population has risen from 6.91 million in 1970 and is projected to rise to 15.94 million by the year 2000 (Figure 1.1 and Table 2.1, UN, 1989). Overcrowding is already a severe problem in Calcutta. Historically, it was the partition of Bengal with the consequent influx of refugees from Bangladesh (then East Pakistan) which had the greatest impact on Calcutta's population. In 1981, 3.24 million (35 per cent) out of 9.16 million residents were classified as slum dwellers.

Climate Calcutta has a tropical savanna climate. The annual mean temperature is 26.8°C, although monthly mean temperatures range from 20°C to 31°C and maximum temperatures in Calcutta often exceed 40°C. The main seasonal influence upon the climate is the monsoon. Maximum rainfall occurs during the monsoon in August (306 mm) and the average annual total is 1,582 mm (WMO, 1971). Moderate north-westerly to north-easterly winds prevail for most of the year with a high frequency of calms. Early morning mists are common. Evening "smog" often occurs due to night-time temperature inversions and mixing heights are generally restricted to below 500 m during the night. Summer is dominated by strong south-westerly monsoon winds. Mean ventilation coefficients are greatest in the pre-monsoon (8,118 $m^2 s^{-1}$) and monsoon (7,410 $m^2 s^{-1}$) periods. Total duration of sunshine is 2,528 hours per annum with maximum insolation occurring in March.

Industry The majority of 187,719 non-agricultural enterprises in Calcutta are wholesale and retail trade (41 per cent); however, manufacturing and repair services do make up 23 per cent of the total number of businesses. Most of the major polluting industries, such as heavy engineering, chemicals, jute, textiles, glass and ceramics, and paper, are listed in Table 11.1. Electricity is provided by two thermal power plants (NEERI, 1988). Many heavy industries are found in the areas surrounding Calcutta and include coal mining, iron and steel, manganese, mica and refining. Tea and jute are the main agricultural products.

Coal is the principal industrial fuel used in Calcutta due to the close proximity of West Bengal's coalfields. Industrial coal consumption for 1989–90 is shown in Table 11.1. In addition to its industrial use, coal is also an important domestic fuel in Calcutta. It is estimated that 90 per cent of slum dwellings burn coal (and charcoal) as fuel which is believed to have an influence upon the high incidence of respiratory disease in these areas (NEERI, 1991a).

Transport As in other Indian cities, it is estimated that at least 25 per cent of all trips are made by walking. Calcutta's public transport system comprises buses, trains, metro, trams, and ferries. In 1980 the bus was the predominant mode of transport (excluding walking), accounting for 67 per cent of all motorized trips compared with only 34 per cent in 1970; para-transit and railway account for a further 24 per cent of trips. Buses and trams are important modes of transport in the city centre as they link the railway stations with people's work places. The State Transport Corporation has a fleet of approximately 650 buses; however, private operators also run services and it is estimated that there are over 5,000 buses plying the roads of Calcutta and a further 4,300 in neighbouring areas. It is assumed that the majority of these buses are diesel-driven as in other Indian

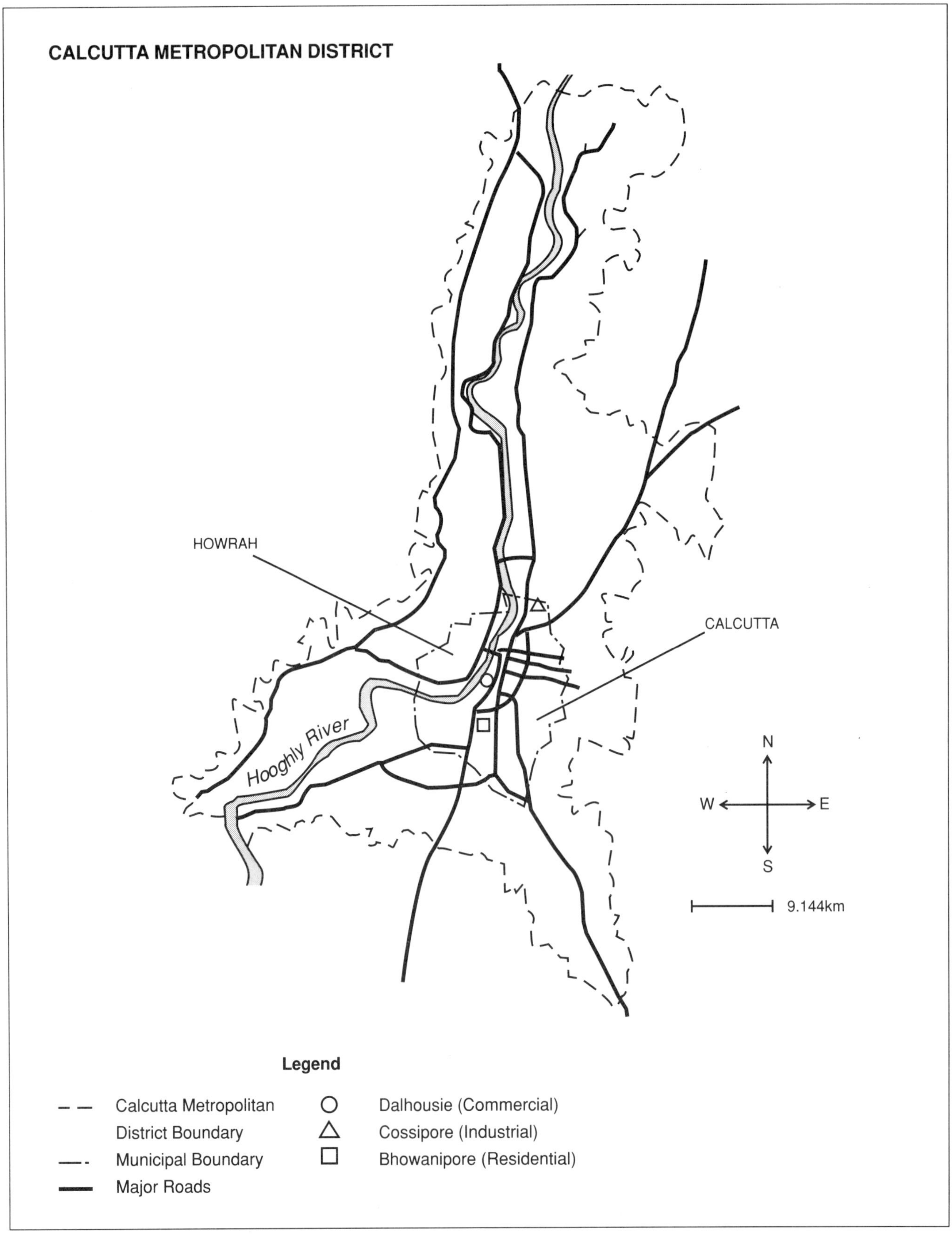

Figure 11.1 *Sketch map of Calcutta Metropolitan District*

Table 11.1 *Industrial coal consumption in Calcutta, 1989–90*

Rank	Industry	No. of units	Coal consumption ($t\ a^{-1}$)
1	Engineering, etc.	277	89,208
2	Forging	24	83,580
3	Rerolling machinery	129	76,428
4	Glass	21	70,680
5	Jute	18	61,776
6	Chemical	86	56,580
7	Ceramics	49	46,812
8	Paper	18	46,296
9	Vanaspathi oil	29	44,424
10	Rubber	110	35,820
11	Refectories	12	13,548
12	Life-saving drugs	14	7,212
13	Others	558	265,596
	Total	1,451	933,960

Total Industrial Coal Supply, 1989–90: 2,100,000 $t\ a^{-1}$
Projected Total Industrial Coal Supply, 2000-01: 3,600,000 $t\ a^{-1}$
After NEERI, 1991a

cities. Local trains mainly transport commuters to the east of the city (approximately 85 per cent of all rail passengers in Calcutta fall in this category). An underground suburban railway (metro) links Esplanade to Tollyganj, a distance of 16 km, and has helped to reduce traffic congestion.

It is estimated that in 1980 10.13 million motorized trips per day were made and that these will rise to 18.3 million by 2000, a slower rate of increase than for other Indian megacities (Faiz et al., 1990). However, total motor vehicle registrations have more than doubled between 1981 and 1989. In 1989 there were approximately 500,000 motor vehicles registered in the Calcutta Metropolitan District. The most significant increase has been in the number of two-wheelers (mainly scooters and mopeds) which generally have two-stroke engines. In 1980–81 the ratio of cars to two-wheelers was approximately 2:1; by 1988-90 the ratio was almost 1:1. In Howrah and other neighbouring districts the number of two-wheelers is increasing at a faster rate, and in these areas motor cycles, scooters and mopeds are the most common motor vehicle type.

Despite an extensive and varied public transport system Calcutta is subject to severe congestion. Slow-moving traffic, such as bullock carts, rickshaws (pulled by people) and bicycles, combined with poor traffic planning, make travelling by road very difficult and have inevitably led to increased vehicular emissions. As in other Indian cities old motor vehicles tend to remain in circulation for many years. The engines of such "old" vehicles (both petrol- and diesel-driven) are often very inefficient in terms of combustion and therefore make a large contribution to overall transport emissions.

11.2 Monitoring

Figure 11.1 is a map of Calcutta Metropolitan District which shows the location of the three GEMS/Air monitoring sites operated by the National Environmental Engineering Research Institute (NEERI). Monitoring of suspended particulate matter (SPM) dates back to 1972 at the Dalhousie (Commercial) and Cossipore (Industrial) stations and to 1973 at the Bhowanipore (Residential) site. Monitoring of gaseous pollutants (sulphur dioxide (SO_2) and nitrogen dioxide (NO_2)) began in 1978 at all three sites. Monitoring was discontinued at the NEERI sites in 1988. Monitoring has since been carried out by the Central Pollution Control Board (CPCB) - West Bengal. However, the results are not comparable with the NEERI data and no indication of methodology, site location or sampling frequency has been provided and therefore these data are not presented here. NEERI recommenced monitoring at the three GEMS sites in April 1990 and these data have been included.

11.3 Air Quality Situation

Sulphur dioxide

Emissions Emissions estimates calculated by NEERI indicate that industry and power generation are the main sources of SO_2 in Calcutta. Figure 11.2 shows that the effects of significant reductions in domestic emissions brought about by a decline in domestic and commercial coal use have, to a large extent, been cancelled by increasing industrial and transport emissions. This trend is projected to continue at least until the year 2000. Overall, it is estimated that SO_2 emissions have remained stable since 1980 at approximately 25,000 tonnes per annum and will remain constant until 2000.

A breakdown of industrial SO_2 emissions conducted by NEERI and CMDA in 1977–78 (Table 11.2) reveals that thermal power plants accounted for 34 per cent. The chemical and engineering industries are responsible for 11 per cent each. Table 11.1 ranks industries (excluding power generation) in terms of

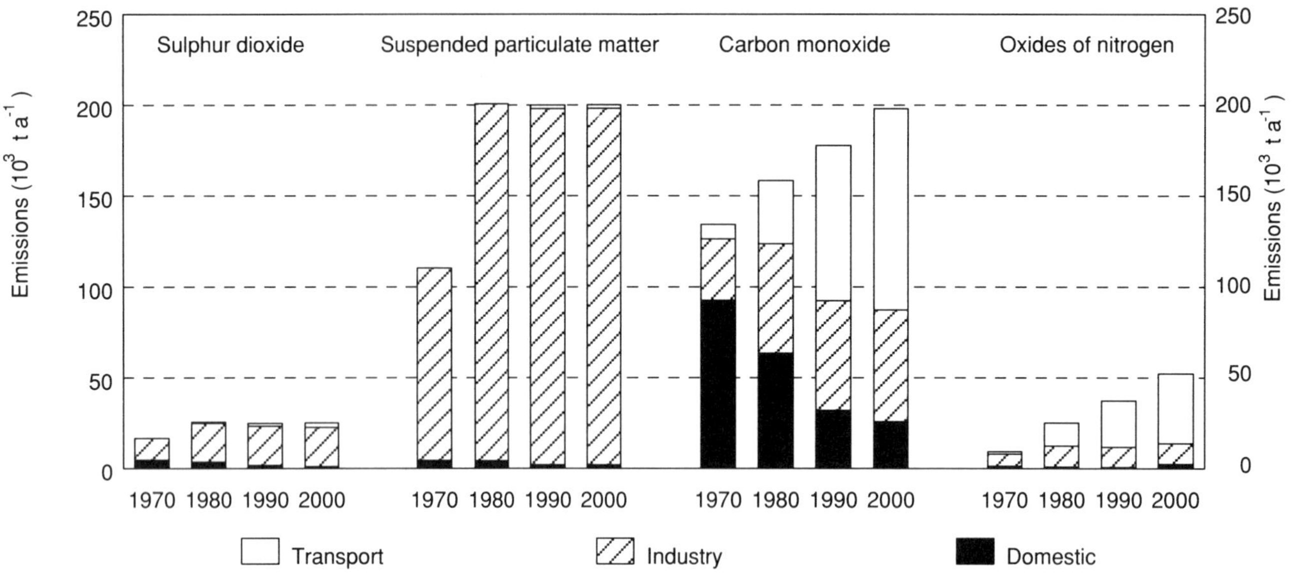

Figure 11.2 *Estimated and projected anthropogenic pollutant emissions by source, 1970–2000*

Source: NEERI, 1991a

coal consumption. In general, the engineering industries consume large quantities of coal and this is reflected in SO_2 emissions. The chemical industry ranks sixth in terms of coal consumption, suggesting that SO_2 emissions may be attributable to actual industrial processes.

The increase in transport SO_2 emissions is attributed to the increase in diesel-driven vehicles. The bus and truck population of Calcutta and Howrah grew by 78 per cent between 1980 and 1989.

Ambient Concentrations Figure 11.3 shows annual mean SO_2 concentrations at the three GEMS/NEERI monitoring sites. The graph shows that concentrations approximately doubled at all sites between 1980 and 1981; a maximum annual mean concentration of 104 μg m^{-3} was observed at the Dalhousie commercial station. It is important to note that this apparent doubling is partly due to missing data for the monsoon months (when concentrations are at a minimum); therefore, the annual values for 1981

Table 11.2 *Industrial emissions in Calcutta, 1977–78*

Industry	No. of units surveyed	Sulphur dioxide		SPM		Carbon monoxide	
		(t a^{-1})	%*	(t a^{-1})	%*	(t a^{-1})	%*
Thermal power plant	2	5,475	34	56,575	44	1,095	5
Chemical	172	1,825	11	20,440	15	6,570	25
Engineering	105	1,825	11	11,315	9	3,650	14
Rubber	80	365	3	9,855	7	3,285	14
Glass and ceramic	36	365	3	6,205	5	2,190	9
Textile	57	365	3	5,840	4	1,825	7
Jute	10	365	3	5,840	4	1,825	7
Paper	11	365	3	1,825	1	730	2
Miscellaneous	135	5,110	29	14,600	11	4,745	18
Total	608	16,060		132,495		25,915	

*Owing to rounding, percentage totals may not equal 100%. After NEERI, 1991a

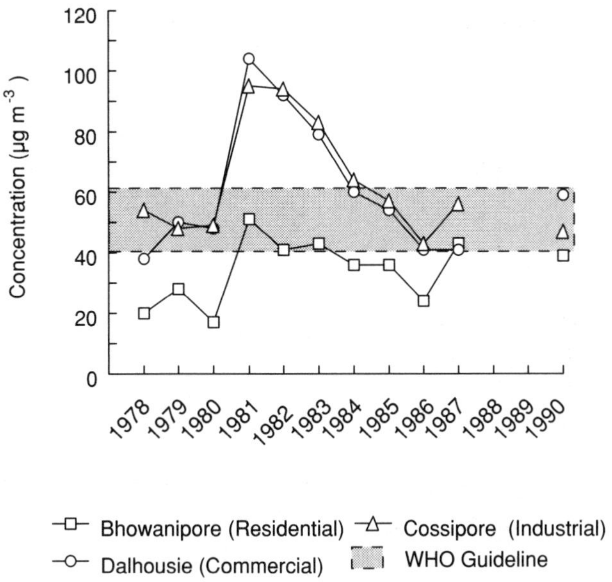

Figure 11.3 *Annual mean concentrations of sulphur dioxide*

Sources: NEERI, 1980, 1983, 1988, 1990, 1991b

cannot be considered representative. However, it is also interesting to note that concentrations in the following year, 1982, were of a similar order when sampling remained consistent throughout the year. Annual 98 percentile concentrations were higher in 1982 than in 1981. Following 1981 and 1982, annual levels fell steadily and by 1985 annual average concentrations at all sites were within or below the WHO guideline range. Concentrations at the Bhowanipore residential station never exceeded the upper limit of the WHO annual guideline range (60 $\mu g\ m^{-3}$) between 1978 and 1987. Monitoring recommenced at the GEMS/NEERI sites in April 1990. The annual arithmetic mean concentrations in 1990 (April to December) were still within the WHO guideline range (NEERI, 1991b).

Ambient SO_2 concentrations peak during the winter (November to February inclusive) with the monthly maximum occurring in November (Figure 11.4). Climatic factors, such as the high percentage of calms and ground-based temperature inversions, are of great importance during the winter. In winter diurnal concentrations are generally higher at night between 2000 hours and 0400 hours due to temperature inversions. Minimum SO_2 concentrations occur in May before the onset of the monsoon season (June to October inclusive). Levels remain relatively low throughout the monsoon.

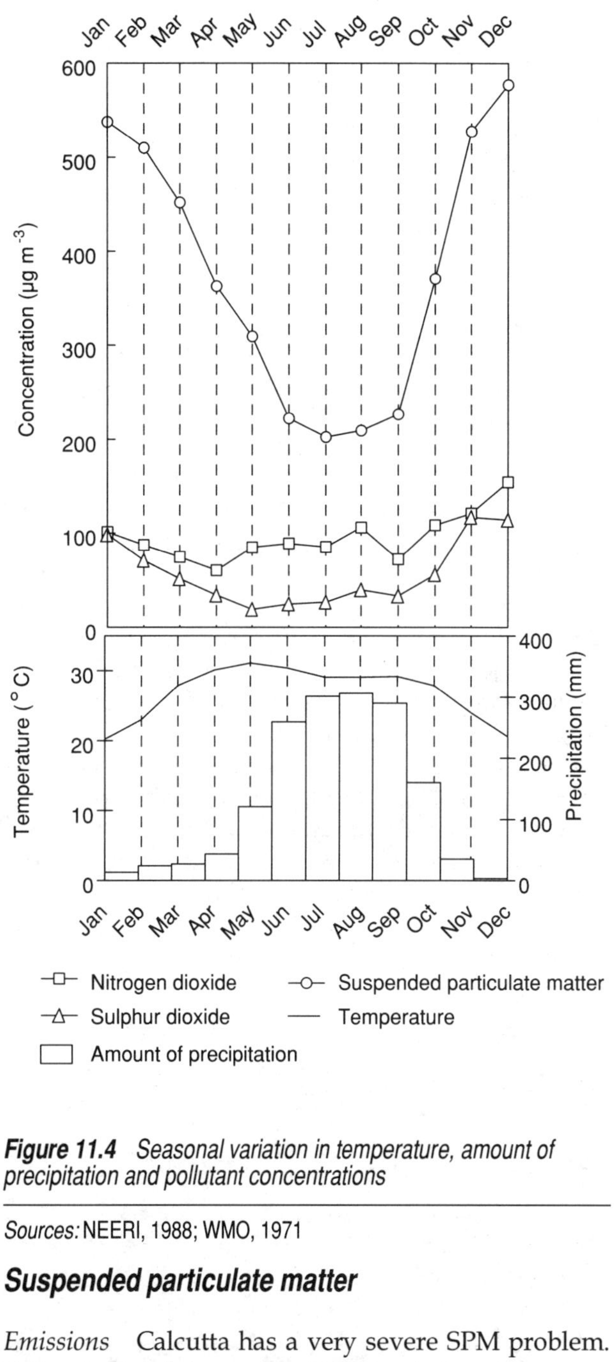

Figure 11.4 *Seasonal variation in temperature, amount of precipitation and pollutant concentrations*

Sources: NEERI, 1988; WMO, 1971

Suspended particulate matter

Emissions Calcutta has a very severe SPM problem. Estimated anthropogenic SPM emissions were approximately 200,000 tonnes per annum in 1990, a similar value to 1980, and it is projected that emissions will remain fairly stable until 2000 (Figure 11.2). Industrial sources account for 98 per cent of the 1990 total. The high emissions and ambient concentrations of SPM (Figure 11.5) result from the high level of coal use by Calcutta's industry, particularly at thermal power plants. Table 11.2 shows that in 1977–78 Calcutta's two thermal power plants accounted for 44 per cent of industrial SPM emissions, followed by the chemical industries which emitted a further 15 per cent. The coal burned in the industrial boilers has a relatively low sulphur content (0.3 per cent) but is high in ash content (24–36 per cent). In 1989-90, 1,451 out of 2,218 registered factories burnt approximately one million tonnes of coal. The influence of the natural dust component of the particulate fraction is not clear as no analysis has been presented.

Ambient Concentrations Historical emissions estimates indicate a 66 per cent increase in SPM emissions between 1970 and 1980. However, monitoring data do not support these estimates, as can be seen in Figure 11.5. Regression of the annual mean SPM concentration between 1972 and 1980 shows no significant overall trend during the 1970s. In fact, a negative trend is observed at the industrial monitoring site at Cossipore for the period 1972–1980. Between 1972 and 1987 all sites exhibited a positive annual mean trend (although not statistically significant). It is possible that SPM from construction activities and the re-entrainment of street dust has declined whereas industrial emissions have increased (Aggarwal, 1991).

Annual mean and 98 percentile concentrations (Figures 11.5 and 11.6) at all stations greatly exceed both WHO guidelines and Indian Air Quality Standards (Appendix I). The overall average concentration in 1987 was 557 μg m^{-3}, over six times the maximum WHO annual

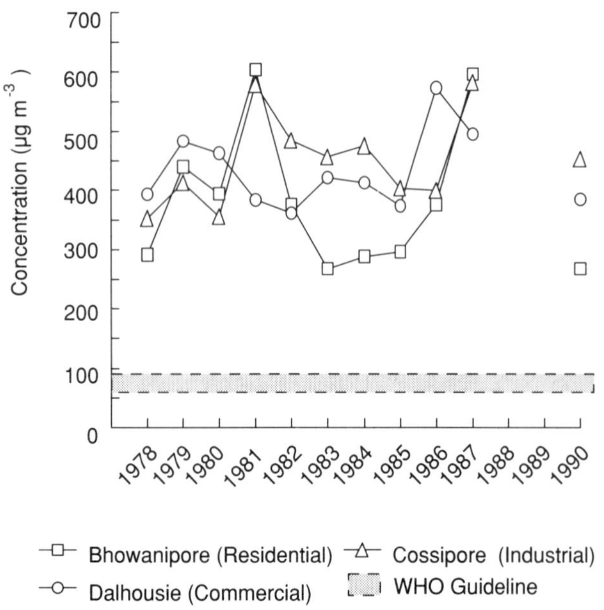

Figure 11.5 *Annual mean concentrations of suspended particulate matter (TSP)*

Sources: NEERI, 1980, 1983, 1988, 1990, 1991b

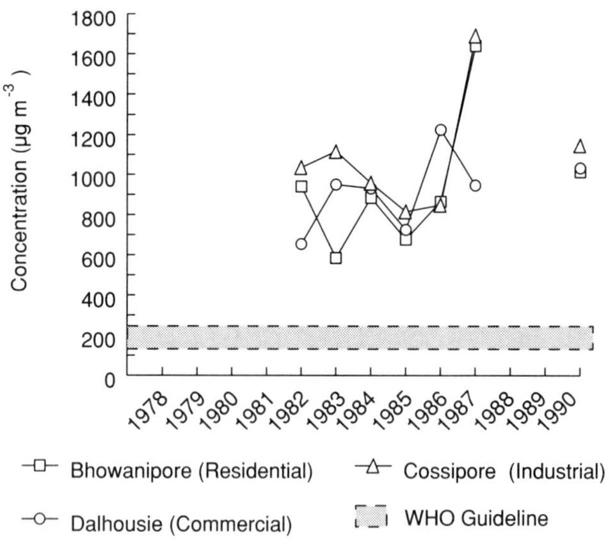

Figure 11.6 *Annual 98 percentile concentrations of SPM (TSP)*

Sources: NEERI, 1980, 1983, 1988, 1990, 1991b

guideline (60–90 μg m^{-3}). The annual 98 percentile concentration of the Cossipore industrial monitoring site reached 1,680 μg m^{-3} in 1987, 14 times the WHO daily guideline and the second highest ever in Calcutta, indicating that episodes of short duration also constitute a problem. There is no significant difference in concentrations between the various sites indicating that high concentrations occur throughout the city. Annual arithmetic mean concentrations in 1990 (April to December) were lower than in 1987 at all three sites (268–453 μg m^{-3}), but were still well above the WHO annual guidelines. Annual 98 percentile concentrations are also extremely high (1,014–1,145 μg m^{-3}).

The contribution of natural dust to overall SPM concentrations is not obvious. However, it is likely to be lower than for other "drier" Indian cities such as Delhi. As Figure 11.4 shows, the monsoon (June to October inclusive) has a pronounced washout effect; concentrations during this period are half those of the winter. Concentrations reach a peak in December and are likely to be influenced by temperature inversions and low wind speeds.

Lead

Ambient Concentrations The lead content of petrol from the Halidia refinery, which supplies Calcutta, is lower than that in Delhi or Bombay at 0.1 g l^{-1}. Despite the relatively low lead content of petrol, annual airborne lead levels, monitored at the three GEMS/NEERI sites, are the highest in India (NEERI,

1991c). Annual concentrations were found to be highest at the residential and commercial sites (0.73 μg m^{-3}), but were below the WHO annual guideline of 1 μg m^{-3}.

Carbon monoxide

Emissions Figure 11.2 shows estimated and projected carbon monoxide (CO) emissions in Calcutta between 1970 and 2000. It is estimated that in 1990, CO emissions totalled approximately 177,000 tonnes per annum. Transport was the greatest source of CO, accounting for 48 per cent of the total, followed by industry at 34 per cent and the remainder classified as domestic emissions - 18 per cent. Figure 11.2 shows the decline of domestic emissions due to changes in fuel use while at the same time motor vehicle emissions have increased by over ten times between 1970 and 1990. Industrial emissions increased from 33,000 tonnes per annum in 1970 to 60,000 tonnes per annum in 1980, but have remained relatively stable since then and are projected to remain at present levels to 2000.

Ambient Concentrations It is not possible to comment on the reliability of emissions estimates as there is no monitoring of ambient CO in the city. It has also not been possible to locate any recent studies referring to CO in Calcutta. It is understood that a detailed CO modelling study is under way in Calcutta (Aggarwal, 1991). Emissions estimates are on a par with those of Bombay, which experiences similar meteorology. However, the contributions of the various sources are very different (industrial and domestic sources are much more important in Calcutta) and it is likely that street-level exposure is low in Calcutta due to lower motor vehicle numbers and emissions and to a relatively "open" urban topography.

Oxides of nitrogen

Emissions Figure 11.2 shows the estimated and projected increase in oxides of nitrogen (NO_x as NO_2) between 1970 and 2000. Transport is now the dominant source of NO_x in Calcutta through the growth in motor vehicle traffic in recent years. It is estimated that in 1970 industry was the major source of NO_x (69 per cent) and that emissions from industrial plant had increased to over 11,000 tonnes per annum by 1980. Estimated industrial emissions have since stabilized and are not projected to increase significantly before 2000. Transport emissions have risen from an estimated 1,825 tonnes per annum in 1970 to 25,550

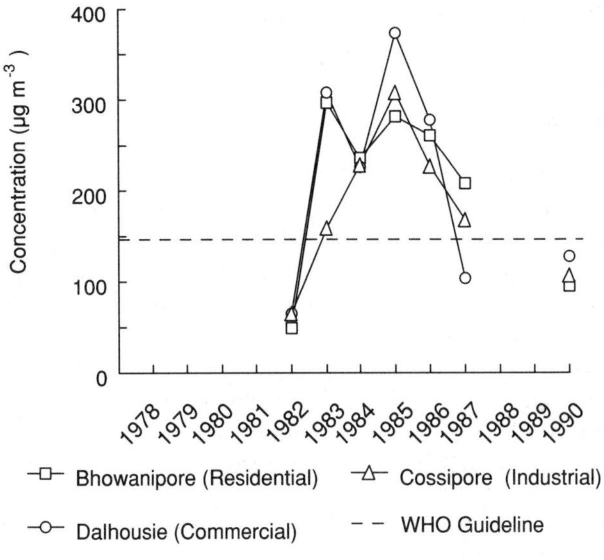

Figure 11.7 *Annual 98 percentile concentrations of nitrogen dioxide in Calcutta*

Sources: NEERI, 1980, 1983, 1988, 1990, 1991b

tonnes per annum in 1990. The main vehicular sources of NO_x are diesel-driven trucks and buses. Although diesel-driven vehicles only account for approximately 10 per cent of Calcutta's motor vehicle population, it is estimated that they are responsible for approximately 90 per cent of motor vehicle NO_x emissions.

Ambient Concentrations Monitoring of NO_2 at the three GEMS/NEERI sites since 1978 has revealed a significant positive trend in annual mean concentrations. Figure 11.7 shows that ambient annual 98 percentile concentrations at all three sites peaked in 1985. In 1986 and 1987 ambient levels decreased significantly, but generally were still above accepted guidelines. Data from 1990 (April to December) suggest that concentrations have fallen further and are now well below accepted guidelines (NEERI, 1991b). No explanation can be given for the decrease in urban concentrations; emissions are believed to be increasing because of increasing traffic. It is possible that meteorological factors such as insolation, the frequency of calms and the frequency of ground-based temperature inversions have influenced annual statistics throughout the 1980s. It is also possible that improved traffic circulation in recent years has helped to reduce NO_2 concentrations in the short term. Maximum concentrations are generally recorded at the Dalhousie commercial site, probably owing to the high traffic densities and congestion in this area.

Figure 11.4 indicates that there is no clear seasonal influence upon monthly NO_2 concentrations. The monthly peak in December is influenced by high insolation and ground-based temperature inversions.

Ozone

Ambient Concentrations Ozone (O_3) is not currently monitored in Calcutta on a regular basis. Monitoring of tropospheric O_3 should be initiated in all major Indian cities to identify whether photochemical smog constitutes a problem. Given the recent rapid increase, and the projected future increase, in precursor emissions in these cities it is likely that O_3 and other photochemical oxidants will increase in importance. Monitoring is needed to quantify the scale of the problem so that remedial measures can be identified.

11.4 Conclusions

Air Pollution Situation It is estimated by NEERI that 60 per cent of Calcutta's residents suffer from some kind of respiratory disease due to air pollution. The burning of coal as an industrial and domestic fuel accounts for a significant proportion of pollutant emissions, especially SPM. Suspended particulate matter from coal combustion is clearly a major problem throughout Calcutta and should be the main focus of immediate control efforts. Surprisingly, SO_2 concentrations are relatively low (within WHO guidelines) which is due to the low sulphur content (0.3 per cent) of the local coal.

It would appear from the data that industrial emissions have, to a large extent, stabilized and in some cases declined. It is not clear what the reasons for these changes are, but it is likely that planning measures restricting industrial development have played an important role. Changes in domestic and commercial fuel use, principally a reduction in coal use, and improvements in burning efficiency will have also helped to reduce emissions.

Carbon monoxide and NO_x emissions from motor vehicles are of increasing concern and probably present the greatest long-term threat to Calcutta's air quality. Calcutta's motor vehicle population doubles every six years, a trend which is likely to continue at least up to 2000. With this rate of growth it is unlikely that even the introduction of the most stringent control measures would reduce overall emissions and ambient concentrations from this source.

The data presented here give only a very limited picture of the air quality situation in Calcutta. A survey of air pollution levels and emissions throughout the Calcutta metropolitan district is required for proper air quality management.

11.5 References

Aggarwal, A. L. 1991 Personal Communication, NEERI, Nagpur.

Faiz, A., Sinha, K., Walsh, M. and Varma, A. 1990 *Automotive Air Pollution: Issues and Options for Developing Countries*, World Bank Policy and Research Working Paper WPS **492**, The World Bank, Washington DC.

NEERI 1980 *Air quality in selected cities in India 1978-1979,* National Environmental Engineering Research Institute, Nagpur.

NEERI 1983 *Air quality in selected cities in India 1980-1981,* National Environmental Engineering Research Institute, Nagpur.

NEERI 1988 *Air quality status in ten cities: India 1982-1985,* National Environmental Engineering Research Institute, Nagpur.

NEERI 1990 *Air quality status in ten cities: India 1986-1987,* National Environmental Engineering Research Institute, Nagpur. (Unpublished report.)

NEERI 1991a *Air pollution aspects of three Indian megacities, Volume III: Calcutta,* National Environmental Engineering Research Institute, Nagpur.

NEERI 1991b *Air quality status 1990,* National Environmental Engineering Research Institute, Nagpur.

NEERI 1991c *Air quality status: Toxic metals, polycyclic hydrocarbons, anionic composition and rain water characteristics (Delhi, Bombay and Calcutta),* National Environmental Engineering Research Institute, Nagpur.

UN 1989 *Prospects of World Urbanization 1988,* Population Studies No. 112, United Nations, New York.

WMO 1971 *Climatological Normals (CLINO) for Climate and Climate Ship Stations for the Period 1931-1960,* No. 117, World Meteorological Organization, Geneva.

12

Delhi

12.1 General Information

Geography The metropolitan area of Delhi consists of two cities: Delhi (or Old Delhi) which was the capital of Muslim India between the twelfth and nineteenth centuries; and New Delhi, located immediately to the south, the capital city of India since 1911. The sprawling city is situated in the territory of Delhi in the north of India (Latitude 28°35′N, Longitude 77°12′E) 160 km south of the Himalayas at an elevation of 216 m above mean sea level. The River Yamuna, a tributary of the Ganges, forms the eastern boundary of the city (Figure 12.1). Delhi is situated between the Great Indian Desert (Thar Desert) of Rajasthan to the west, the central hot plains to the south, and the cooler, hilly region to the north and east.

Demography UN estimates indicate that Delhi's population has more than doubled between 1970 (3.53 million) and 1990 (8.62 million) and is projected to increase to 12.77 million by the year 2000 (UN, 1989). In 1981 the city covered an area of 591.9 km^2 with a population density of 9,647 persons per km^2. Overcrowding is not as great a problem as in Calcutta or Bombay but the encroachment of the city into rural areas is of great concern. Rural-urban and inter-urban migration are the main reasons for Delhi's population growth. Employment opportunities in government offices, business and commerce and, more recently, industry all influence population growth. In the past there seems to have been a lack of co-ordination in discouraging employment opportunities in Delhi. For example, there has been no major decentralization of government.

Climate The region has a tropical steppe climate. The general prevalence of continental air leads to relatively dry conditions with extremely hot summers. Monthly mean temperatures range from 14.3°C in January (minimum 3°C) to 34.5°C in June (maximum 47°C). The annual mean temperature is 25.3°C (WMO, 1971). The main seasonal climatic influence is the monsoon, typically from June to October. The mean annual rainfall total is 715 mm. Maximum rainfall occurs in July (211 mm). The heavy rains of the monsoon act as a "scrubber". North-westerly winds usually prevail; however, in June and July south-easterlies predominate. Wind speeds are typically higher in the summer and monsoon periods; in winter, calms are frequent (20 per cent of the time).

A regular pre-monsoon feature is the Andhi dust storm where westerly winds from the Great Indian Desert deposit large concentrations of suspended particulate matter (SPM) into Delhi's atmosphere. Alternatively, pre-monsoon calms often lead to increased pollution loads due to lack of mixing/dilution. Ground-based temperature inversions are a regular feature. These restrict mixing height to low levels thus limiting pollutant dispersal.

Industry As India's capital city Delhi has become a nucleus of trade, commerce and industry in the northern region. Government office complexes are a major source of employment and the city is also home to important medical, agricultural and educational institutions. Delhi is often categorized as a "service town"; however, industry is rapidly expanding. There was a 57 per cent increase in industrial units from 26,000 in 1971 to 41,000 in 1981, despite the introduction of planning restrictions on large industry.

Engineering, clothing and chemicals predominate, although electronics and electrical goods are gaining in importance. Most industries are located in the west, south and south-east of the city. North-westerly winds often drag pollution from western industrial areas across the city. It has been recommended that all future industrial development is limited to the south-east to avoid such problems (Murty and Tangirala, 1990) and the Okhla industrial estate has recently been built in this area. Increasing industrial productivity combined with rapid urbanization mean that there is a greater demand for energy in Delhi than can be supplied; therefore, industry cannot generate at full capacity. Delhi has two major thermal power plants (Baderpur and Indraprastha).

Trade, banking and commerce are also assuming increasing importance, although not on the scale of Bombay. This change from "service town"to commercial centre will continue to have an important influence over land-use changes and population.

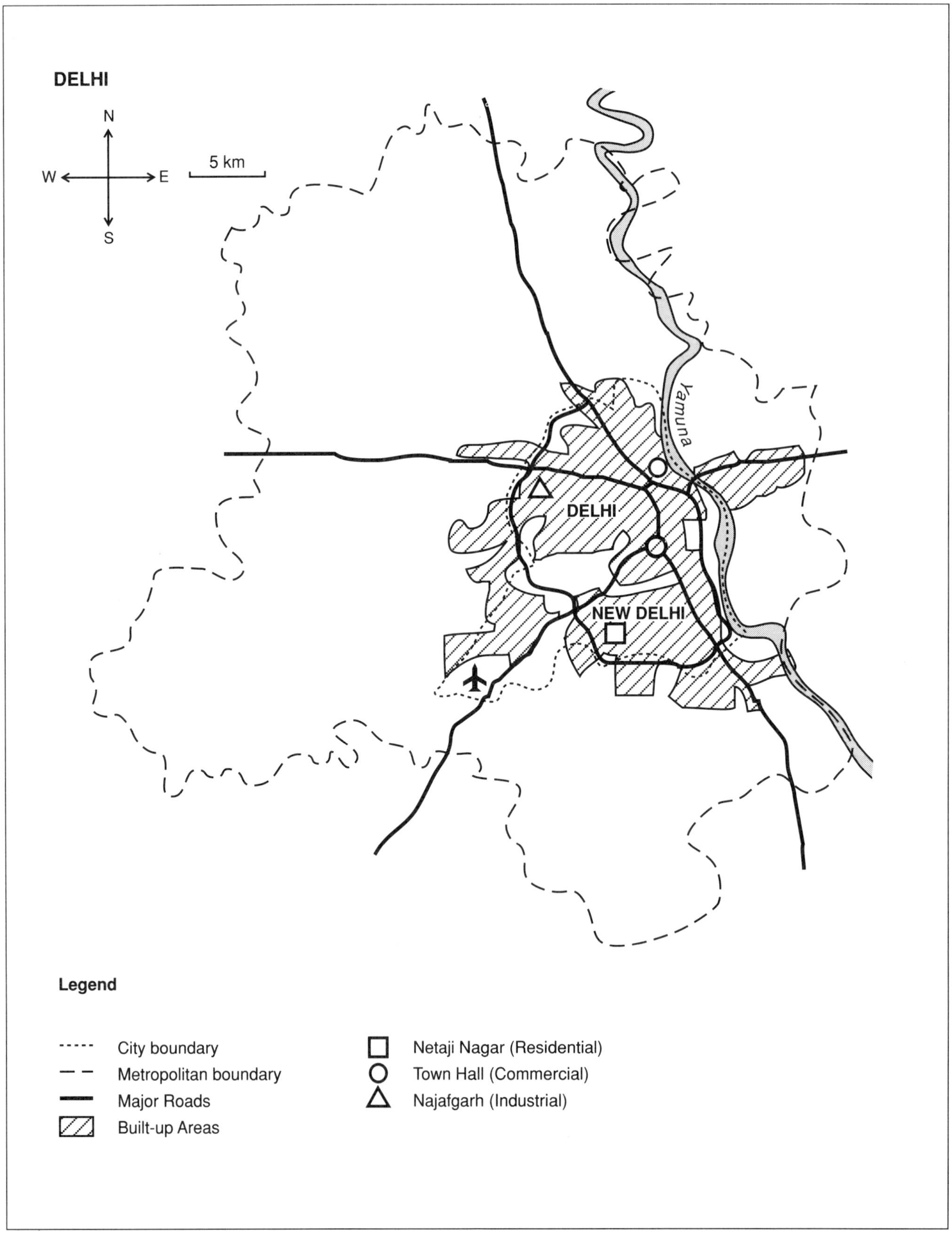

Figure 12.1 *Sketch map of Delhi*

Transport Delhi has long been the goods distribution centre of India because of its position at the communications/transport hub of the sub-continent. Modal split studies reveal that in the 1980s the bus was the dominant form of transport in Delhi (Faiz et al., 1990). Motor vehicle registrations between 1971–81 rose by 142 per cent compared with a population growth rate of 57 per cent over the same period. It is estimated that in 1989 there were currently 1.66 million vehicles registered in Delhi (10 per cent of the national total) and between 170,000 new ones are registered each year (500 per day). It is also estimated that 70,000 vehicles enter Delhi from neighbouring states every day. Figure 12.2 shows that scooters and motor cycles are the most common form of motorized vehicle; in 1981 they accounted for 11 per cent of all trips and were the third most used mode of transport after buses and bicycles. It is estimated that in 1989–90 two-wheelers represented 76 per cent (1,115,000) of motor vehicles in Delhi and were responsible for a large proportion of motor vehicle emissions. With an annual growth rate of 13 per cent the number of two-wheelers is projected to rise to approximately 300,000 by 2000–2001 (85 per cent of total motor vehicles). Cars and jeeps will more than double in numbers over the same period from an estimated 207,405 in 1989–90 to approximately 535,000 in 2000–2001.

The city road system consists of old, convoluted roads in the old city and of modern, diagonal roads in New Delhi. The traffic problem is also compounded by poor traffic planning and a disregard for regulations. Low driving speeds, poor engine maintenance and the great age of the motor vehicle fleet result in high pollutant emissions. Mass transport facilities are inadequate although the establishment of a ring railway and ring bus service have helped to mitigate part of the overall problem.

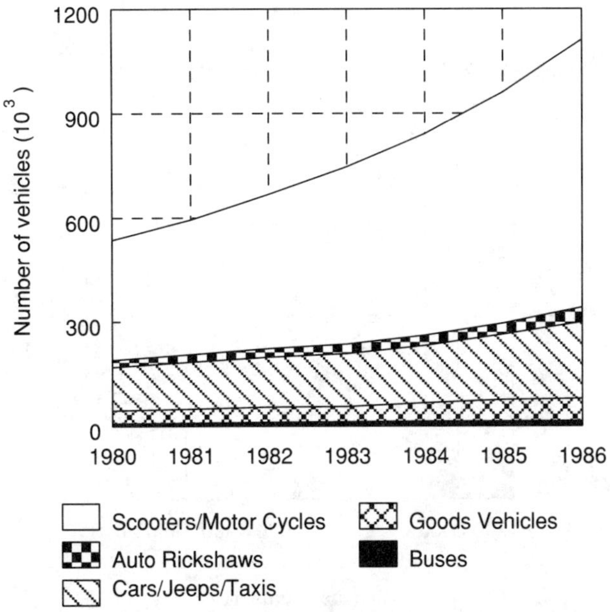

Figure 12.2 *Motor vehicle registrations by type, 1980–1986*

Source: NEERI, 1991a

12.2 Monitoring

Ambient air quality in Delhi has been monitored by the National Environmental Engineering Research Institute (NEERI) as part of the National Air Quality Monitoring Network (NAQMN). Sulphur dioxide (SO_2), SPM and nitrogen dioxide (NO_2) have been measured at three sites (Figure 12.1) representing commercial (Town Hall), industrial (Najafgarh) and residential (Netaji Nagar) areas between 1978 and 1987. Monitoring was suspended in 1988 before recommencing in 1990. Results were published in a series of *Air Quality in Selected Cities* reports. Annual mean airborne lead (Pb) concentrations were determined at the three NEERI sites in 1990. The Central Pollution Control Board (CPCB) commenced air quality monitoring at five stations in 1987; however, these results are not directly comparable with those from the NEERI sites because of differences in siting criteria and methodologies. NEERI have not undertaken any monitoring of carbon monoxide (CO) or ozone (O_3) in Delhi. Reference has been made to research studies on these pollutants.

12.3 Air Quality Situation

Sulphur dioxide

Emissions Figures 12.3 and 12.4 demonstrate that industrial sources, and power stations in particular, are responsible for the majority of SO_2 emissions in Delhi. It is estimated that the two power stations, Baderpur and Indraprastha, produce around 25,550 tonnes of SO_2 per annum. Indraprastha alone produces 30 per cent of total SO_2 emissions. Estimates produced by the CPCB (NEERI, 1991a) put total anthropogenic SO_2 emissions at approximately 59,000 tonnes per annum in 1987. NEERI estimates are lower; their 1990 estimate is approximately 45,000 tonnes per annum with a projected increase to 49,000 tonnes per annum by 2000. There are major differences in the two inventories. NEERI suggest that domestic emissions have remained relatively stable throughout the 1980s

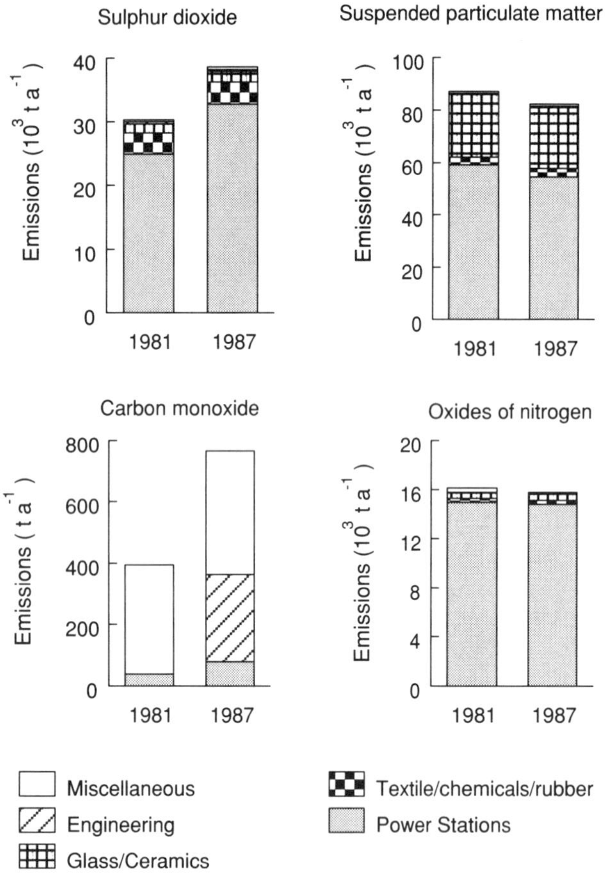

Figure 12.3 *Estimated industrial pollutant emissions by source, 1981 and 1987*

Source: NEERI 1991a

through changes in fuel use. However, in the inventory produced by the CPCB, domestic and commercial emissions estimates are higher and increase significantly from 8,395 tonnes per annum in 1981 to 12,400 tonnes per annum in 1987, a period of just six years. Ambient air quality data displayed in Figures 12.5 and 12.6 support NEERI's suggestion that industrial emissions have increased during the 1980s. The CPCB inventory (Figure 12.3) also points to an increase in most industrial SO_2 emissions, including power stations, between 1981 and 1987.

It is agreed that SO_2 emissions from transport sources have increased and will continue to increase due to the increasing motor vehicle population. The number of diesel-driven vehicles (the main vehicular source of SO_2) increased from 16,658 in 1971 to 75,709 in 1987. Delhi is a major goods distribution centre and therefore it is likely that many public and private carriers will be based in Delhi. Many lorries will also travel into Delhi every day; however, it is unlikely that the number of lorries registered in Delhi are ever all operational in the city on any single day.

Ambient Concentrations Figure 12.5 shows increasing SO_2 concentrations at the three GEMS/NEERI stations. Concentrations are consistently higher at the commercial and industrial sites. Annual mean concentrations at the commercial and industrial sites have exceeded the WHO guideline range in each year since 1984 until monitoring was discontinued in 1988. However, data for 1990 suggest that concentrations

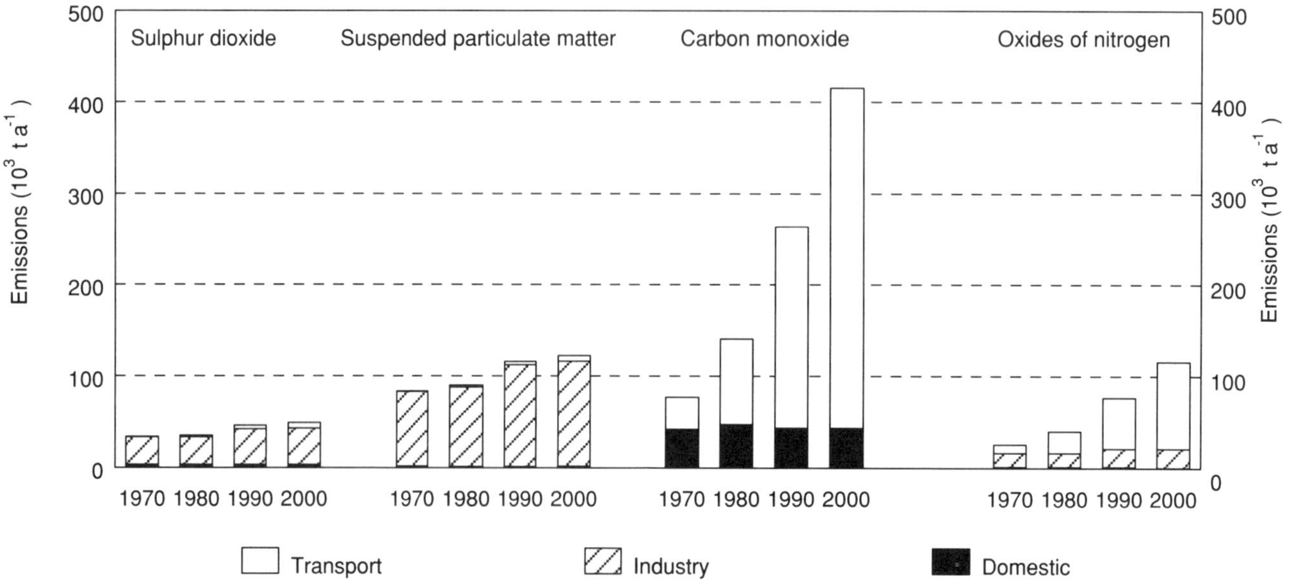

Figure 12.4 *Estimated and projected anthropogenic emissions by source, 1970 –2000*

Source: NEERI 1991a

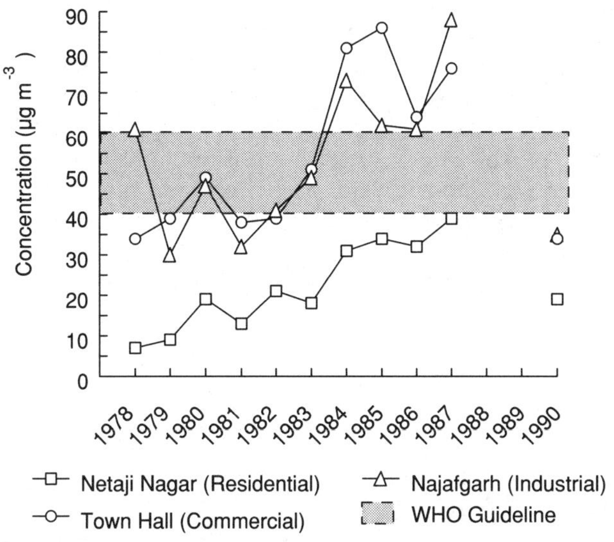

Figure 12.5 *Annual mean sulphur dioxide concentrations*

Sources: NEERI, 1980, 1983, 1988, 1990, 1991b

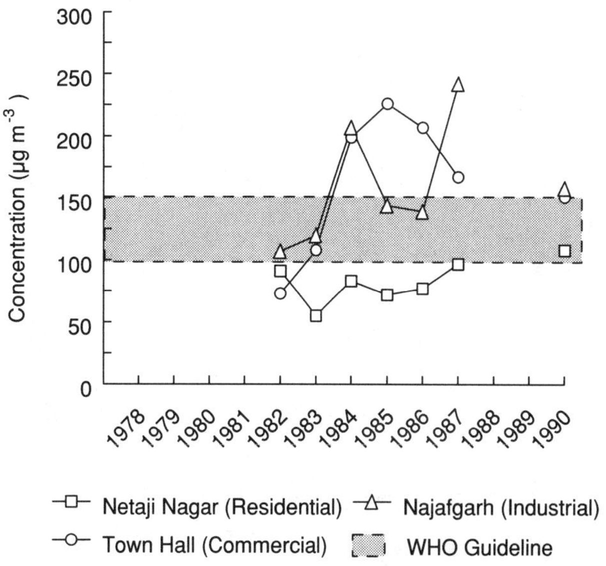

Figure 12.6 *Annual 98 percentile sulphur dioxide concentrations*

Sources: NEERI, 1980, 1983, 1988, 1990, 1991b

have fallen below the WHO annual guideline range at all three sites. Figure 12.6 shows a similar trend in 98 percentile concentrations. At the residential site 98 percentile levels were also increasing and approaching the WHO daily guideline (NEERI, 1983, 1988, 1990, 1991b).

Sulphur dioxide concentrations peak in winter (November to February), probably as a result of lower temperatures, especially at night. These cause demand for space heating and therefore the burning of coal. More importantly, winter calms and ground-based temperature inversions are likely to hinder dispersion of emissions.

Suspended particulate matter

Emissions Estimated emissions of SPM follow a very similar pattern to SO_2 emissions. It is estimated by NEERI that total SPM emissions were around 115,700 tonnes per annum in 1990 and will increase to 122,600 tonnes per annum in 2000 (NEERI, 1991a). NEERI attribute increasing emissions throughout the 1980s mainly to industry (Figure 12.4). However, inventories conducted on behalf of CPCB indicate that industrial and domestic SPM emissions decreased between 1981 and 1987 and that the overall increase in SPM was attributable to transport. Figure 12.3 shows that the glass and ceramic (bricks) industries are the major sources of SPM after power stations. It is estimated that the two power stations, Baderpur and Indraprastha, produce 25,500 tonnes of fly ash per annum, causing a major soot problem (settled particulates) around these places. Electrostatic precipitators at the plants are old and poorly maintained, although a drop in SPM emissions from this source between 1981 and 1987 would suggest that their efficiency is improving.

Domestic emissions have remained, and will continue to remain, relatively stable mainly due to rapid urban population growth which counters the reductions achieved by changes in fuel use. Delhi has seen a doubling of kerosene and liquid petroleum gas (LPG) use between 1981 and 1987 and this trend is likely to continue. The use of soft coke, firewood and charcoal reduced substantially over the same period, as did that of dung cakes (NEERI, 1991a).

Anthropogenic emissions are not the only source of SPM in Delhi. Natural dust is blown in from the surrounding arid areas, such as the Great Indian Desert (Thar Desert). The Andhi dust storms are a regular climatic feature in June, preceding the monsoon rains and depositing large amounts of dust and SPM into the atmosphere. The monthly mean SPM concentration in June is over 600 $\mu g\ m^{-3}$; the WHO 24-hour guideline is 150–230 $\mu g\ m^{-3}$. This natural dustfall will remain in circulation for long periods. Settled particulates and those "washed out" of the atmosphere by the monsoon rains are likely to be resuspended during drier conditions. Wind speeds also increase during the summer leading to further resuspension.

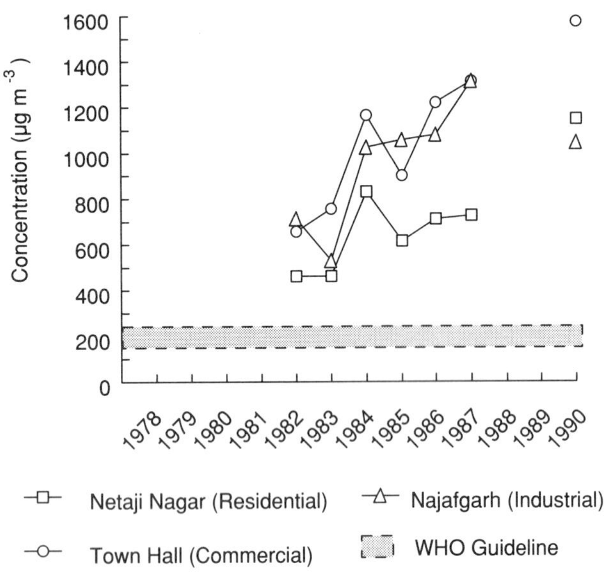

Figure 12.7 *Annual 98 percentile SPM (TSP) concentrations*

Sources: NEERI, 1980, 1983, 1988, 1990, 1991b

Some estimates attribute one-third of all anthropogenic SPM in Delhi's atmosphere to road transport. A survey of over 50 roadside sites in 1987 on the influence of motor vehicles on the spatial distribution of SPM and other pollutants indicated that there is a strong correlation between motor vehicle densities (i.e., at junctions) and SPM concentrations (Mathur, 1988). Besides direct motor vehicle emissions, resuspension of natural dust and settled particulates is also likely to be an important factor. Spot SPM samples taken at Connaught Place (the commercial centre of New Delhi) were found to exceed those around the Indraprastha power plant which has a widely recognized smoke and soot problem. Transport SPM are probably of greatest concern in terms of health effects due to the carcinogenic nature of certain diesel smoke constituents (e.g., polynuclear aromatic hydrocarbons (PAHs)). Buses in particular are a major source of diesel smoke. Measurement of smoke emissions from three bus companies showed that 82 per cent of buses have a smoke density higher than the permissible limit of 65 Hartridge Smoke Units (HSU). Around 46 per cent gave a smoke density of 90 HSU and above which is extremely high and indicative of jet black smoke (Maximum 100 HSU). Trucks and "Tempos" are also particularly bad emitters. It has been suggested that buses and public goods carriers should be inspected by an independent organization for smoke levels. Such an organization would require legal powers to enforce emissions limits on vehicles. Maintenance schedules and guidelines for operators would be required.

Ambient Concentrations Analysis of annual mean ambient SPM concentrations shows no significant trend. However, 98 percentile concentrations at all stations (Figure 12.7) have increased and at the residential and commercial sites concentrations have more than doubled between 1982 and 1990 (NEERI, 1983, 1988, 1990, 1991b). These concentrations are far in excess of all the recognized WHO guidelines and also exceed the Indian air quality standards (Appendix I). Even taking into account the huge natural component of SPM in Delhi it is clear that these levels constitute a substantial risk to human health.

Lead

Emissions The Pb content of petrol distributed from the Mathura refinery which serves Delhi is relatively high at 1.8 g l^{-1}. Given this Pb content and the number of petrol-driven motor vehicles registered in Delhi it is estimated that Pb emissions from this source are of the order of 600 tonnes per annum.

Ambient Concentrations Annual Pb concentrations in Delhi in 1990 at the three GEMS/NEERI sites ranged from 0.37 μg m^{-3} at the residential site to 0.67 μg m^{-3} at the industrial site (NEERI, 1991c). Results of spot samples taken at over 50 traffic junctions in 1987 show a wide range from 0.129 μg m^{-3} at Janak Cinema to 4.6 μg m^{-3} at Palika Bazar (Mathur, 1988). Concentrations at the vast majority of junctions were below 1 μg m^{-3}, the WHO annual guideline.

Carbon monoxide

Emissions Emissions inventories for CO in Delhi suggest that the major source, as in most other cities of the world, is transport and in particular motor vehicles. It is estimated by NEERI (1991a) that emissions have increased from 140,000 tonnes per annum in 1980 to 265,000 tonnes per annum in 1990 (Figure 12.4). Total CO emissions are projected to reach over 400,000 tonnes per annum by the year 2000. Domestic emissions were around 42,000 tonnes per annum in 1990, similar to those estimated for 1970 and the projected figure for 2000. As with other pollutants, changes in fuel use account for the reduction in per capita domestic emissions.

The increase in motor vehicles (Figure 12.2) throughout the 1980s accounts for the huge increase in CO emissions. By the year 2000 it is estimated that CO emissions from two-wheelers will be approximately equal to those for all motor vehicles in 1991-92.

Modelling studies have been conducted on the concentrations of traffic CO in Delhi. As expected, maximum CO concentrations are reached at peak traffic hours. These are 0900 hours to 1100 hours and 1700 hours and 1900 hours (Singh et al., 1990).

Ambient Concentrations Carbon monoxide is not routinely monitored by NEERI or by CPCB in Delhi. However, spot surveys found that concentrations around heavy traffic junctions such as Delhi Gate, Seelampur, Okhla and Vikas Minar were all around 5 mg m^{-3} over eight hours; the WHO eight-hourly guideline value is 10 mg m^{-3} (Mathur, 1988).

Oxides of nitrogen

Emissions Total emissions of oxides of nitrogen (NO_x) in 1990 were around 73,000 tonnes per annum and follow a similar trend to CO emissions (Figure 12.4). Industrial emissions accounted for around 15,000 tonnes per annum during the 1980s and it is estimated that Indraprastha power station produces 50 per cent of industrial NO_x alone. Over 90 per cent of industrial NO_x emissions are attributable to power stations. Transport NO_x emissions have increased significantly over the past 20 years and projections for the rest of the century suggest that NO_x emissions will continue to rise due to increasing motor vehicle traffic. Diesel-driven goods vehicles and buses are by far the most important source of transport NO_x. It is estimated that goods vehicles and buses, which constitute 7 per cent of vehicle numbers, contribute 85.5 per cent of vehicular NO_x emissions (NEERI, 1991a). It is estimated that by the year 2000 diesel-driven vehicles will account for 96 per cent of vehicular NO_x emissions (excluding two- and three-wheelers).

Ambient Concentrations Figure 12.8 shows the increasing trend in average NO_2 concentrations through the 1980s, although annual concentrations in 1990 were below those of 1987. Levels at the commercial Town Hall site are greatest owing to high motor vehicle densities in this area. Levels at the Najafgarh industrial site are higher than those at the residential station. Ninety-eight percentile values at all sites were still well below the WHO 24-hour guideline of 150 µg m^{-3}. However, the increasing trend in NO_2 concentration gives cause for concern because of the forecast increase in motor vehicle numbers and NO_x emissions (NEERI, 1983, 1988, 1990, 1991b).

A survey of busy traffic junctions revealed eight-hourly concentrations of NO_x in excess of 500 µg m^{-3}. It is not clear whether measurements represent nitric oxide (NO) or NO_2 concentrations or both. Even if these results are for NO it can be assumed that, given Delhi's meteorology, NO_2 levels will also be high (Mathur, 1988). Such concentrations cannot be considered representative of Delhi as a whole. However, they are important when considering personal exposure of individuals at these locations (i.e., drivers, traffic police, street vendors, etc.).

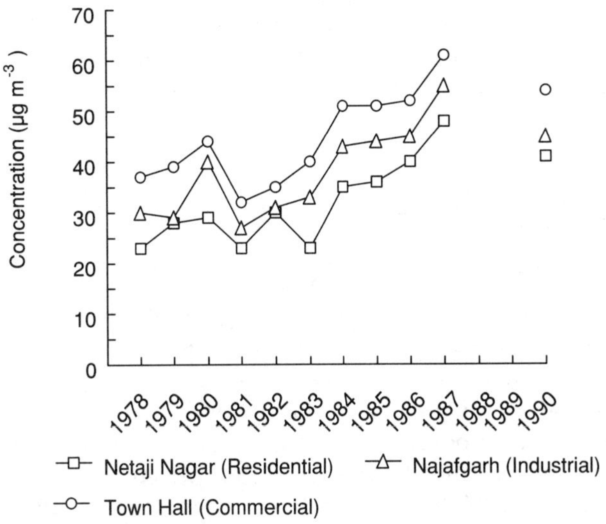

Figure 12.8 *Annual mean nitrogen dioxide concentrations*

Sources: NEERI, 1980, 1983, 1988, 1990, 1991b

Levels of NO_2 peak in November following the monsoon. Maximal insolation occurs in October (288 hours) and November (285 hours) immediately after the monsoon. Levels remain relatively constant for the rest of the year. The lowest concentrations occur during the monsoons when precipitation is at a maximum and insolation is at a minimum.

Ozone

Ambient concentrations Ozone is not routinely monitored by NEERI or by CPCB in Delhi. No reference can be found with regard to any long-term monitoring of O_3 in Delhi. Several reports refer to an increasing incidence of smog in New Delhi over the years (CSE, 1985). However, it is not clear whether these are referring to photochemical smog or simply a reduction in visibility brought about by the increase in SPM and other pollutants. As with other Indian cities, Delhi's climate is favourable in terms of ozone formation, especially during the winter. Oxides of nitrogen and hydrocarbon emissions are forecast to increase dramatically by the year 2000 and it is possible that O_3 will become a serious air quality problem over the next ten years. Continuous long-term monitoring of O_3 should be regarded as a major priority in Delhi in order to identify if and when O_3 levels pose a significant risk to health in the city.

12.4 Conclusions

Air Pollution Situation The trend in air pollution in Delhi is upward through increasing urbanization and associated motorization and industrialization.

The suspension and resuspension of natural dust is one of the main air quality problems in Delhi. Reducing exposure, for example through education, would probably be the most cost-effective way of dealing with this problem. Continuing urbanization and increases in motor vehicle traffic mean that suspended dust is likely to increase even if steps are taken to reduce man-made SPM emissions.

Although Indian crude oil and coal are low in sulphur, SO_2 levels repeatedly exceed WHO guidelines in Delhi. The increase in motor vehicle SO_2 and SPM emissions is particularly disturbing because of the potential numbers of people exposed to this source. Some form of emission test for both petrol- and diesel-driven vehicles should be introduced. Delhi still has a relatively small motor vehicle population, especially cars, when compared with cities in high income countries. This is borne out by NO_2 concentrations which are still within accepted guidelines.

The degree of pollution attributed to Delhi's power stations gives cause for concern. It is reported that there is a high incidence of tuberculosis in workers inside the Indrapastha site and the high levels of SPM are believed to be an important cofactor (NEERI, 1991a). This plant is associated with high SPM, SO_2 and NO_x concentrations not only in the immediate vicinity of the plant but throughout Delhi. Pollution control measures should be introduced to all urban power generation facilities. Alternatively these plants should be replaced with modern power stations located away from the city to the south-east.

12.5 References

CSE 1985 *The State of India's Environment 1984-85, A Second Citizens' Report*, Centre for Science and Environment, New Delhi.

Faiz, A., Sinha, K., Walsh, M. and Varma, A. 1990 *Automotive Air Pollution: Issues and Options for Developing Countries*, World Bank Policy and Research Working Paper. WPS 492, The World Bank, Washington DC.

Mathur, H. B. 1988 *Nature and Effects of Vehicular Pollution*, Indian Institute of Technology, New Delhi.

Murty, B. P. and Tangirala, R. S. 1990 An Assessment of the Assimilative Capacity of the Atmosphere in Delhi, *Atmospheric Emvironment*, **24A**, 845-848.

NEERI 1980 *Air quality in selected cities in India 1978-1979,* National Environmental Engineering Research Institute, Nagpur.

NEERI 1983 *Air quality in selected cities in India 1980-1981,* National Environmental Engineering Research Institute, Nagpur.

NEERI 1988 *Air quality status in ten cities: India 1982-1985,* National Environmental Engineering Research Institute, Nagpur.

NEERI 1990 *Air quality status in ten cities: India 1986-1987,* National Environmental Engineering Research Institute, Nagpur. (Unpublished report.)

NEERI 1991a *Air pollution aspects of three Indian megacities, Volume I: Delhi,* National Environmental Engineering Research Institute, Nagpur.

NEERI 1991b *Air quality status 1990,* National Environmental Engineering Research Institute, Nagpur.

NEERI 1991c *Air quality status: Toxic metals, polycyclic hydrocarbons, anionic composition and rain water characteristics (Delhi, Bombay and Calcutta),* National Environmental Engineering Research Institute, Nagpur.

Singh, M. P., Goyal, P., Basu, S., Agarwal, P., Nigam, S., Kumari, M. and Panwar, T. S. 1990 Predicted and measured concentrations of traffic carbon monoxide over Delhi, *Atmospheric Environment,* **24A**, 801-810.

UN 1989 *Prospects of World Urbanization 1988,* Population Studies No. 112, United Nations, New York.

WMO 1971 *Climatological Normals (CLINO) for Climate and Climate Ship Stations for the Period 1931-1960,* No. 117, World Meteorological Organization, Geneva.

13

Jakarta

13.1 General Information

Geography Jakarta, the capital of Indonesia, is located in the tropics (Latitude 6°08′S, Longitude 106°45′E). The city is on the north-west coast of the island of Java at the mouth of the Ciliwung River (Figure 13.1). Metropolitan Jakarta occupies 590 km^2 on a level alluvial plain with a mean elevation of 7 m above sea level. There are no natural topographical barriers near Jakarta.

Demography The population of Jakarta is growing rapidly. The population was 6.5 million in 1980 and 7.9 million in 1985. To ease the population density problems, Jakarta expanded into 2,620 km^2 of surrounding land between 1980 and 1985. Much of this was agricultural land. The population was projected to grow to 9.42 million in 1990 (UN, 1989) and is projected to be 13.23 million by the year 2000 and thus Jakarta's population will have more than doubled between 1980 and 2000. The population densities in the various parts of the urban area range from more than 30,000 per km^2 in central Jakarta to less than 10,000 per km^2 in most of the other parts of Jakarta.

Climate Jakarta has a tropical rain forest climate. The annual mean daily minimum and maximum temperatures are 22°C and 33°C, respectively. The annual rainfall is high and totals 1,760 mm, most of which falls during the rainy season. The monthly mean relative humidity varies between 75 and 83 per cent. The annual mean wind speeds are very low (of the order of 1 m s^{-1}), which indicates the potential for a high air pollutant buildup.

Industry Jakarta is a major trade, financial and industrial centre. The main industries are breweries, soap and margarine factories, and iron foundries. The per capita income in Indonesia was US$430 in 1988. There are no figures available to assess the importance of various industries in Jakarta. If the GNP contributions of various sectors for Indonesia are taken, services/ commerce, agriculture /fishing/ forestry, and manufacturing/industry are the most important sectors.

Coal, crude oil and natural gas are important primary energy sources consumed in Indonesia. Total coal production in Indonesia was approximately 6 million tonnes in 1989–90, and it is projected to more than double by 1993–94.

In metropolitan Jakarta a total of approximately 11 million tonnes of oil and gas were sold during 1987, of which approximately 6 million tonnes (or 55 per cent) were petrol, 2.1 million tonnes (or 19 per cent) were liquid petroleum gas (LPG), and 0.7 million tonnes (or 6 per cent) were diesel (Central Bureau of Statistics, 1988).

Transport Motor vehicle registrations (e.g., cars, buses, trucks, and motorcycles) totalled about 760,000 in 1980, but grew to 1.12 million in 1983, and to 1.34 million in 1986. In 1987, the most recent year for which information was available, motor vehicle registrations totalled 1.38 million. The number of motor vehicles is growing at a rate of 3-4 per cent per annum. Jakarta has no developed mass transit rail system. Public transportation is mainly provided by diesel buses with no emissions controls.

13.2 Monitoring

Ambient air pollution has been measured at 19 different monitoring sites since 1980, as shown in Figure 13.1. Four of the stations are part of the GEMS/Air programme. GEMS/Air data were originally collected by the Ministry of Health, and since 1980 all stations have been operated by the Jakarta Municipal Government (JMG). It is important to note that the locations of the GEMS/Air sites were changed by JMG in 1986. This hinders the comparison of data obtained before and after 1986. The four GEMS/Air stations are located close together in an area of high population density and low air pollution and therefore the GEMS/Air stations do not monitor worst case situations and do not give results representative of different sections of Jakarta.

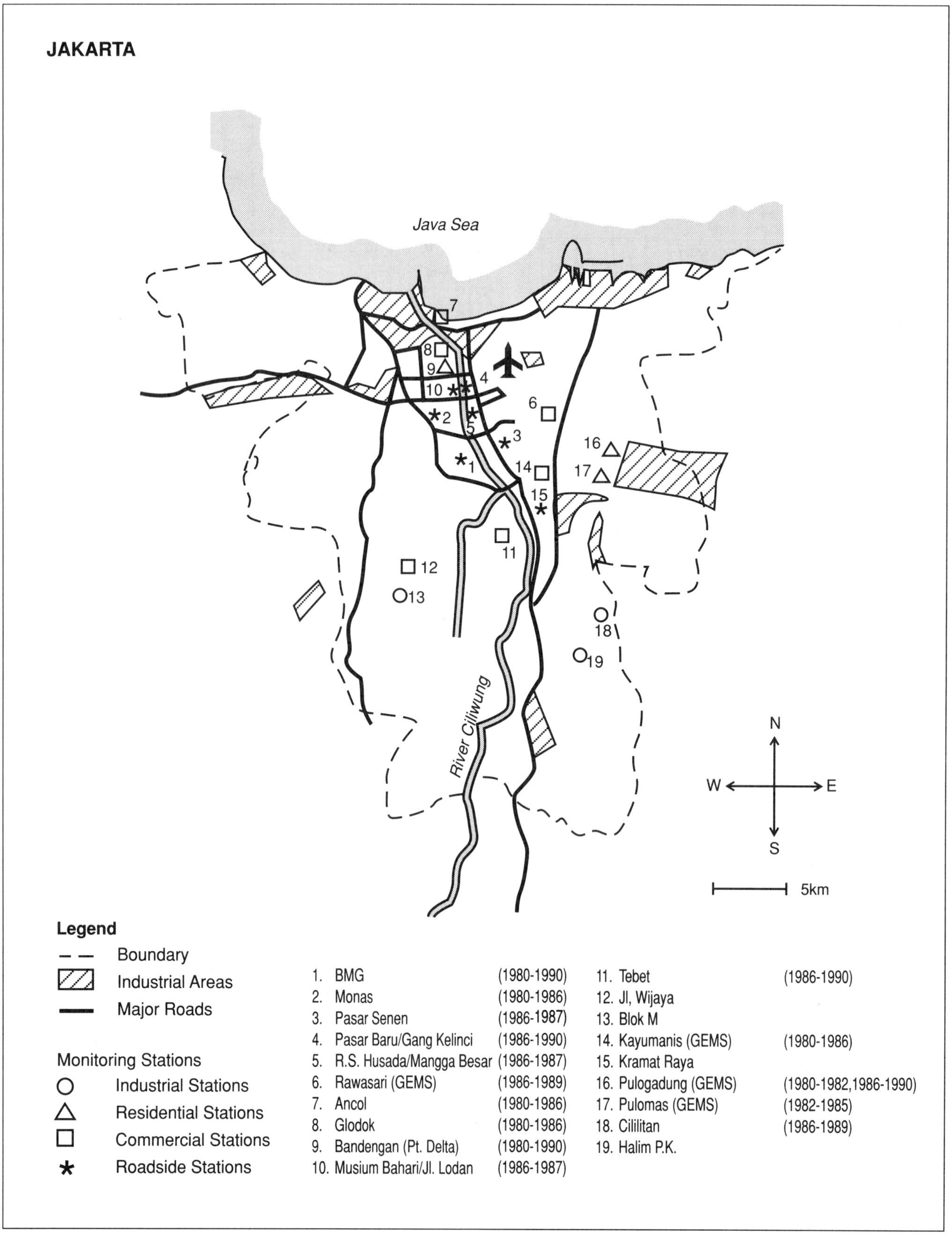
JAKARTA
Java Sea
River Ciliwung
N
S
W
E
5km
Legend
Boundary
Industrial Areas
Major Roads
Monitoring Stations
Industrial Stations
Residential Stations
Commercial Stations
Roadside Stations
1. BMG (1980-1990)
2. Monas (1980-1986)
3. Pasar Senen (1986-1987)
4. Pasar Baru/Gang Kelinci (1986-1990)
5. R.S. Husada/Mangga Besar (1986-1987)
6. Rawasari (GEMS) (1986-1989)
7. Ancol (1980-1986)
8. Glodok (1980-1986)
9. Bandengan (Pt. Delta) (1980-1990)
10. Musium Bahari/Jl. Lodan (1986-1987)
11. Tebet (1986-1990)
12. Jl, Wijaya
13. Blok M
14. Kayumanis (GEMS) (1980-1986)
15. Kramat Raya
16. Pulogadung (GEMS) (1980-1982,1986-1990)
17. Pulomas (GEMS) (1982-1985)
18. Cililitan (1986-1989)
19. Halim P.K.

Figure 13.1 *Sketch map of Jakarta*

The GEMS/Air monitoring stations operate on a once-every-sixth day schedule for suspended particulate matter (SPM) and sulphur dioxide (SO_2) and the other JMG stations operate once-every-eight days from August to February.

13.3 Air Quality Situation

Sulphur dioxide

Emissions A detailed air pollutant emissions inventory for Jakarta was not available. An attempt was made to estimate the emissions from industry, domestic fuel use and transport in metropolitan Jakarta for the base year 1989 (Winarto, 1989). This estimate is summarized in Table 13.1.

Total SO_2 emissions in 1989 were estimated to be 24,700 tonnes per annum. This figure seems rather low compared with the population number and the number of industrial activities, even if it is taken into account that there is little heavy industry in Jakarta. The majority of SO_2 emissions (63 per cent) come from the "industry" sector which includes power stations as well as industrial fuel combustion and industrial processes. According to the above mentioned estimates, domestic fuel combustion contributes about 2,600 tonnes per annum and transport (mainly diesel vehicle traffic) contributes another 6,500 tonnes per annum. It is not possible to assess the validity of the emissions data given. However, if compared with other cities of similar size, climate, and industrial characteristics, the SO_2 emissions could well be up to 5, or even 10 times, higher than the figures given in the available estimate.

Ambient Concentrations Sulphur dioxide is routinely monitored at the GEMS/Air stations as well as at the JMG stations. Sulphur dioxide levels at the GEMS/Air sites are generally below the minimum detectable level of the pararosanaline method (5 $\mu g\ m^{-3}$). Similarly, the JMG reported annual means consistently less than 29 $\mu g\ m^{-3}$ (0.01 ppm) from 1983 to 1990. Those values are well below the WHO annual mean guideline range of 40 - 60 $\mu g\ m^{-3}$.

Maximum 24-hour mean concentrations of SO_2 were reported to be around 240 $\mu g\ m^{-3}$ (0.09 ppm) in 1983, but maximum daily means decreased to 8 $\mu g\ m^{-3}$ (0.003 ppm) in 1986–1989. This remarkable sudden drop to about 3 per cent of the previous values cannot be explained at this time. In 1990, the maximum monthly SO_2 means concentration at the Badan Meterologi dan Geofisika (BMG or Department of Transportation, Meteorology and Geophysics Office) site was around 20 $\mu g\ m^{-3}$.

The available data suggest that SO_2 concentrations in Jakarta are probably not a serious problem. The reason for this is that there is no fuel combustion for domestic heating and there is little heavy industry in the urban area. Furthermore, the climatic conditions may favour rapid dispersion of SO_2.

Table 13.1 *Estimated annual emissions ($t\ a^{-1}$) in metropolitan Jakarta (base year 1989)*

	SO_2	NO_x	CO
Industry	15,500	3,300	300
Domestic	2,600	2,000	300
Transport	6,500	15,000	321,700
Others	100	200	3,300
Total	24,700	20,500	325,600

After Winarto, 1989

Suspended particulate matter

Emissions There are natural emissions sources of SPM (such as wind-blown dusts) as well as anthropogenic sources (e.g., motor vehicle exhaust; and particulates from industrial activities). Natural sources of SPM are not quantifiable and cannot be incorporated in emissions inventories. There is only one preliminary estimate of anthropogenic SPM emissions in Jakarta available (Winarto, 1989). According to this estimate, total anthropogenic SPM emissions in Jakarta would be 70 tonnes per annum. This number is too low, probably by a factor of 50–100. For instance, in Karachi, with a total motor vehicle population of 650,000 in 1989, SPM emissions from motor vehicle traffic were estimated to be more than 7,000 tonnes per annum. In Jakarta, with a motor vehicle registration of more than 1.38 million in 1987, SPM emissions from traffic would be expected to be 10,000–18,000 tonnes per annum. Furthermore, some important sources of SPM, like open burning of waste, have not been included in the inventory.

Ambient Concentrations Suspended particulate matter is monitored at the four GEMS/Air stations as well as at the JMG sites. Annual means at the GEMS/Air sites were about 150–300 $\mu g\ m^{-3}$ from 1980 to 1990 (Figure 13.2), well above the WHO annual mean guideline of 60–90 $\mu g\ m^{-3}$, at all monitoring stations in all the years covered. The SPM concentrations increased by about 50 per cent from 1986 and 1990 in

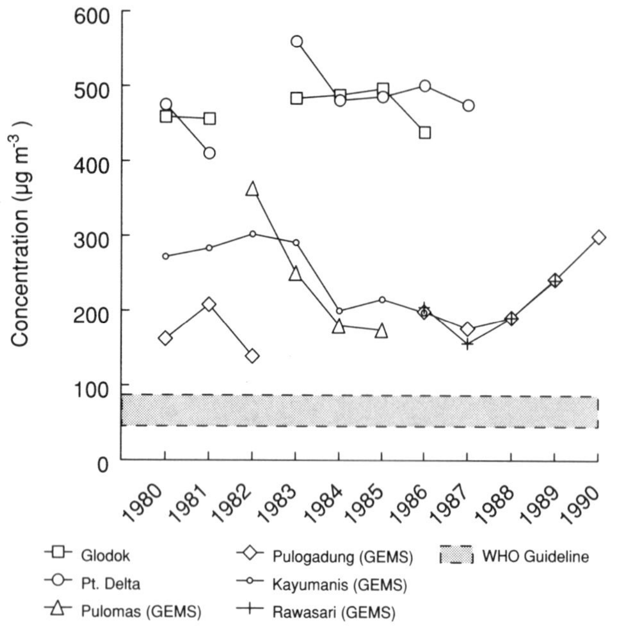

Figure 13.2 *Annual mean suspended particulate matter (TSP) concentrations*

Sources: GEMS data; Office of State Ministry of Population and Environment, 1990

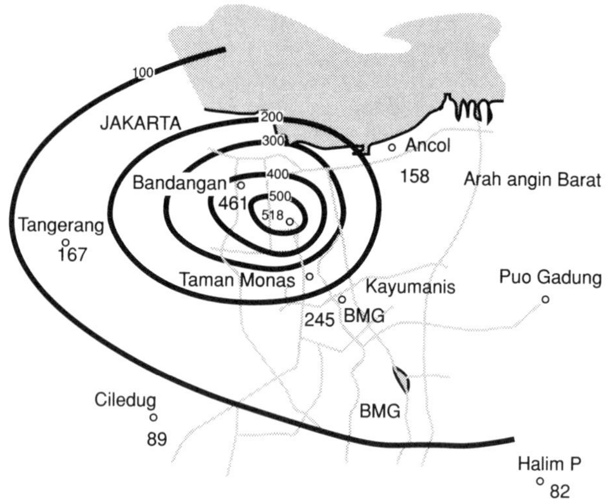

Figure 13.3 *Suspended particulate matter isopleths 1980–1985*

After Office of State Ministry of Population and Environment, 1990

the industrial site at Pulogadung as well as in the residential site at Rawasari. For example, the 1990 annual average at the GEMS station at Pulogadung was 304 μg m^{-3} which means even the WHO daily guideline of 230 μg m^{-3} (not to be exceeded on more than 2 per cent of days) is exceeded. The high level of SPM pollution and the continuing upward trend of annual mean levels give cause for concern.

Suspended particulate matter levels at two heavily polluted JMG sites (commercial site at Glodok, industrial site at Bandengan (Pt. Delta)), which are included in Figure 13.2, are even higher than at the GEMS/Air sites. Mean concentrations were about 400–550 μg m^{-3} at these two monitoring stations from 1980 to 1987. The reason for the relatively lower values reported by the GEMS/Air stations is that the latter are located in areas outside the main SPM pollution zones. This is demonstrated by the 1980–1985 SPM isopleth map in Figure 13.3 as calculated by the Office of State Ministry for Population and Environment (1990). The areas of highest SPM levels are the city centre and the eastern parts of western Jakarta where SPM annual means exceed 400 μg m^{-3}. The SPM levels are much lower in the eastern parts of the city where the GEMS/Air sites are located.

There is no information available on the chemical composition and physical size distribution of SPM in Jakarta. Thus it is not possible to assess the toxic potential or the health risk associated with particulates in Jakarta.

Lead

Emissions No emissions estimates for lead (Pb) were available for this report.

It has been reported that tetraethyl Pb was added to petrol used in Jakarta until 1982, and since then alkylbenzene compounds have been added to improve octane ratings (Winarto, 1989). However, it was indicated in another report (Tri-Tugaswati et al., 1987) that in 1987 Pb levels in petrol were still the same as in 1981: equivalent to 2.5 ml tetraethyl Pb per gallon of "premium" (which is about 0.6 g Pb per litre) and 3.0 ml per gallon of "super" (which is about 0.73 g Pb per litre). If it is assumed that Pb levels in petrol are as high as reported by Tri-Tugaswati et al. (1987), and taking into account a petrol consumption of 6 million tonnes per annum, then Pb emissions would be approximately 3,000 tonnes per annum, which would be extremely high. Yet, it should be noted that the present Pb levels in petrol sold in Jakarta are not known, so emissions of Pb in Jakarta cannot be estimated with any accuracy.

Ambient Concentrations There are very few data on Pb levels in Jakarta. Lead is not routinely reported by GEMS/Air sites, although the filters from SPM sampling are archived for Pb analyses at a future date. A report on the environment in Indonesia (Office of State Ministry for Population and Environment, 1989) cites concentrations of a number of air pollutants in

Table 13.2 *Traffic volume, suspended particulate matter concentrations and lead concentrations, July 1985*

Sampling site	Traffic volume (Motor vehicles per hour)	SPM concentration ($\mu g\ m^{-3}$)	Pb concentration ($\mu g\ m^{-3}$)
Salemba State Road, Central Jakarta	6,570	263	3.59
Percetakan Negara Street, Central Jakarta	1,284	161	1.68
Ciganjur District, South Jakarta	40	8	0.31

After Tri-Tugaswati et al., 1987

Jakarta, but Pb is absent from the discussion of air pollutants.

A 1979 study revealed that short-term ambient air Pb levels in Jakarta were as high as 90 $\mu g\ m^{-3}$ at the kerbside of the busiest street (Tri-Tugaswati et al., 1987). Although the averaging period is not known, those values are far above the WHO annual mean guidelines of 0.5-1 $\mu g\ m^{-3}$ and even above the already high city of Jakarta's 1981 standard of a maximum allowable Pb concentration of 60 $\mu g\ m^{-3}$.

A study in July 1985 showed that Pb concentrations at three monitoring sites were 0.3-3.6 $\mu g\ m^{-3}$, with the Pb levels strongly correlated to the respective traffic volumes (Tri-Tugaswati et al., 1987). The results of this study are summarized in Table 13.2. Those levels were considerably lower than the 1979 values. The sampling period of this study was not reported, but the Pb concentrations were still significantly above WHO annual mean guidelines.

Carbon monoxide

Emissions A detailed air pollutant emissions inventory for Jakarta is not available. Preliminary estimates for carbon monoxide (CO) emissions from industry, domestic fuel use and transport in metropolitan Jakarta in 1989 (Winarto, 1989) are summarized in Table 13.1.

According to the available estimates total CO emissions in 1989 were 325,600 tonnes per annum. This value seems low considering the high number of registered vehicles. More than 99 per cent of the CO emissions are attributed to the traffic sector. All other emissions sectors were estimated to be very small. However, it has to be taken into account that some important emissions sources (e.g., open burning of waste) have not been considered in the emissions inventory.

Ambient Concentrations There are at present no CO data from the GEMS/Air network, but there are data from the JMG sites. The eight-hour mean CO levels were found to be around 3.5 mg m^{-3} (3 ppm) in a residential area and at a bus terminal (Cililitan site), but were up to 27 mg m^{-3} (23 ppm) at the Glodok station in a city centre commercial area (Office of State Ministry of Population and Environment, 1990). This CO level is well above the WHO guideline of 10 mg m^{-3} as an eight-hour average, indicating that CO is a problem at areas congested by traffic in Jakarta. Badan Meterologi dan Geofisika reports a strong bimodal CO pattern on weekdays, with highest values occurring in the peak traffic hours between 0700–0900 hours and 1800–2000 hours.

Oxides of nitrogen

Emissions Oxides of nitrogen (NO_x) were covered in the Jakarta emissions estimates for the year 1989 (Winarto, 1989). Total NO_x emissions were estimated to be around 20,500 tonnes per annum (Table 13.1). Oxides of nitrogen in Jakarta are emitted primarily by motor vehicles (75 per cent) and by industry (15 per cent). The estimated contribution of industrial sources seems to be relatively small, particularly as power plants and large industrial boilers are usually important NO_x sources.

Ambient Concentrations GEMS/Air stations reported annual mean NO_x concentrations of 2-4 $\mu g\ m^{-3}$, and maximum 24-hour concentrations of 5-10 $\mu g\ m^{-3}$ during 1986–1989. These values are considerably below the WHO 24-hour mean guideline of 150 $\mu g\ m^{-3}$. However, as noted above, GEMS stations are located away from the city centre and thus reflect primarily suburban ambient air pollution.

Jakarta Municipal Government data indicate that the annual mean NO_x concentrations fell from 113 $\mu g\ m^{-3}$ to 9.4 $\mu g\ m^{-3}$ from 1983 to 1986 and similarly, maximum 24-hour values fell from 395 $\mu g\ m^{-3}$ to 15 $\mu g\ m^{-3}$. This sudden drop in NO_x concentrations cannot be explained with the available information, but it seems likely that besides a possible improvement in air quality, the siting, sampling or instrumentation of the monitoring stations must have had a major influence.

Measurements performed in 1987 and 1988 at industrial areas in Jakarta gave annual mean NO_x concentrations of 17-36 μg m^{-3}. During 1989 and 1990, the average NO_x concentration at the Bandengan station in the city centre was 28 μg m^{-3}. Even if all the NO_x were NO_2, WHO guidelines would be met.

Ozone

Emissions As shown above, emissions of NO_x were estimated to be about 20,500 tonnes per annum. A preliminary estimate for "hydrocarbon" emissions gave 14,600 tonnes per annum, of which 90 per cent were attributed to traffic (Winarto, 1989). This estimate is certainly much too low. In addition to traffic, there are many other volatile organic compound sources like industrial processes (e.g., refineries), petrol and solvent evaporation, and natural sources (e.g., vegetation) which are not considered in the available estimates.

Ambient Concentrations Ozone (O_3) is measured at eight stations in the JMG area network. Results of O_3 monitoring in 1986 -1987 are summarized in Table 13.3. Annual mean O_3 concentrations range from 2 μg m^{-3}–15 μg m^{-3}, which is relatively low. Thus it might be concluded that the annual or daily mean concentrations of O_3 are also low. The highest one-hour maximum of 85.8 μg m^{-3} occurred at the Pasar Senen site. This value indicates that the WHO one-hour mean guidelines of 150–200 μg m^{-3}, as well as the national one-hour ozone standard of 160 μg m^{-3}, were met at all sites during 1986 -1987.

***Table 13.3** Ozone concentrations (μg m^{-3}) 1986–1987*

Station name	1 h maxima	Annual mean
Bandengan	8.2	2
Pasar Senen	85.8	15
Cililitan	41.6	6.2
Tebet	16.4	4.2
Musium Bahari	8.2	3.4
Pasar Baru	42	5.2
Pulogadung	23.6	4.0
R.S.Husada	13.8	2.8

After Bureau of Population and Environment, 1988

13.4 Conclusions

Air Pollution Situation From the available data it can be concluded that levels of SPM are generally very high in Jakarta. The applicable WHO annual mean guidelines for SPM were exceeded in all monitoring stations during the whole monitoring period, and there was an increasing trend from 1986–1990.

In the city centre, and especially in areas heavily influenced by traffic, there are also very high concentrations of Pb and of CO. Although there are only a very few data for 1985 and there are no data available for the last few years, it seems that Pb levels exceed the WHO guidelines by up to factor of three in areas with heavy traffic.

Sulphur dioxide and NO_x are probably of minor importance for the air pollution situation in Jakarta. Sulphur dioxide annual mean concentrations are about half the WHO guidelines, and NO_x levels are even lower. For both pollutants, there was a remarkable sudden drop of the measured concentrations between 1983 and 1986 which at present cannot be explained.

For O_3, only data for 1985 and 1986 were available. In those years, long-term average O_3 levels were low. Short-term O_3 concentrations, however, on some days of the year were very high in certain parts of the city (e.g., at the two bus terminal monitoring sites). On those days, the WHO one-hour guideline was exceeded by a factor of four in the most polluted sites.

Generally, low windspeeds contribute to the buildup of pollutants in Jakarta. Observed high O_3 levels may be due to local photochemical formation of O_3 from reactive organic compounds under favourable meteorological circumstances. It is anticipated that NO_2 and O_3 levels will increase as the city expands and the number of vehicles increases. These O_3 episodes will also have an impact on regions downwind from Jakarta.

Main Problems From the available monitoring data and from the preliminary emissions inventory, it must be concluded that the transportation sector (particularly motor vehicle emissions) is the primary source of air pollutants in Jakarta (Achmadi, 1990a). In addition, the high levels of SPM might also be mainly or partly caused by industrial processes and natural dusts.

Logically, air pollution is worse in areas which are heavily influenced by traffic. This is particularly the case for Pb pollution, where traffic is the main emissions source. There are no recent data on Pb levels in petrol but, because of the high toxicity of Pb, regular monitoring should be performed and Pb levels in petrol sold in Jakarta should be evaluated.

At present the main causes of air pollution in Jakarta cannot be sufficiently identified as only preliminary emissions estimates have been established. An up-to-date, detailed and comprehensive emissions inventory would greatly facilitate the analysis of the major pollution problems, such an inventory being the basis for an effective air quality management plan.

Health Effects It is difficult to relate air pollution problems in Jakarta to mortality, partly because accurate acute respiratory infection (ARI) data are not available (Ministry of Health, 1989). Since ARI "alone or in combination with acute diarrhoeal diseases and immunizable diseases are the biggest portion of preventable deaths among children below the age of five" (Ministry of Health, 1989), other diseases confound the picture. A 1986 health survey showed that 14.4 per cent of infant deaths and 8.4 per cent of deaths between one and four years of age were due to ARI. A 1986 National Household Survey in Indonesia revealed that ARI accounts for more than 42.4 per cent and 40.6 per cent of total morbidity of infants and of those in the one-four year age group, respectively. Among hospitalized patients in Jakarta, ARI accounts for 31 per cent of patients. Although risk factors for morbidity and mortality due to ARI include malnutrition and low birth weight, preventive strategies recommended by the government include reducing "indoor air pollution (smoke from biomass and tobacco) and urban air pollution" (Ministry of Health, 1989).

Lead concentrations in Jakarta air reported in the 1980s have been much greater than WHO guideline values. Since Pb in air is usually associated with the respirable particulate size fraction (less than 10 μm in aerodynamic diameter), high Pb levels in air are of particular concern (WHO, 1987). In 1980 high levels of Pb were noted in Jakarta city bus drivers (Achmadi, 1990b). In 1985, levels of Pb in blood were 18 $\mu g\ dl^{-1}$ in Jakarta public transit drivers, but only 9 $\mu g\ dl^{-1}$ in an agricultural group (Tri-Tugaswati et al., 1987). These results confirm that motor vehicle traffic is the major source of lead air pollution in Jakarta.

13.5 References

Achmadi, U. F. 1990a *Current State of Air Pollution and its Public Health Significance in Indonesia,* University of Jakarta (Unpublished report.)

Achmadi, U. F. 1990b *Impact of Current Pollutants on Public Health,* University of Jakarta (Unpublished report.)

Bureau of Population and Environment 1988 *The balance of regional population with the environment,* Appendix **3**, Pemerintah Daerah Khusus Ibukota, Jakarta.

Central Bureau of Statistics 1988 *Statistik Indonesia,* Government of Republic of Indonesia, Jakarta.

Ministry of Health 1989 *Programme for the Control of Acute Respiratory Infections in Children, Plan of Operation 1989–1994,* Government of Republic of Indonesia, Jakarta.

Office of State Ministry of Population and Environment 1989 *Population and Environment: An Overview,* Government of Republic of Indonesia, Jakarta.

Office of State Ministry of Population and Environment 1990 *Environmental Quality in Indonesia,* Government of Republic of Indonesia, Jakarta.

Tri-Tugaswati, A., Suzuki, S., Koyama, H. and Kawada, T. 1987 Health effects of air pollution due to automotive lead in Jakarta, *Asia-Pacific Journal of Public Health* **1**, 23–27.

WHO 1987 *Air Quality Guidelines for Europe,* World Health Organization, Copenhagen.

Winarto, W. 1989 *Contributions to emissions of gases and particulates by the combustion of fuels by industry, motor vehicles, households and waste incineration in Jakarta,* Institut Teknologi Bandung, Bandung.

UN 1989 *Prospects of World Urbanization 1988,* Population Studies No. 112, United Nations, New York.

14

Karachi

14.1 General Information

Geography Karachi (Latitude 24°51′N, Longitude 67°02′E) is located on the coast of the Arabian Sea immediately north-west of the Indus River delta (Figure 14.1). It is the largest city in Pakistan and capital of the province of Sind.

Karachi is situated on the shores of a natural harbour. Moving away from the coast the ground rises gently forming a large plain to the north and east on which the city is built. The city is between 1.5 and 37 m above mean sea level. Two seasonal rivers run through Karachi, the Malir River which passes through the eastern part of the city and the Layari River which runs through the densely populated northern parts of the city. A few isolated hills and ridges are found to the north and east; the highest point is Manghopir Hill, which has an elevation of 178 m.

Demography Karachi, as the biggest industrial and commercial centre in Pakistan, offers considerable employment and business opportunities. Consequently, an increasing number of people migrate from rural areas into the city. The Karachi Development Authority (KDA) has estimated that the annual growth rate of the city is about 6 per cent, of which approximately one-half are newcomers from rural communities. In 1990 Karachi had an estimated population of 7.67 million, which is expected to grow to 11.57 million by the year 2000 (UN, 1989).

The city authorities will have to cope with the increasing growth of urban population and provide housing and other civic amenities which are already under heavy strain. The urban area has expanded from 2,500 km^2 in 1972 to 3,530 km^2 in 1982. Such a rate of expansion means that conventional urban planning is exceptionally difficult.

Climate Karachi's climate can be classified as desert. The dry season extends from November to February; annual mean precipitation in Karachi is around 200 mm with the maximum falling in July. Monthly mean temperatures range from 20°C in January to approximately 30°C in June. Average wind velocity is 12 m s^{-1} during June and July, and 3.5 m s^{-1} from January to March. During the south-west monsoon season winds blow from the sea towards the coast, whereas during the north-east monsoon their direction is reversed. Therefore, pollutants are pushed inland during the south-west monsoon season and are blown out to sea during the north-east monsoons. In the summer, winds are strong, thus resulting in low ambient pollutant concentrations.

Industry Textiles and footwear are the main manufacturing goods produced in Karachi, followed by metal products, food and beverages, paper and printing, wood and furniture, machinery, chemicals and petroleum, leather and rubber and electrical goods.

Karachi handles the entire sea-borne trade of Pakistan and that of neighbouring Afghanistan which is land-locked. Much of Karachi's industry is located in industrial estates. The most important industrial areas are the Landhi, Korangi and Sind Industrial Trading Estates (referred to hereafter as LITE, KITE, and SITE), the latter being the one with the largest number of industrial units.

Pakistan is self-sufficient in natural gas and resources of low-sulphur coal, but there are major imports of crude oil and petroleum products (IUAPPA, 1991).

Oil-fired power plants in the Port Qasim area consumed about 950,000 tonnes of furnace oil in 1988–89, which was about 80 per cent of the city's total furnace oil consumption in power generation. Other power stations, such as the Karachi Electrical Supply Company (KESC) and Pakistan Steel Mills power plants, consume approximately 650 million m^3 of natural gas per annum. Pakistan Steel Mills utilizes approximately 136,000 tonnes of coal per annum for the production of coke (Ghauri et al., 1988).

Kerosene, firewood, charcoal and cow dung account for approximately 50 per cent of domestic fuel use, mainly in those dwellings where natural gas is not available. It is estimated that 32 per cent of dwellings use kerosene for lighting and 66 per cent use electricity. Emissions from kerosene stoves are over twice those fuelled by natural gas. An estimated

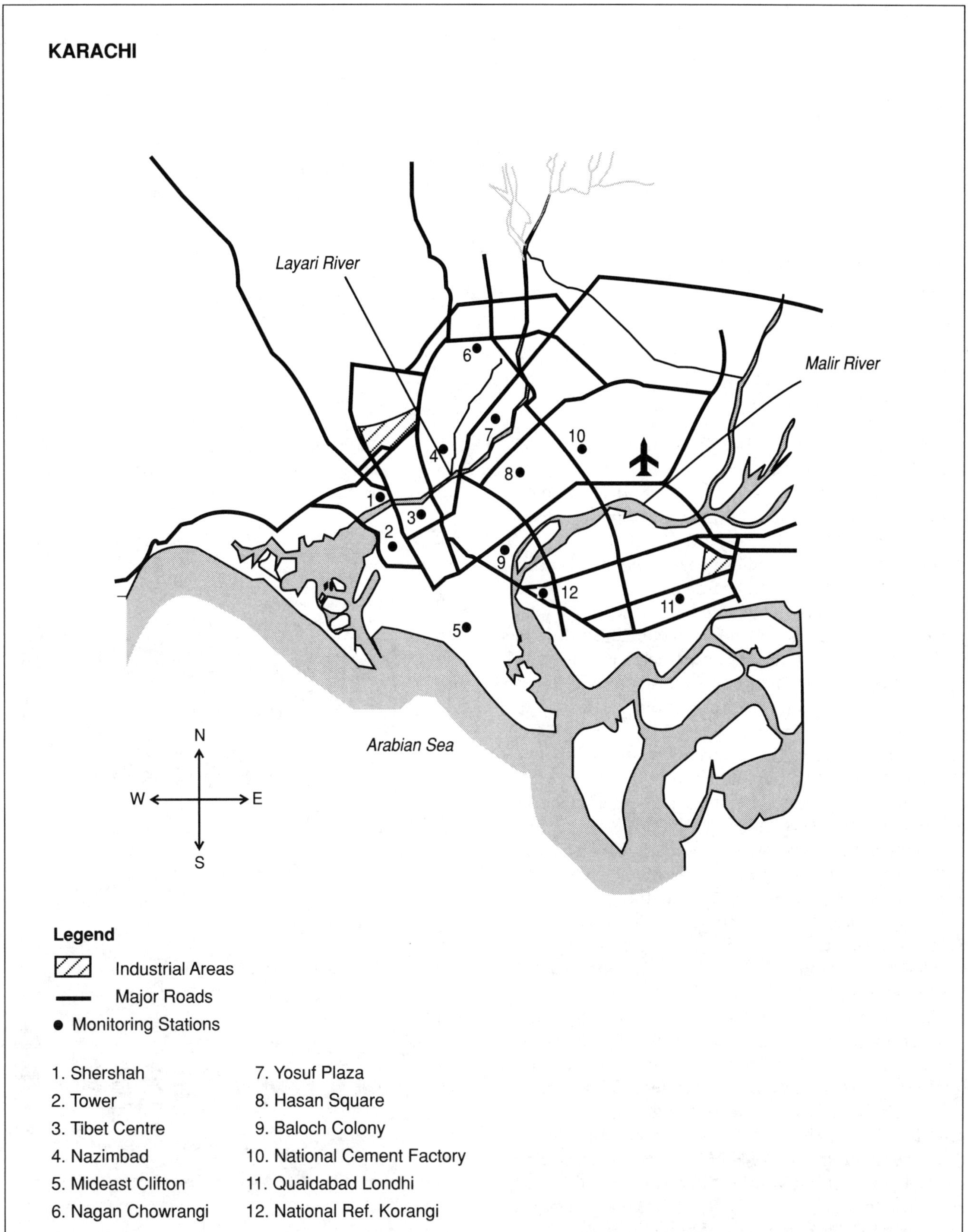
KARACHI
Layari River
Malir River
1
2
3
4
5
6
7
8
9
10
11
12
Arabian Sea
N
S
W
E
Legend
Industrial Areas
Major Roads
Monitoring Stations
1. Shershah
2. Tower
3. Tibet Centre
4. Nazimbad
5. Mideast Clifton
6. Nagan Chowrangi
7. Yosuf Plaza
8. Hasan Square
9. Baloch Colony
10. National Cement Factory
11. Quaidabad Londhi
12. National Ref. Korangi

Figure 14.1 *Sketch map of Karachi*

Table 14.1 *Motor vehicle registrations*

	1980	1989	2000*
Cars, jeeps/station wagons	113,100	277,500	465,200
Motorcycles/scooters	138,600	278,200	460,000
Taxis	8,500	10,600	16,300
Motor rickshaws	15,600	16,900	21,900
Buses/minibuses	7,500	12,000	20,700
Trucks	8,000	13,300	21,600
Other	16,500	38,200	69,300
Total	307,800	646,700	1,075,000

* Projected
After Beg, 1990

17,000 dwellings use biomass fuels such as fire-wood and cow dung.

It is estimated that each year around 2.9 million tonnes of waste are burned uncontrolled in the open.

Transport Road transport has become the major mode of transportation for both passengers and goods in Karachi. Twenty-seven per cent of Pakistan's total vehicle population is registered in Karachi. While in 1980 the number of registered vehicles was around 300,000, this number had grown to about 650,000 in 1989. The number of vehicles is projected to rise to about 1.1 million by the year 2000. The annual growth rate between 1980 and 1989 was 12.5 per cent, 2.5 times higher than population growth rate. In 1989, cars and motorcycles each made up about 43 per cent of the total vehicle population (Table 14.1).

Modal split studies undertaken in 1987 showed that 57 per cent of trips were made by public transport and 43 per cent by private transport. However, the share of public transport is declining. A Karachi mass transit study forecasts that public transport would only account for 45 per cent by the year 2000 (Beg, 1990).

Automotive air pollution problems in Karachi are compounded by the high number of old and poorly maintained vehicles and the pronounced "canyon effect" of city streets. Certain areas (e.g., Mohammad Ali Jinnah Road, Liaqat Road, Chundrigar Road, Zeb-un-Nisa Street, Abdullah Haroon Road and also in the Federal Area and the SITE) are severely affected by motor vehicle pollution. Motor vehicle pollutant emission trends from 1980 to 1989 and projected emissions for the year 2000 are shown in Figure 14.2.

Karachi also has a large airport and a large naval port which are additional sources of traffic-related pollutants. In 1989–90, 4,053 arrivals and sailings and 843 port "shiftings" were recorded, totalling about 14,000 hours of movement in port with a consumption of more than 17,000 tonnes of fuel oil.

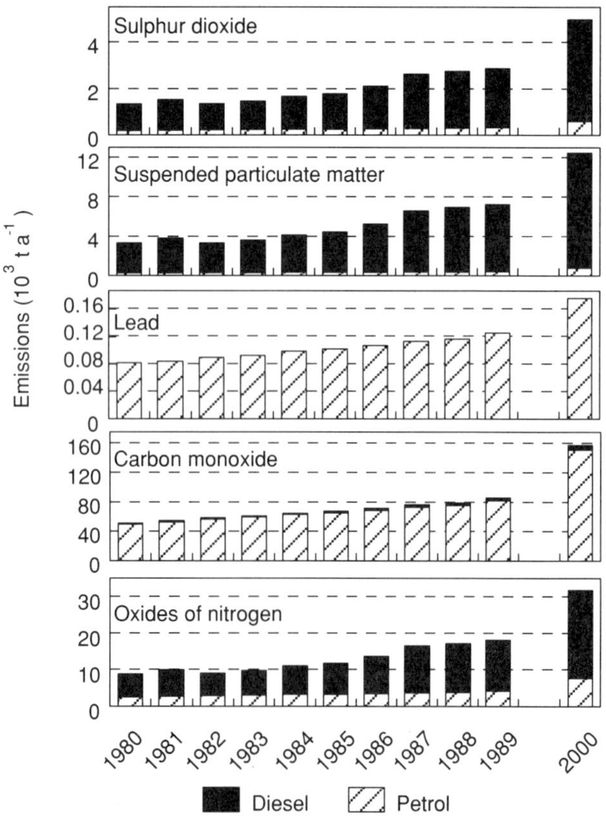

Figure 14.2 *Motor vehicle emissions trends*

Source: Beg, 1990

Table 14.2 Daily mean concentrations of major air pollutants, 12–13 June 1988

Location	SO_2 ($\mu g\ m^{-3}$)	CO ($mg\ m^{-3}$)	NO_2 ($\mu g\ m^{-3}$)	O_3 ($\mu g\ m^{-3}$)
Area A				
Tower	134	7	38	n.d.
Tibet centre	100	7	51	n.d.
Grumandir		5	63	n.d.
Hasan Square	67	5	111	n.d.
Area B				
Shershah	67	3	60	n.d.
Nazimabad	67	7	48	n.d.
Nagan Chowrangi	50	7	70	n.d.
Area C				
Yousuf Plaza	42	3	179	n.d.
National Cement Factory	50	2	544	n.d.
Balouch Colony	33	4	43	50
Mid-east Clifton	25	5	53	46
Area D				
Quaidabad Landhi	33	4	50	n.d.
National Refinery Korangi	67	6	48	36

n.d. = not detectable

After Ghauri et al., 1988

14.2 Monitoring

Karachi has not, as yet, submitted any data to GEMS/Air and does not appear at this stage to have a co-ordinated monitoring network. The majority of the air quality data presented here are from a research study by Pakistan Space and Upper Atmosphere Research Commission (SUPARCO) on assessment of air pollution in metropolitan Karachi (Ghauri et al., 1988). This study has been extended in recent years to observe the trends of the major urban air pollutants, such as sulphur dioxide (SO_2), suspended particulate matter (SPM), carbon monoxide (CO), oxides of nitrogen (NO_x) and ozone (O_3) in Karachi. Aerosol studies have also been undertaken to analyse ambient aerosols for up to 40 different elements. At present, measurements of these pollutants are made regularly at three different sites in and around Karachi. An assessment survey has also been carried out at 12 different sites (Figure 14.1) simultaneously from 0600 to 2100 hours for four consecutive days. The SUPARCO study focused on the areas of the Karachi metropolitan area shown in Table 14.2.

These areas were selected after a study of prevailing wind patterns in Karachi and thus of the transport of pollutants. The mean levels of major air pollutants measured at two days in the various sites are given in Table 14.2.

14.3 Air Quality Situation

Sulphur dioxide

Emissions Most SO_2 emissions are a result of the combustion of sulphur-containing fuels. An emissions inventory for Karachi was prepared by using 1989 estimates provided for this report (Beg, 1990), and by using information from other studies or from industries. There are several sectors missing in the inventory (e.g., industrial and commercial furnace oil combustion), but the data are more complete than for many other megacities. The estimates for the year 1989 are summarized in Table 14.3.

Total SO_2 emissions amount to about 77,000 tonnes per annum, about 85 per cent of which is emitted by fuel oil-fired power stations (Figure 14.3). Domestic fuel combustion contributes relatively low SO_2 emissions (3,000 tonnes per annum), because low-sulphur fuels like natural gas, kerosene, charcoal and wood are most frequently used. Sulphur dioxide emissions from industrial processes (oil refineries, coke production, metal and steel production) are low compared with total emissions, but their impact on ambient air pollution levels in the areas surrounding those industries is important. For instance, coke production at Pakistan Steel Mills caused SO_2 emissions of about 1,300 tonnes per annum, and the two refineries at KITE emit 3,000 tonnes per annum.

Table 14.3 *Emissions inventory (t a^{-1})*

	SO_2	SPM	NO_x	CO
Power generation				
Furnace oil	65,000	1,000	12,000	600
Natural gas	100	200	7,000	200
Industrial and commercial fuel consumption				
Furnace oil				
Natural gas	50	360	600	340
Other				
Domestic fuel consumption				
Fuel oil				
Natural gas	150	230	1,500	140
Other (Kerosene/charcoal, wood)	3,000	20,000	1,000	30,000
Industrial processes				
Metal/steel/coke	1,300	21,000	10	27,200
Refinery	3,000	3,000		100
Port activities		4,500		
Food		6,000		
Textile		500		
Chemical	10	500		
Construction		18,500		
Traffic and transport				
Motor vehicles	2,800	7,200	18,200	84,000
Ships	1,200	50	150	10
Aircraft	200	200	1,000	5,000
Waste				
Open burning	200	23,000	9,000	123,000
Total	77,010	106,000	50,460	270,590

Sources: Beg, 1990; Ghauri et al., 1988

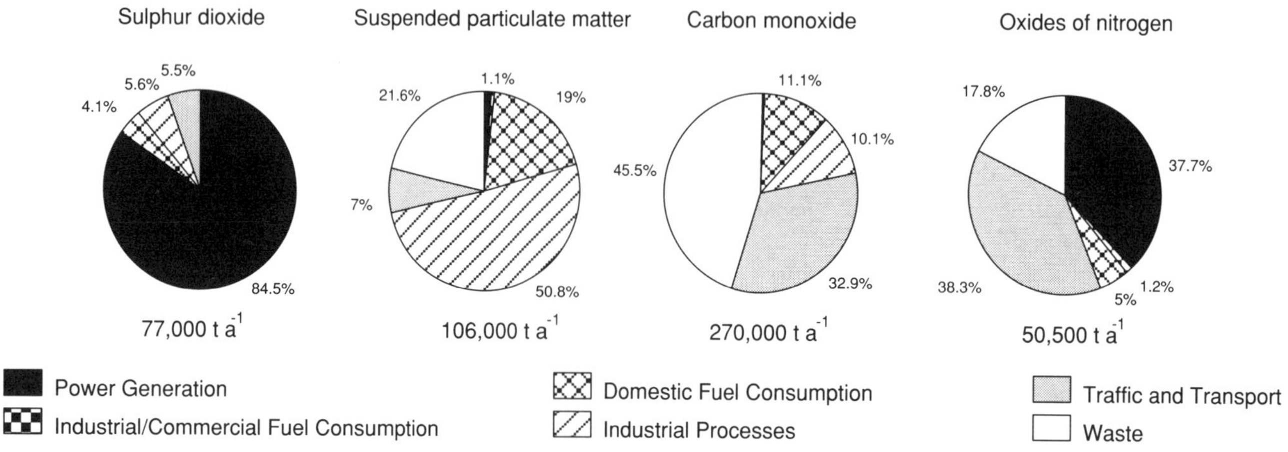

Figure 14.3 *Emissions by sector, 1989*

Sources: Beg, 1990; Ghauri et al. ,1988

Traffic SO_2 emissions are thought to be relatively small. Motor vehicles (mainly diesel-powered vehicles) emit 2,800 tonnes per annum, ships 1,200 tonnes per annum and aircraft 200 tonnes per annum.

Ambient Concentrations Ambient concentrations in Karachi as monitored in the SUPARCO study on 12–13 June 1988 are summarized in Table 14.2.

Daily mean SO_2 levels were 67-134 µg m^{-3} in the city centre and 25–67 µg m^{-3} in the other areas. Those values are below the WHO daily guideline (150 µg m^{-3}). However, when assessing the SUPARCO data it must be taken into account that June is generally a month with low air pollution due to heavy rainfall and dominant winds from the sea.

Suspended particulate matter

Emissions In Karachi there are natural sources of particulate matter (such as wind-blown dusts from surrounding desert areas) as well as anthropogenic sources (e.g., motor vehicle exhaust; and particulates from industrial activities like iron and steel production, asphalt mixing plants, cement kilns). Most natural sources of SPM are not quantifiable and cannot be incorporated in emissions inventories. An overview of anthropogenic SPM emissions in the year 1989 for Karachi was prepared by estimates from Beg (1990), and by incorporating additional information

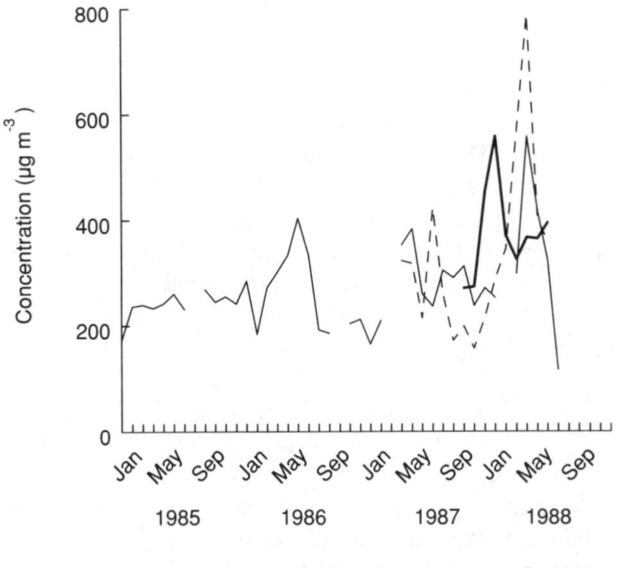

***Figure 14.4** Monthly mean suspended particulate matter (TSP) concentrations*

Source: Ghauri et al., 1988

***Table 14.4** Annual mean* total suspended particulate concentrations (µg m^{-3})*

Year	SPARCENT	SITE	Saddar
1985	239	n.a.	n.a.
1986	265	n.a.	n.a.
1987	275	254	333
1988	328[a]	459[b]	397[b]

[a] January–July [b] January–June
* calculated from monthly mean concentrations
n.a. = no data available
Source: Ghauri et al., 1988

from other studies or from the industries themselves. The estimates for the year 1989 are summarized in Table 14.3.

Total anthropogenic SPM emissions in Karachi have been estimated to be 106,000 tonnes per annum, with about half of this amount coming from industrial activities (mainly iron/steel/coke production and the construction industry). Pakistan Steel Mills is the largest single source of industrial SPM emitting about 21,000 tonnes per annum. Construction activity, especially dust escaping from vehicles engaged in construction activities (4,000 lorries enter and leave Karachi every day carrying building materials) is another important source of fugitive dust emissions. Similarly, port activity such as the offloading of coal, iron ore, grain and fertilizer gives rise to large quantities of SPM emissions. It is estimated that Karachi's two port areas (Karachi Port Trust and Port Qasim) emit a total of 4,500 tonnes per annum SPM. Domestic combustion of solid fuels is another large source and accounts for 20,000 tonnes per annum of SPM emissions (or 19 per cent of the total). Power plants and industrial fuel combustions contribute only small amounts of SPM because the combustion process in large boilers is more efficient.

In the inventory, solid refuse burning is estimated to be the greatest single source of SPM. It accounted for 23,000 tonnes per annum or 22 per cent of total emissions. Traffic contributes about 7,000 tonnes per annum SPM, with by far the most of it from motor vehicles.

Ambient Concentrations Daily total suspended particulate (TSP) measurements were made on glass fibre filters using high-volume samplers at three different sites (Space Science Division (SPARCENT), SITE, and Saddar Karachi) between 1985 and 1988 (Ghauri et al., 1988). Relative standard deviations were 20–30 per cent thus indicating that the TSP level did not drastically change from day to day.

Table 14.5 Trace constituents in <1 μm and 1–10 μm aerosols during the sampling period from 27 February – 6 March 1987

	Sonmiani Beach		SITE Karachi		SPARCENT		Pakistan Steel Mill	
	<1 μm	1–10 μm	<1 μm	1–10 μm	<1 μm	1–10 μm	<1 μm	1-10 μm
Sodium[a]	0.9	4.9	0.3	0.76	0.25	0.8	0.28	0.54
Chlorine[a]	0.04	8.6	0.6	1.24	0.58	1.5	0.4	1.23
Aluminium[a]	0.24	0.64	2.2	2.5	1	1.6	2.4	2.6
Scandium[b]	0.05	0.17	0.2	0.37	0.14	0.29	0.33	0.6
Bromine[b]	11	5	11	3	24	7	19	6
Lead[b]	4.8	0.2	214	44	287	56	65	16
Antimony[b]	<0.1	<0.1	5.4	1	8	1.3	0.73	0.23
Iron[a]	0.2	0.55	1	1.5	0.56	1.2	3.7	7.5
Manganese[b]	3.8	12	56	38	26	29	280	350
Vanadium[b]	<2	<2	7.3	4.9	4.1	3.7	14	7.3
Nickel[b]	1.9	1.5	14	3.8	3.3	2.8	162	5.5
Chromium[b]	<2	<2	11	12	6	9	11	11
Selenium[b]	0.5	0.4	0.7	0.4	0.5	0.3	1	0.6

[a] Concentrations given as μg m^{-3}　　[b] Concentrations given as ng m^{-3}
After Beg, 1990

Monthly mean concentrations of TSP as observed by this SUPARCO study are shown in Figure 14.4. Monthly means were between 150 μg m^{-3} and 250 μg m^{-3} at the SPARCENT in most months. In three months, however, SPM concentrations reached more than 400 μg m^{-3} at the SPARCENT site (June 1986, April 1988, May 1988). Levels of SPM at the industrial SITE station were generally higher. At this station, four of the fifteen months exceeded 400 μg m^{-3} and the maximum was 788 μg m^{-3} in April 1988.

Annual mean concentrations calculated from the monthly mean data show the clearly increasing trend of SPM (TSP) ambient levels (Table 14.4). In the SPARCENT site, annual means were 239 μg m^{-3} in 1985, 265 μg m^{-3} in 1986, and 275 μg m^{-3} in 1987. In 1988, the mean value for the January-July period was as high as 328 μg m^{-3}. In the SITE and the Saddar Karachi sites, the 1988 January-June mean values were 459 μg m^{-3} and 397 μg m^{-3}, respectively. All these concentrations are higher than the WHO daily guideline range of 150–230 μg m^{-3}. These data are consistent with a 1983 survey, where mean "smoke" levels of 400–460 μg m^{-3} were found during the January–June period (Beg, 1990).

Table 14.5 summarizes the concentrations of trace constituents in aerosols sampled between 27 February and 6 March 1987. The trace element analyses show that in Karachi there are five major source components: soil/limestone (aluminium, calcium, selenium, manganese, iron, cobalt, thorium), sea spray (sodium, chlorine), fossil fuels (selenium and non-marine sulphates), vehicular traffic (bromine, lead) and metal plating/air conditioning (bromine). It was calculated that soil and limestone account for 64 per cent of total aerosols, a large part of which probably come from a cement plant in Gulshan-e-Iqbal. The remaining 36 per cent consist of soot from fossil-fuel combustion, vehicular traffic aerosols, other anthropogenic sources (zinc, antimony and lead), water vapour, and undetermined organic and inorganic constituents.

Lead

Emissions The lead (Pb) concentrations in petrol in Pakistan are between 1.5–2 g l^{-1} which is relatively high on a world scale. Nevertheless, emissions of Pb from petrol-driven motor vehicles in Karachi were estimated to be about 125 tonnes per annum in 1989 (Beg, 1990), which seems very low considering the high Pb levels in petrol.

Ambient Concentrations Concentrations of inorganic Pb in Karachi are very high. Mean concentrations in urban areas were about 1-3 μg m^{-3} with maximum values of 7-9 μg m^{-3} in areas of heavy traffic (Beg, 1990). Those values are far above the WHO annual mean guideline range of 0.5–1.0 μg m^{-3}. Lead levels measured in aerosols (<10 μm) sampled between 27 February and 6 March 1987 were 0.05–0.34 μg m^{-3} (Beg, 1990). During that study it was shown that 80-96 per cent of Pb is in respirable particles, which underlines the threat of Pb pollution to human health.

Carbon monoxide

Emissions An inventory of CO emissions in the year 1989 for Karachi was prepared by using estimates from Beg (1990), and by incorporating additional information from other studies or from the industries themselves. The estimates for the year 1989 are summarized in Table 14.3.

Total CO emissions were estimated to be 271,000 tonnes per annum. It is estimated that 123,000 tonnes per annum (more than 45 per cent of total emissions) arise from open burning of 2.9 million tonnes of waste. The emissions from waste burning exceed even emissions from motor vehicles and other traffic sources (89,000 tonnes per annum or 33 per cent).

Domestic fuel combustion was estimated to emit another 30,000 tonnes per annum or 11 per cent. The major industrial source for CO is metal, steel and coke production which contributes nearly as much CO as all domestic sources. It should be noted that the local emission density at these point sources is frequently extremely high and the industries may have a great impact on the local air quality.

The contribution from the other (known) sources is relatively small, especially in the power generation sector, where very efficient boilers with low CO emission rates are used.

Ambient Concentrations Monitoring of CO has been undertaken in several studies since 1969. During a survey in 1969, CO concentrations of 6–23 mg m^{-3} (5–20 ppm) were recorded near 26 roadside locations, with values of 12–41 mg m^{-3} (10–35 ppm) in the centre of the roads during traffic congestion (Beg, 1990). The averaging time of these measurements was not reported, so the results are difficult to assess.

High CO values were found during a 1983 study: mean CO concentrations in various parts of the city ranged from 12–23 mg m^{-3} (10–20 ppm) during the sampling period from January to June (Beg, 1990).

A survey in 1988 showed that the means from 0800 to 1800 hours during one to three days ranged from 2–57 mg m^{-3} (2–49 ppm) (Beg, 1990). The highest daily means were found in city centre locations like Tibet Centre, Plaza Cinema, Nazimabad, or Nagan Chowrangi. The lowest value was recorded at a site 100 m off the road. The maximum short-term CO concentrations during this study ranged from 6 mg m^{-3} (5 ppm) at the off-the-road site up to 107 mg m^{-3} (92 ppm) at the sites where there was heavy traffic.

Surprisingly, data reported from a survey on 12–13 June 1988 are much lower (see Table 14.2). Here, the daily mean concentrations are between 2 and 7 mg m^{-3}. During that study, the diurnal variation of CO concentration ranged from 2–10 mg m^{-3} with the peak values in the morning and afternoon rush hours.

It is not clear how to interpret the available CO data. While a series of measurements from 1969 to 1987 found CO concentrations in the range of 10–30 mg m^{-3} in many monitoring sites throughout the city, which exceed the WHO eight-hour mean guideline value of 10 mg m^{-3}, the 1988 data are lower by a factor of four. The 1988 data would suggest that the WHO guidelines on CO would not be violated in Karachi. It seems very unlikely that CO concentrations decreased that much between 1987 and 1988. Thus one cause for the contradictory monitoring data could be that the locations of the sampling sites were not comparable between the various studies.

Oxides of nitrogen

Emissions Oxides of nitrogen were covered in the Karachi emissions data for the year 1989 which were prepared by Beg (1990). The estimates for the year 1989 are summarized in Table 14.3.

Total NO_x emissions were estimated to be around 50,000 tonnes per annum. Power generation and traffic together make up 76 per cent of total emissions. Power generation from oil and natural gas account for 19,000 tonnes per annum or 38 per cent of the total NO_x emissions. This is because power stations normally have a very high NO_x emission rate because of high-temperature combustion processes. This negative effect of efficient combustion processes can also be seen in the domestic sector (2,500 tonnes per annum), where the otherwise relatively "clean" natural gas causes the majority of emissions. Transport is the second large emissions sector. About 18,000 tonnes per annum are emitted by motor vehicles, and another 1,000 tonnes per annum by aircraft. Most of the motor vehicle emissions come from diesel-powered engines and, due to the growing traffic, their emissions are projected to exceed 30,000 tonnes per annum by the year 2000. Open burning of wastes was estimated to contribute about 9,000 tonnes per annum of the 18 per cent total NO_x emissions. Industrial processes are not considered as major NO_x emissions sources in the available inventory. However, considerable NO_x emissions can certainly be expected from fertilizer production industries.

Ambient Concentrations Little long-term monitoring of NO_x has been undertaken in Karachi. A summary of daily averages from a SUPARCO survey on 12–13 June 1988 throughout Karachi is presented in Table 14.2. The reported daily average nitrogen dioxide (NO_2) concentrations were between 38 μg m^{-3} and 544 μg m^{-3}. All but two stations reported daily mean values below the WHO daily guideline of 150 µg m^{-3}.

It should be noted that by far the highest concentrations were measured in the vicinity of the National Cement Factory.

During that NO_x monitoring study, the diurnal variation of NO_x concentration in city centre locations between 0700 and 2100 hours ranged from 28–95 µg m^{-3}). No clear diurnal trend could be found.

Ozone

Emissions As shown above, emissions of NO_x have been estimated to be about 50,000 tonnes per annum. There is no emissions inventory available for volatile organic compounds (VOCs). Sources of VOC emissions are incomplete combustion, industrial processes (e.g., refineries), petrol and solvent evaporation, and natural sources (e.g., vegetation).

Ambient Concentrations Ozone concentrations were included in the SUPARCO study on air pollution in Karachi. Three monitoring sites were chosen, bearing in mind the south-westerly prevailing wind direction during most of the year. One station was located in the SITE area, located to the south-west of the city (site 1); another station was located in the city centre with heavy vehicular traffic (site 2); and another station represented a suburban site in the north of the city (site 3). This arrangement allowed identification of O_3 resulting from precursor emissions generated in the city and comparison of O_3 levels up-wind and down-wind of the central urban area (Ghauri et al., 1992).

Measurements of surface O_3 made between 1986 and 1988 have shown that Karachi has a marked afternoon maximum O_3 concentration between 1200 and 1500 hours, irrespective of climatic conditions and site location. This O_3 maximum is delayed at the downwind sites 2 and 3 which demonstrates the contribution of the city's polluted air to O_3 formation. While O_3 concentrations at the upwind site were as low as 2–50 µg m^{-3}, O_3 maxima at the downwind sites 2 and 3 reached levels of 80 µg m^{-3} and 100 µg m^{-3}, respectively. However, the WHO one-hour mean guidelines of 150–200 µg m^{-3} were not exceeded during the monitoring period (Ghauri et al., 1992).

14.4 Conclusions

Air Pollution Situation The air pollution situation cannot be sufficiently characterized as there is no long-term monitoring providing statistically valid ambient air quality information. From a few case studies during 1969–1988 it seems that levels of SPM, Pb, CO and NO_x are extremely high. Short-term or long-term averages exceeded the applicable WHO guidelines in many parts of the urban area.

Suspended particulate matter exceeded WHO annual guidelines in all monitoring stations during the whole monitoring period, and there was an increasing trend from 1985–1988. Mean concentrations of Pb exceeded the WHO guidelines by a factor of three, and in areas of heavy traffic by a factor of up to nine. This is even more severe as most of the Pb-containing particles are respirable.

Carbon monoxide levels were reported to be extremely high at traffic-related monitoring sites, exceeding WHO guidelines by a factor of up to five. However, a study in 1988 showed much lower CO concentrations below the WHO guidelines. Similarly, SO_2 concentrations monitored during this 1988 case study are relatively low. One of the causes for this discrepancy might be the fact that CO levels are actually very high in traffic-related areas, but rather low in locations away from the main roads.

The high concentrations of NO_2 reported during the case study in 1988 cannot be explained at present.

Main Problems One of the main problems of Karachi is the lack of long-term routine ambient air quality monitoring. Thus it is not possible to analyse the major air pollutants, to identify their sources, and to assess their impact. If pollutants are monitored during single case studies only, the location of the monitoring sites and the exposure of the sampling devices are normally not comparable, leading to confusion in the interpretation of data. Furthermore, no trends in air pollution can be identified.

From the preliminary emissions inventory, it must be concluded that the main pollution problems in Karachi stem from motor vehicle traffic, from several industrial sources which produce large amounts of air pollutants and which cause severe problems in the surrounding urban areas, and from the open burning of waste. Power stations within the city using fuel oil are the main source of SO_2 emissions, but the ambient levels of SO_2 seem to be low.

Motor vehicle traffic is responsible in large part for NO_x and CO pollution and for nearly all the Pb pollution. In addition, many vehicles emit black smoke as a result of improper burning of fuel which is usually due to lack of maintenance. The black smoke normally contains toxic compounds such as unburnt hydrocarbons, paraffins, polynuclear aromatics and aldehydes. The auto-rickshaw and other two-stroke vehicles produce large quantities of smoke.

Two-stroke vehicles use a lubricant together with petrol. Drivers of auto-rickshaws use more lubricant than recommended and use "tinny" exhaust pipes for unretarded flow of exhaust gases in order to obtain high speeds without overheating the engine.

The open burning of refuse in and around Karachi also has great impact upon air quality and is a problem which needs to be addressed as a matter of urgency. The main problem associated with refuse disposal is the lack of collection vehicles; this results in only 25 per cent of refuse being collected. The rubbish which is collected is simply burned at sites on the outskirts of the city. Open waste burning accounts for almost half of all CO emissions and for more than 20 per cent of particulate emissions.

Control Measures In Pakistan, ambient air is considered acceptable if maximum levels are below the WHO guidelines. In 1983, the Pakistan Environmental Protection Agency (PEPA) was established which has powers to establish regulations for the control of pollutants from stationary and mobile sources, and to declare clean air zones (Commins, 1990; IUAPPA, 1991).

Industrial processes are subject to the Environmental Protection Ordinance which came into force in 1983. For the pollution caused by the National Cement Industries in Karachi, the PEPA has initiated action which resulted in the installation of dust control equipment in 1991.

For mobile sources there are effectively no regulations on exhaust emissions. Further work is being carried out before the introduction of an unleaded regular grade petrol which would help to reduce Pb levels in urban areas in the long term.

For the control of domestic sources and smaller industrial sources, the local or territorial government is responsible. Since 1989, the responsibility of ambient air pollution monitoring and direct control of large industrial pollution sources in Karachi was given to the Department of Health and its Pollution Control Society of Sind (PCSS) (IUAPPA, 1991).

Health Effects The lack of any long-term monitoring in Karachi makes proper assessment of the urban air quality very difficult. Apart from a small study on blood Pb levels, little epidemiological work has been undertaken in the city or in Pakistan as a whole. Air pollution problems in Karachi have in no way been properly quantified and therefore the options for air quality management cannot be evaluated fully.

Of great importance to health is the effect of pollutants on communities around areas of high emission densities (e.g., in Mohammad Ali Jinnah Road, Sadar and Korangi). Air pollution in these highly populated localities comes mainly from motor vehicles, power plants, tanneries and textile mills. Particulate emissions from cement factories are also of concern. High concentrations of respirable silicate particles are known to cause pulmonary diseases such as silicosis.

During the burning of wastes many toxic compounds are released which pose a considerable health risk to the population affected by the smoke.

14.5 References

Beg, M. A. A. 1990 *Report on Status of Air Pollution in Karachi, Past, Present and Future,* Pakistan Council of Scientific and Industrial Research, Karachi.

Commins, B. 1990 *Air Pollution in Pakistan,* WHO Assignment Report. (Unpublished report.)

Ghauri, B.M.K., Salam, M. and Mirza, M.I. 1988 *A Report on Assessment of Air Pollution in the Metropolitan Karachi,* Space Science Division (SPARCENT), Pakistan Space and Upper Atmosphere Research Commission (SUPARCO), Karachi.

Ghauri, B.M.K., Salam, M. and Mirza, M.I. 1992 Surface Ozone in Karachi. In: *Ozone depletion: implications for the tropics,* M. Ilyas (Ed.), University of Science Malaysia, Penang, Malaysia/United Nations Environment Programme, Nairobi, Kenya, 169–177.

IUAPPA 1991 *Clean Air Around the World,* Second edition, International Union of Air Pollution Prevention Associations, Brighton.

UN 1989 *Prospects of World Urbanization 1988,* Population Studies No. 112, United Nations, New York.

15

London

15.1 General Information

Geography London is the capital of the United Kingdom (UK) and is the most populous city in Europe. It is situated in the south-east of England on the banks of the River Thames, 65 km west of its estuary on the North Sea, 5 m above mean sea level (Latitude 51°30′N, Longitude 0°10′W).

Demography Like most cities, London's population is classified in a number of ways: the urban agglomeration, as used in UN estimates (Figure 1.1 and Table 2.1), had an estimated population of 10.57 million in 1990 (UN, 1989). This classification covers a wider area than that of Greater London, an amalgamation of the 33 London boroughs (districts), shown in Figure 15.1 (area 1,579 km^2), which had an estimated population of 6.7 million in 1991 (LRC, 1990). In this document London is defined as the Greater London area, as this is the area to which most of the UK source material refers. The population density of Greater London in 1987 was 4,287 persons per km^2 and central London's working population is approximately 1.25 million. London's population decreased slightly during the 1970s before starting to rise again in the 1980s. Projections indicate that the population will increase by approximately 200,000 to 10.79 million over the next 10 years (UN, 1989). A steady increase in population has been observed in London's socially deprived areas in recent years (LRC, 1990). It is these deprived areas which often have the worst air quality because of their proximity to industry and to transport corridors. The main factors governing population change in London are employment, housing supply and trends in the housing market. London has a low birth and death rate in comparison with many of the other megacities.

Climate London has a marine west coast climate. The average annual mean temperature at the London Weather Centre in Central London is 11.4°C with a January minimum of 5.5°C and a July maximum of 18.1°C. London causes a pronounced urban heat island effect – city temperatures are typically 2–3°C higher than surrounding rural areas. Rainfall is irregular and the annual mean of 597 mm may vary by ± 30 per cent per annum (LRC, 1990; WMO, 1971).

Autumn (September–November) usually induces the highest pollution levels in London. High barometric pressure over the south-east of England and continental Europe results in light winds thus reducing the dilution of air pollution. The concurrent cooling of the ground causes a temperature inversion (a few tens of metres above the surface); traffic pollution is trapped near the ground and power station emissions from the east descend on London before they can disperse in the upper atmosphere. Such conditions predominate particularly in November.

Industry London is one of main banking and insurance centres of the world. Until the 1960s London was also an important port; the docks and associated industry to the east of the city were a major source of employment. Changes in working practices and a decrease in Britain's manufacturing base led to the closure of most of the docks, although a few wharves remain in operation. Most of the London docks are now undergoing intensive commercial and residential redevelopment. The industries associated with the docks have also tended to relocate out of London. Major industrial areas still occur in the east, along the Thames in Greenwich, Woolwich and Dagenham. Other industrial areas are located along the River Lea in north London, at the railway sidings at Willesden to the north-west, along the Grand Union Canal to the west and in Croydon to the south.

Transport In the past, London has relied heavily upon its radial rail and underground system to move people about the city. Recent changes in the nature and location of employment have not been accommodated by the existing public transport network. This, combined with other social and economic factors, has led to an increase in motor vehicle ownership and use (OECD, 1988). In 1988 Greater London had a motor vehicle population of 2.7 million (LRC, 1990). This means that there is one car for every three people in London. The south-east of England, including London and the outlying metropolitan districts, accounts for ownership of

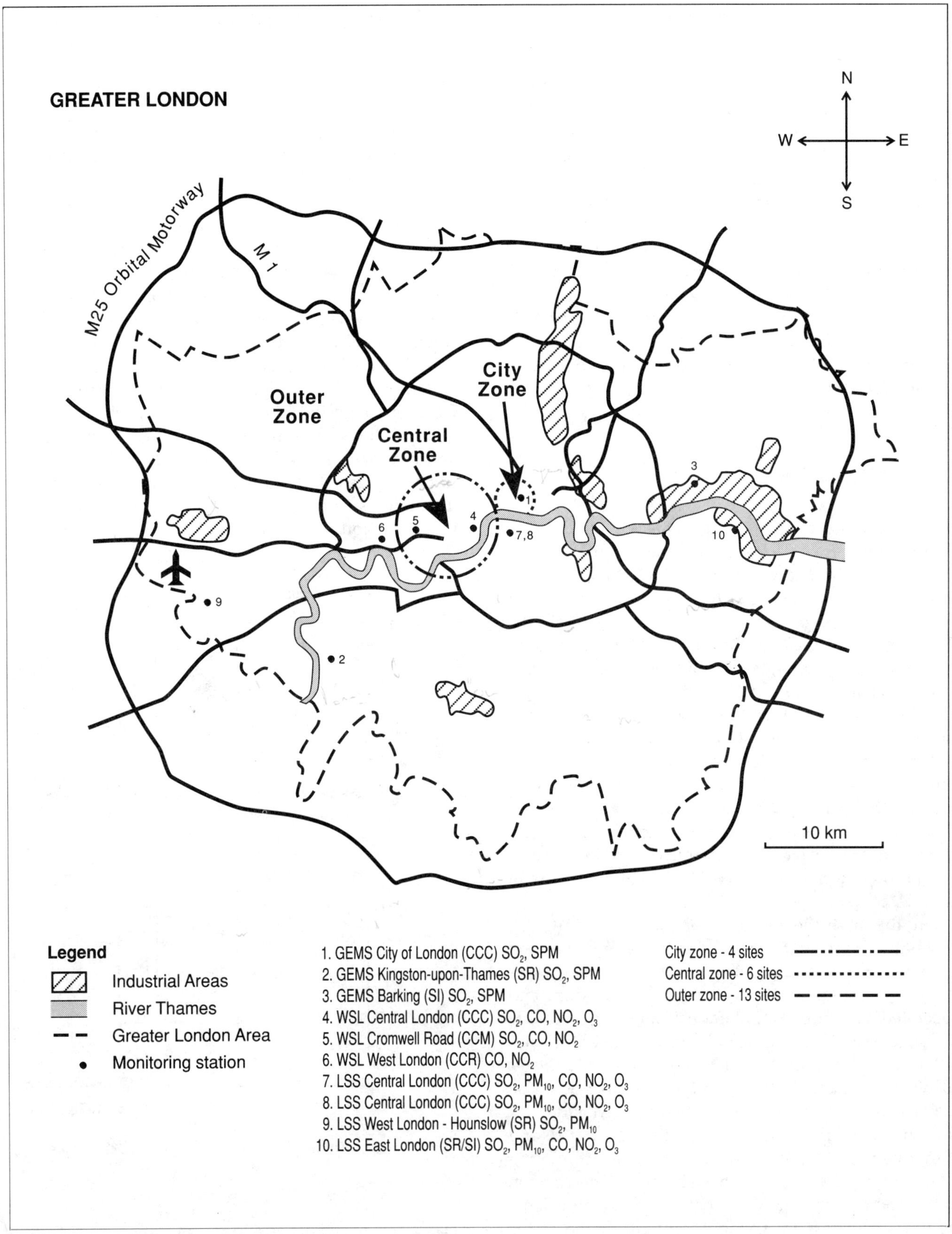
GREATER LONDON
N
W
E
S
M25 Orbital Motorway
M 1
Outer Zone
City Zone
Central Zone
1
2
3
4
5
6
7,8
9
10
10 km
Legend
Industrial Areas
River Thames
Greater London Area
Monitoring station
1. GEMS City of London (CCC) SO2, SPM
2. GEMS Kingston-upon-Thames (SR) SO2, SPM
3. GEMS Barking (SI) SO2, SPM
4. WSL Central London (CCC) SO2, CO, NO2, O3
5. WSL Cromwell Road (CCM) SO2, CO, NO2
6. WSL West London (CCR) CO, NO2
7. LSS Central London (CCC) SO2, PM10, CO, NO2, O3
8. LSS Central London (CCC) SO2, PM10, CO, NO2, O3
9. LSS West London - Hounslow (SR) SO2, PM10
10. LSS East London (SR/SI) SO2, PM10, CO, NO2, O3
City zone - 4 sites
Central zone - 6 sites
Outer zone - 13 sites

Figure 15.1 *Sketch map of London*

over 6 million cars, 35 per cent of all cars in the UK. It is estimated that road traffic in London has increased by 70 per cent over the past 20 years, yet the road area has increased by only 10 per cent over the same period. The resulting congestion has meant that the average speed of traffic during the morning peak period has dropped from 29.12 km h^{-1} in 1968 to 27.19 km h^{-1} in 1986 (LRC, 1990; OECD, 1988). Changes in employment location have led to a shift in journey patterns. The number of car journeys into central London peaked in 1984 and have declined since, mainly owing to the introduction of cheap travel passes on public transport, stricter parking restrictions and the completion of London's orbital motorway, the M25. The M25 has accelerated the relocation of employment to the outskirts of London, thus increasing traffic movements in this area.

15.2 Monitoring

The air quality data presented here for London are from a number of sources. The locations of monitoring stations are shown in Figure 15.1. Monitoring is undertaken by three organizations in London. On behalf of the Department of the Environment (DoE) the Department of Trade and Industry (DTI) delegate to Warren Spring Laboratory (WSL) the co-ordination of the national networks which monitor urban smoke, sulphur dioxide (SO_2), nitrogen dioxide (NO_2), carbon monoxide (CO), ozone (O_3) and airborne lead (Pb). Warren Spring Laboratory operate three urban continuous monitoring stations in London and co-ordinate the smoke/SO_2 monitoring undertaken by London boroughs as part of the national survey. The three GEMS/Air sites, all of which have now ceased operation, were part of this national urban smoke/SO_2 survey.

In the past the Greater London Council (GLC), Scientific Services Branch, later to become London Scientific Services (LSS), also operated a similar, overlapping, smoke and SO_2 survey in London and operated a London Air Pollution Monitoring Network. This consisted of four continuous monitoring sites measuring SO_2, NO_2, CO, airborne Pb and particulates between 1986 and 1990. London Scientific Services also co-ordinated the London-wide Nitrogen Dioxide Diffusion Tube Survey (now run by Rendel Science and Environment), the London-wide Ozone Monitoring Programme and various other surveys in the London area. These surveys and the network were, to a large extent, sponsored by London boroughs. Some of the boroughs undertake their own monitoring; however, few of these data are ever published.

15.3 Air Quality Situation

Sulphur dioxide

Emissions Three detailed emissions inventories have been carried out in London (Figure 15.2) in 1975/76 (hereafter referred to as 1975), 1978 and 1983/84 (hereafter referred to as 1983). Total emissions of SO_2 fell from 179,000 tonnes per annum in 1975 to 49,000 tonnes per annum in 1983. Commercial, industrial and institutional sources predominate and accounted for 83 per cent of total emissions in 1983; the remainder was split between domestic and transport sources. Fuel oil is the dominant source of SO_2, followed by coal and gas oil (Munday et al., 1989). Table 15.1 shows SO_2 emissions estimates produced by the GLC for the period 1965–1980 (OECD, 1988).

There has been a marked shift away from fuel oil and coal, with its high sulphur content (3.5 per cent), to natural gas. The shut-down of electricity generating stations in London has also helped to accelerate the reduction in SO_2 levels. In 1975 there were 15

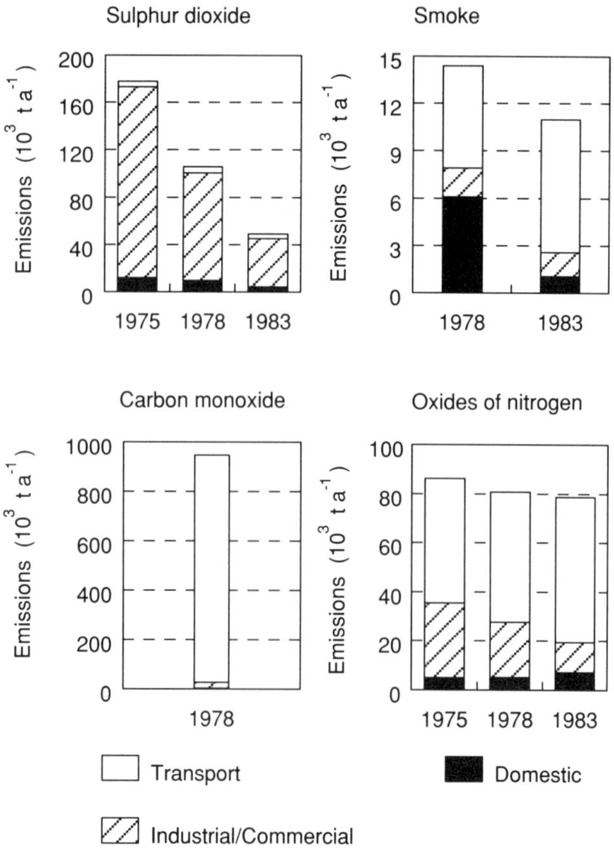

***Figure 15.2** Estimated emissions in Greater London by source*

Sources: Munday et al., 1989; OECD, 1988; Ball and Armorgie, 1983

Table 15.1 Estimated sulphur dioxide emissions 1965–1980 (10^3 t a^{-1})

	Domestic	Industrial	Commercial	Power Stations	Total
1965	47	122	56	170	395
1970	30	111	58	159	358
1975	10	45	57	58	170
1980	7	25	32	17	81

Note: Road transport contributes approximately 1% of SO_2 emissions   After OECD,1988

major power stations in operation in London; by 1988 there were only three. Much of London's generating capacity is now provided by power stations outside London on the Thames estuary (Ball and Armorgie, 1983). Specific legislation restricting the sulphur content of fuel (coal and oil) to below 1 per cent has been in force since 1972 in the City of London (central business district). The decline of manufacturing industry in the area is also a major causal factor in the reduction of SO_2 emissions.

Ambient Concentrations Monitoring of SO_2 at County Hall in London dates back to 1931. Annual mean concentrations of between 300–400 μg m^{-3} were typical until the mid-1960s when a steady decline began. Annual mean concentrations today are around 20–30 μg m^{-3}. A similar trend has been observed throughout the UK (Laxen and Thompson, 1987; LSS, 1990a). Annual mean GEMS/Air data (Figure 15.3) and long-term trends

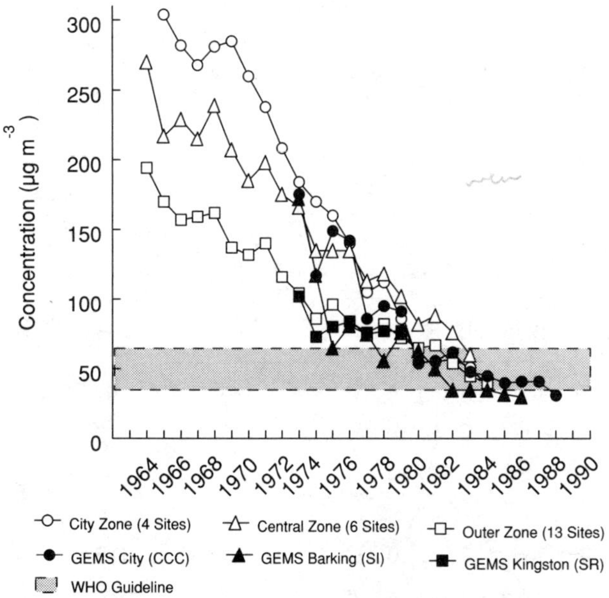

Figure 15.3 *Annual mean sulphur dioxide concentrations*

Sources: Laxen and Thompson, 1987; GEMS data

from WSL and LSS monitoring networks (Laxen and Thompson, 1987) show that concentrations are now well below the EC annual limit values and WHO long-term guidelines (40–60 μg m^{-3}). A number of factors are responsible for this fall in ambient concentrations over the past 30 years, the early introduction and enforcement of "Smoke Control Orders" in London as a result of the 1956 Clean Air Act (amended in 1964 and 1968) being the most important. This had a major effect on fuel use, as described above.

Although annual mean SO_2 concentrations have now fallen below WHO guideline levels, high pollution episodes still pose a significant risk to health on certain days in each year. Exceedences of short-term guidelines (500 μg m^{-3} 10-minute mean; 350 μg m^{-3} one-hour mean) are still a regular occurrence. In 1989/90 the one-hour guideline value was exceeded for 19 hours in total over five days at the WSL Central London site, compared with only one hour in the whole of 1988/89, 13 hours in 1987/88 and 37 hours in 1985/86 (Broughton et al., 1990a, 1990b, 1991). London Scientific Services reported exceedences on seven days in 1989 at their central London station, with a maximum recorded value of 1,640 μg m^{-3}, double any level recorded by LSS since 1986. Such episodes occur when meteorological conditions (easterly winds) force pollution from remote sources, such as power stations in the Thames estuary, into London (LSS, 1990a).

Suspended particulate matter

Emissions Suspended particulate matter (SPM) from stationary sources has followed a very similar trend to SO_2 in London over the past 25 years owing to the effects of the Clean Air Act (1956, 1964 and 1968). The reduction in solid fuel use, together with the introduction of solid, smokeless fuels, caused a steady decline in emissions throughout the 1970s and 1980s. In 1983 smoke emissions were estimated to total almost 11,000 tonnes per annum, compared with 14,000 tonnes (excluding power stations) in 1978. Motor

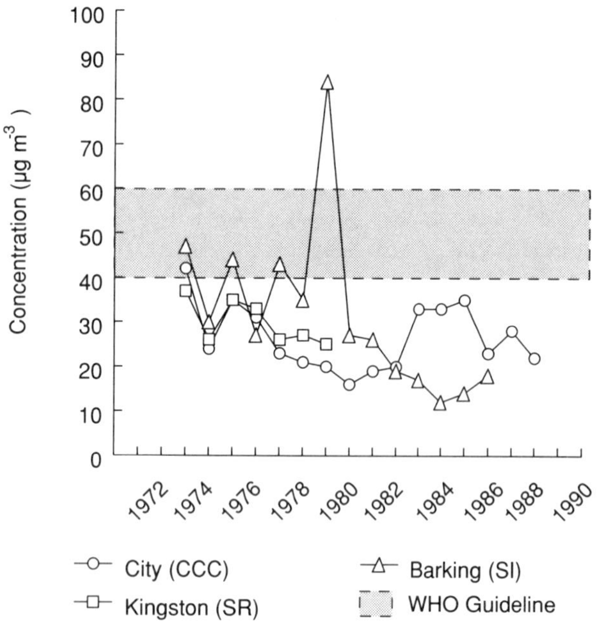

Figure 15.4 *Annual mean suspended particulate matter (smoke) concentrations*

Source: GEMS data

vehicles emissions are an increasingly important source of smoke in London; transport is estimated to contribute 76 per cent to total anthropogenic emissions. Smoke emissions from diesel-engined vehicles were estimated to account for 66 per cent of total smoke emissions in 1984. Diesel smoke has been of concern in London for a number of years, especially where people are at direct risk from exposure. At present, smoke emissions are defined for production-line vehicles in European Directive 72/306/EEC and are based upon the "acceptability" to a panel of observers of the visible appearance of the emitted smoke. It has been suggested by the UK Government that tighter emissions standards, based upon the US 1994 standards, should be introduced and enforced.

Ambient Concentrations Annual mean smoke concentrations have now levelled out generally below 30 μg m^{-3}. The GEMS/Air data (Figure 15.4) demonstrate how levels at the commercial city centre site exceed those at the suburban industrial site.

In 1988 LSS presented results of a baseline metal-in-dust sampling survey in Greater London. The study revealed that industrial areas (e.g., Lea Valley) and areas with high traffic densities (e.g., central London) had the highest concentrations of cadmium, copper, iron, lead and zinc (Schwar et al., 1988).

Lead

Emissions Airborne Pb, principally emitted from motor vehicle exhausts, has been a major political and scientific issue in the UK for over 20 years. The reduction in permissible Pb levels in petrol from 0.4 to 0.15 g l^{-1}, which came into force on 1 January 1986, was the subject of many studies throughout the UK. It is estimated that Pb emissions from motor vehicles are in the order of 350-700 tonnes per annum (10–20 per cent of the national total).

Ambient Concentrations The reduction in the Pb content of petrol, approximately 63 per cent, was not immediately realized in terms of urban airborne Pb levels in London. A study conducted in 1986 (Jensen and Laxen, 1987) on environmental lead levels during the phase down of lead in petrol observed decreases in lead-in-air concentrations of 34–55 per cent. The results suggest that a residual component exists

Table 15.2 *Mean airborne lead concentrations 1984–1987*

			Lead concentration (μg m^{-3})				1987 concentration as a percentage of concentrations	
Survey	Sampling Period	Site	1984	1985	1986	1987	1985	1986
Multi-Element (WSL)	Weekly	Brent	0.570	0.640	0.300	0.290	45	97
	Monthly	Victoria	0.520	0.480	0.270	0.280	58	104
Continuous Monitoring (WSL)	Weekly	Cromwell Road	1.410	1.450	0.660	n.a.		
LSS	Monthly	County Hall	0.420	0.380	0.200	0.230	61	115
Overall mean for UK – 20 sites				0.443	0.212	0.229	52	108

n.a. No data available

After McInnes, 1988

comprising industrial Pb emissions, Pb transported over large distances and re-suspension of Pb in dust. These results correspond with those presented by McInnes (1988) for the UK as a whole which include data from several different monitoring surveys in London (Table 15.2).

Continuous monitoring by LSS at four sites in London between 1986 and 1989 showed that ambient Pb levels have decreased still further, probably through the increased availability and use of unleaded petrol (LSS, 1990a). Annual concentrations are now well below WHO guidelines at background sites. Roadside concentrations at the central London site do not exceed the upper limit of the guide range (1 $\mu g\ m^{-3}$). Airborne Pb levels are likely to fall further as the proportion of vehicles using unleaded petrol increases. The UK market share of unleaded petrol increased from nil at the beginning of 1987 to 33 per cent in May 1990. Unleaded petrol is available at 92 per cent of outlets in the UK and, since October 1990, all new cars in the EC must be capable of using unleaded petrol.

Carbon monoxide

Emissions No detailed CO emissions estimates or inventories have been produced for the London area since 1978. In 1978 total anthropogenic emissions (not including power stations) were around 950,000 tonnes per annum. Petrol-driven vehicles were estimated to contribute 95 per cent of the total (OECD, 1988). Theoretically, by the end of the century CO emissions from motor vehicles should be approximately half those of today because of increasing engine efficiencies and the introduction of three-way catalytic converters to all new petrol-driven cars in 1993.

Ambient Concentrations Carbon monoxide is monitored at three central London sites by WSL and was monitored until recently at the four LSS stations. Measurements are made on a continuous basis using infra-red absorption techniques. Several of the London boroughs have also undertaken localized surveys to determine ambient concentrations and the personal exposure of residents to CO.

The data presented generally refer to "ambient" stations which are chosen to be representative of large areas (e.g., city centre). By necessity these stations have to be located away from any dominant source which may interfere with the sites representativeness. However, as CO is chiefly emitted by motor vehicles, it is important to make some measurements at the roadside in order to give an indication of the levels to which many people are exposed.

Improvements in motor vehicle engine efficiency and a reduction in domestic emissions have helped to reduce ambient levels over the past 20 years. The highest eight-hour mean concentrations between 0900 and 1700 hours at the four LSS monitoring sites between 1986 and 1989 are presented in Figure 15.5. In 1989 the WHO eight-hour (at any time) guideline was exceeded on 27 days at LSS roadside site (LSS, 1990a). At the WSL sites In 1989/90 the WHO eight-hour guideline was exceeded on eight occasions over seven days at the Cromwell Road (roadside) site, but on only two occasions on one day at the nearby west London site (Broughton et al., 1991). Maximum daily eight-hour mean CO levels at a roadside monitoring station operated by the London Borough of Southwark on the busy Old Kent Road (one of the main arterial routes into London) indicated 86 exceedences of the WHO eight-hour guideline between October 1985 and September 1986 (Woodbridge, 1989). High localized concentrations such as these are of major concern.

Carbon monoxide concentrations peak in winter during high pollution episodes. These episodes tend to correspond with ground-based temperature inversions which trap vehicle emissions close to the ground. The winter of 1988/89 was a good example (Table 15.3) as it was much drier than average and low wind speeds and high insolation were dominant factors in the buildup of pollution.

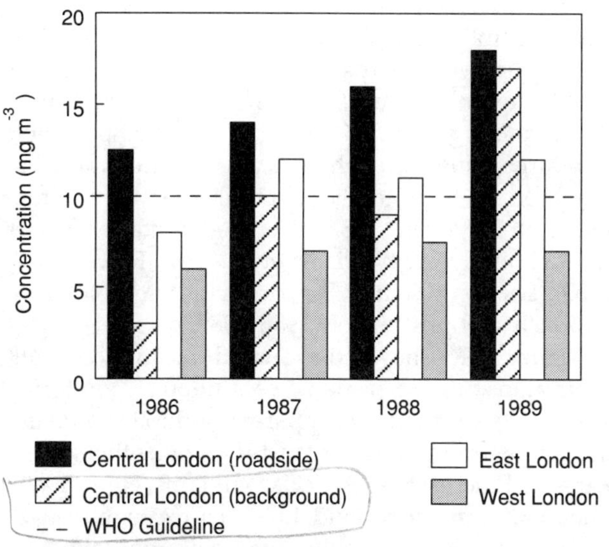

Figure 15.5 *Highest eight-hour mean (0900–1700 hours) concentration of carbon monoxide*

After LSS, 1990a

Table 15.3 Weather conditions and carbon monoxide levels, October 1988 – January 1989

	Weather related to long-term means [a]			
Month	Mean Temperature	Sunshine hours	Rainfall	Days with 8 hour CO > 10 mg m^{-3} [b]
October	+0.4 °C	+4%	+10%	3
November	- 0.6 °C	+48%	-68%	13
December	+2.9 °C	-27%	-78%	8
January	+2.5 °C	+44%	-10%	3

[a] London Weather Centre data [b] Central London Roadside Station After Laxen, 1989

Oxides of nitrogen

Emissions Anthropogenic oxides of nitrogen (NO_x) emissions totalled an estimated 79,000 tonnes in 1983. Transport, and in particular motor vehicles, is the major anthropogenic source of NO_x in London (Figure 15.2). In 1984 transport accounted for about 75 per cent of total NO_x emissions compared with only 57 per cent in 1975. Industrial/institutional emissions fell from 30,500 tonnes (34 per cent) in 1975 to 12,200 tonnes (15 per cent) in 1983 (Munday et al., 1989).

Ambient Concentrations Nitrogen dioxide and nitric oxide (NO) concentrations are monitored continuously (chemiluminescent gas analysers) at several sites in London. WSL has monitored NO and NO_2 on a continuous basis at its central London and Cromwell Road sites since 1973 and 1975 respectively. These sites recently became part of the national NO_2 network, set up in compliance with the EC Directive on NO_2. The EC air quality standards have set guide values for NO_2 at 135 µg m^{-3} for a 98 percentile of hourly values throughout a year and at 50 µg m^{-3} for a 50 percentile of hourly values throughout a year. A limit value has also been set, to protect human health and the environment, at 200 µg m^{-3}, 98 percentile of hourly values throughout a year (Williams et al., 1988; Bower et al., 1989). London Scientific Services monitored NO_x continuously at four sites from 1986–1989.

Figure 15.6 shows the annual mean NO_2 concentrations at the various continuous monitoring sites. No clear trends are apparent, some sites remaining stable throughout their history while others show a marked increase (LSS, 1990a; Williams, et al., 1988). The WHO one-hour guideline was exceeded on 25 hours over nine days (plus nine daily guideline exceedences) at the WSL west London site in 1989/90 compared with 11 hours on two days (15 daily exceedences) at Cromwell Road and one hour at the central London site (Broughton et al., 1991). It is interesting to note that, during the second week of December 1991, London experienced the highest NO_2 concentrations since records began (a maximum hourly average of 867 µg m^{-3} (423 ppb) at the central London site). During the episode (12–15 December) NO_2 levels were in excess of 205 µg m^{-3} (100 ppb) for an average of 72 hours and eight hours over 600 µg m^{-3} (300 ppb). The episode is believed to have been caused by motor vehicle emissions trapped during a period of cold calm weather (WSL, 1992).

Since 1986 LSS has co-ordinated the London-wide NO_2 diffusion tube survey, which consists of 64 sites throughout the Greater London area (LSS, 1990b). Diffusion tube surveys have been used to produce concentration distribution maps for London (Figure 15.7) (Munday et al., 1989). Districts displaying high NO_2 concentrations tend to correspond with high

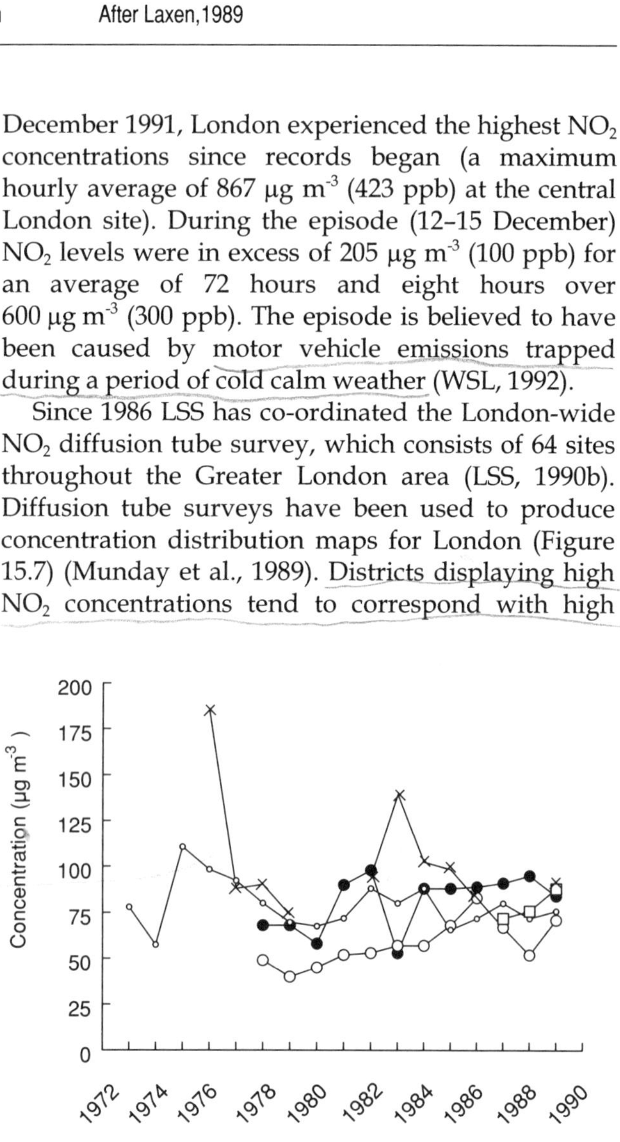

Figure 15.6 *Annual mean nitrogen dioxide concentrations*

Sources: Broughton et al., 1990a, 1990b, 1991; LSS, 1990a; Parker et al., 1990; Williams et al., 1988

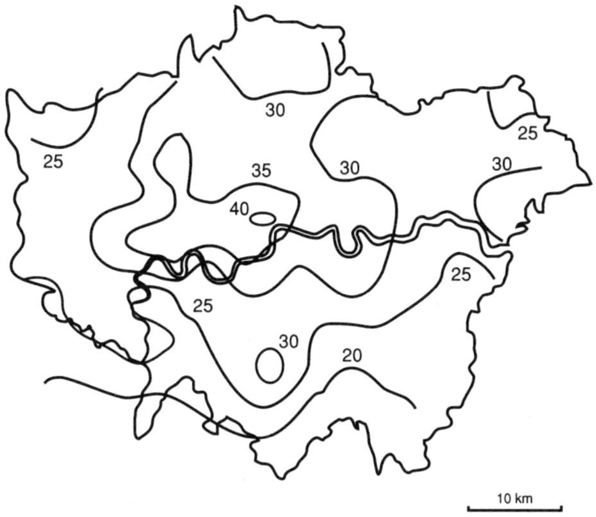

Figure 15.7 *Annual mean nitrogen dioxide concentration in ppb at non-roadside sites from the 1984–1985 diffusion tube survey*

After United Kingdom Photochemical Oxidants Review Group, 1990

traffic densities. The most obvious examples of this are central and west London. Results from the London-wide NO_2 diffusion tube survey reveal a 2 per cent drop in NO_2 concentrations across 50 sites between 1986 and 1989; this is insufficient to determine any significant trends. During 1989, 16 sites in six central London boroughs had an estimated 98 percentile greater than the EC limit value of 200 μg m^{-3}; all but two of 64 sites throughout London had an estimated 50 or 98 percentile greater than the Directive Guide Values of 50 or 135 μg m^{-3} respectively. It should be noted that the majority of these sites are urban "roadside" sites (LSS, 1990b).

Modelling studies by WSL forecast that significant reductions in ambient NO_2 are unlikely before the turn of the century because of increasing traffic (Munday et al., 1989).

Ozone

Ambient Concentrations Ambient tropospheric O_3 measurements were first carried out in London in the summer of 1972 at a single site in the city centre. Many studies throughout the 1970s, both in and around London, did much to identify the scale of the O_3 problem throughout western Europe. The most important finding of these initial studies was that elevated concentrations of O_3 were much more frequent and widespread than originally supposed.

In 1982 a major collaborative investigation, the "London Ozone Survey" was initiated. The study lasted four years and examined O_3 generation in the London region at a maximum of 20 ground-based monitoring stations. Monitoring was conducted over a limited number of intensive "campaigns" each summer on selected days when meteorological conditions were conducive to oxidant formation. The objective was to determine the behaviour of O_3 throughout the atmospheric boundary layer and to quantify locally generated O_3 fluxes (Ball, 1987; Varey et al., 1988). The survey revealed that precursor emissions generated in London can increase downwind O_3 concentrations by between 38–154 μg m^{-3} (18–72 ppb) within a few hours (up to 10 hours). This "urban plume" effect increases rural concentrations which may have adverse effects on forests and commercial crops; in addition it may contribute to urban O_3 in cities and towns downwind of London. A summary of exceedence statistics for sites in central, west and east London are presented in Table 15.4 for the period 1975–1985. There is no evidence of a trend in UK O_3 concentrations since monitoring began. Figure 15.8 shows the number of days when hourly O_3 concentrations exceeded 171 μg m^{-3} (80 ppb) (WHO one-hour guideline range is 150–200 μg m^{-3} (76–100 ppb)) at the LSS central London monitoring site. Peaks in O_3 levels correspond with exceptional meteorological conditions; the summers of 1975, 1976 and 1989 were among the sunniest since records began.

In 1989 there were 11 occasions at the LSS central London site when the hourly mean value exceeded 214 μg m^{-3} (100 ppb), the upper limit of the WHO guideline range. Five major elevated O_3 incidents occurred in 1989. On three of these occasions a significant proportion of the O_3 had its origins in continental Europe. Such episodes are characterized by anticyclonic weather conditions giving clear skies, low wind speeds and high insolation – ideal conditions for photochemical oxidation (LSS, 1990c). Monitoring at the WSL central London site revealed that in 1988/89 the annual average O_3 concentration was lower than at any other station in the UK NO_x/O_3 network. At this site the WHO one-hour guideline was exceeded on 16 hours over six days and the eight-hour guideline on 10 hours over eight days (Broughton et al., 1991). Low levels at urban monitoring stations are attributed to NO scavenging as a result of local primary NO_x emissions (Bower et al., 1988).

Table 15.4 Ozone statistics from 1975 to 1985

Year	South-west Suburbs 80*	100*	120*	maximum (ppb)	Central London 80*	100*	120*	maximum (ppb)	North-east Suburbs 80*	100*	120*	maximum (ppb)
1975	130	50	11	138	79	33	12	150	81	28	3	122
1976	346	184	98	211	134	68	44	212	247	126	57	178
1977	n.a.	n.a.	n.a.	n.a.	3	0	0	87	n.a.	n.a.	n.a.	n.a.
1978	95	36	11	157	6	1	0	103	n.a.	n.a.	n.a.	n.a.
1979	21	8	5	156	17	3	2	153	13	3	1	128
1980	n.a.	n.a.	n.a.	n.a.	8	3	0	116	n.a.	n.a.	n.a.	n.a.
1981	n.a.	n.a.	n.a.	n.a.	24	5	0	112	n.a.	n.a.	n.a.	n.a.
1982	25	4	2	123	2	0	0	91	14	2	0	119
1983	66	15	2	137	29	0	0	99	10	6	0	114
1984	28	3	0	117	10	0	0	88	22	6	5	160
1985	21	2	0	109	8	0	0	98	4	0	0	88

n.a. = no data available

* For each site, the first three columns give the number of hours above 80 ppb (171 $\mu g\ m^{-3}$), 100 ppb (214 $\mu g\ m^{-3}$) and 120 ppb (257 $\mu g\ m^{-3}$). The fourth column is the maximum one-hour mean for the year.

Note: South-west data are from Teddington or Kew, north-east data are from Hainault or Chigwell, central London data are from County Hall roof top site. Data relate broadly to the summer period, i.e., May to September, but complete coverage was not obtained and only qualitative comparisons should be attempted between sites and between different years. WHO one-hour guideline is 150-200 $\mu g\ m^{-3}$ (76-100 ppb).

After Ball, 1987

15.4 Conclusions

Air Pollution Situation London is a good example of how legislation combined with economic alternatives to coal have reduced "traditional" pollutant emissions from stationary sources. The main source of SO_2 pollution in London is now power stations outside the city on the Thames estuary. These SO_2 emissions often descend on London when light prevailing winds are from the east. Such conditions are usually responsible for the worst SO_2 episodes in the metropolis.

Despite London's relatively stable population, the number of vehicle registrations is increasing. Road traffic increased by 7 per cent in 1989 alone. Such growth in traffic can only result in increased emissions of CO, NO and secondary pollutants such as NO_2 and O_3. Automotive pollution levels in central London will probably remain relatively constant and possibly even decrease owing to the limited available road area. However, concentrations in the suburbs are likely to increase through the increasing volume of traffic and the subsequent congestion. Potentially this could lead to an increase in emissions in suburban residential areas and thus increase the exposure of the population in these areas. In 1993 EC legislation will instigate stringent emissions standards requiring all new cars to have catalytic converters. In theory this should curb NO_x emissions from petrol-engined cars significantly. However, modelling undertaken by WSL projects only

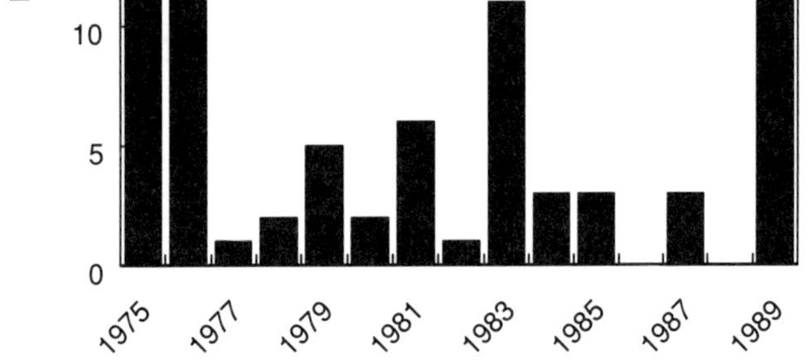

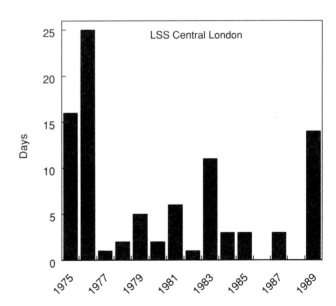

Figure 15.8 Number of days when hourly ozone concentrations exceeding 80 ppb (171 $\mu g\ m^{-3}$) at LSS Central London site

After LSS, 1990c

a 5 per cent decline in NO_2 levels in central London by the year 2000 because of the expected increase in traffic (Munday et al., 1989).

Main Problems There is a definite need for strategic co-ordinated monitoring in Greater London. At present local authorities lack localized and London-wide information which would be of use in terms of activating emergency control procedures, evaluating risks to human health and providing a useful data base for land use planning purposes. These should be the primary objectives of air quality monitoring. At present monitoring simply aims to judge compliance with air quality standards.

Control Measures The UK, through the UN Economic Commission for Europe (ECE) Convention on Long-range Transboundary Air Pollution - NO_x Protocol, is committed to stabilizing national NO_x emissions at 1987 levels by 1994. It is hoped to achieve this mainly through implementation of the EC Large Combustion Plants Directive, which commits the UK to reducing NO_x emissions from existing plant by 15 per cent by 1993 and 30 per cent by 1998. Oxides of nitrogen emissions from motor vehicles should be reduced through the introduction of new EC standards for passenger cars which come into force at the end of 1992. This requires that NO_x emissions from all new cars will be reduced by 75 per cent, achieved by fitting all new cars with catalysts. However, it is estimated that approximately half UK road transport emissions are attributable to diesel vehicles and at present there is no imminent legislation planned to curb these emissions. In addition, the UN ECE Convention on Long-range Transboundary Air Pollution is currently preparing a Protocol on VOCs aimed at controlling O_3 precursors. The UN ECE Protocol is likely to be backed up with an EC Directive which will dictate the control measures to be adopted in Britain.

15.5 References

Ball, D. J. 1987 A History of Secondary Pollutant Measurements in the London Region 1971–1986, *Science of the Total Environment* **59**, 181–206.

Ball, D. J. and Armorgie, C. J. 1983 Implications for London of European Air Quality Standards for Sulphur Dioxide and Suspended Particulates, *Environmental Pollution (Series B)* **5**, 207–232.

Bower, J. S., Broughton, G. F. J., Dando, M. T. 1988 *Ozone monitoring in the UK: A review of 1987/88 data from monitoring sites operated by Warren Spring Laboratory*, Report No. LR 678 (AP)M, Warren Spring Laboratory, Stevenage.

Bower, J. S., Broughton, G. F. J., Dando, M. T. and Lees, A. J., 1989 *Monitoring for the European Community Nitrogen Dioxide Directive in the United Kingdom: A Review of Data for 1987*, Warren Spring Laboratory, Stevenage.

Broughton, G. F. J., Bower, J. J., Laight, S. E. and Driver, G. S. 1990a *Air quality in the UK: A summary of results from Instrumental Air Monitoring Networks in 1987/88*, Warren Spring Laboratory, Stevenage.

Broughton, G. F. J., Bower, J. J., Laight, S. E. and Driver, G. S. 1990b *Air quality in the UK: A summary of results from Instrumental Air Monitoring Networks in 1988/89*, Warren Spring Laboratory, Stevenage.

Broughton, G. F. J., Willis, P. G. and Bower, J. J. 1991 *Air quality in the UK: A summary of results from Instrumental Air Monitoring Networks in 1989/90*, Warren Spring Laboratory, Stevenage.

Jensen, R. A. and Laxen, D. P. H. 1987 The Effect of Phase-Down of Lead in Petrol on Levels of Lead in Air, *Science of the Total Environment* **59**, 1–8.

Laxen, D. P. H. and Thompson, M. A. 1987 Sulphur Dioxide in Greater London, 1931-1985, *Environmental Pollution* **43**, 103–114.

Laxen, D. P. H. 1989 Winter smogs return to London, *London Environmental Bulletin*, London Scientific Services, Autumn, London.

LRC 1990 *Population and Statistics*, London Research Centre, London.

LSS 1990a *London Air Pollution Monitoring Network – Fourth Report 1989*, LSS/LWMP/120, London Scientific Services, London.

LSS 1990b *Nitrogen Dioxide Survey 1989: Results and Report*, LSS/LWMP/121, London Scientific Services, London.

LSS 1990c *London-Wide Ozone Monitoring Programme: Summary of Results for 1989*, LSS/LWMP/109, London Scientific Services, London.

McInnes, G. 1988 *Airborne Lead Concentrations in the United Kingdom 1984–1987,* Warren Spring Laboratory, Stevenage.

Munday, P. K., Timmis, R. J., Walker, C. A. 1989 *A dispersion modelling study of present air quality and future oxides of nitrogen concentrations in Greater London,* Warren Spring Laboratory, Stevenage.

OECD 1988 *Cities and Transport,* Organisation for Economic Co-operation and Development, Paris.

Parker, V. J., Sweeney, B. P., Bower, J. S., Broughton, G. F. J., Warwick, A. J., Lilley, K., Powell, K. and Stevenson, K. J. 1990 *Survey of Gaseous Air Pollutants at Selected UK Sites: XXII Data Digest for Cromwell Road, Central London, Stevenage and Sibton, April 1985 to March 1987,* Warren Spring Laboratory, Stevenage.

Schwar, M. J. E., Moorcroft, J. S., Laxen, D. P. H., Thompson, M. and Armorgie, C. 1988 Baseline Metal-in-Dust Concentrations in Greater London, *Science of the Total Environment* **68**, 25–43.

United Kingdom Photochemical Oxidants Review Group 1990 *Oxides of nitrogen in the United Kingdom,* Department of the Environment, London.

UN 1989 *Prospects of World Urbanization 1988,* Population Studies No. 112, United Nations, New York.

Varey, R. H., Ball, D. J., Crane, A. J., Laxen, D. P. H. and Sandalls, F. J. 1988 Ozone Formation in the London Plume, *Atmospheric Environment* **22,** 1335–1346.

Williams, M. L., Broughton, G. F. J., Bower, J. S., Drury, V. J., Lilley, K., Powell, K., Rogers, F. S. M. and Stevenson, K. J. 1988 Ambient NO_x Concentrations in the UK 1976–1984: A Summary, *Atmospheric Environment* **22**, 2819–2840.

WMO 1971 *Climatological Normals (CLINO) for Climate and Climate Ship Stations for the Period 1931–1960,* No. 117, World Meteorological Organization, Geneva.

Woodbridge, R. H. 1989 *A Graphical Presentation of Roadside Air Pollution Data,* Brunel University, London.

WSL 1992 *Initial analysis of NO_2 pollution episode, December 1991,* Warren Spring Laboratory, Stevenage.

16

Los Angeles

16.1 General Information

Geography The Los Angeles metropolitan area is situated on the west coast of the United States of America (Latitude 34°00′N, Longitude 118°15′W) and covers an area of approximately 16,600 km^2. This area is bounded by the Pacific Ocean to the west and south, and the San Gabriel, San Bernardino and San Jacinto mountain ranges to the north and east. Mean altitude is 113m, ranging from sea level to 900m. The Los Angeles metropolitan area includes all of Orange County and the non-desert parts of Los Angeles, Riverside and San Bernardino counties.

Demography The Los Angeles metropolitan area (hereafter referred to as the South Coast Air Basin or "LA Basin") is the second largest population centre in the US. Since the early 1940s the LA Basin has been one of the most rapidly growing areas of the USA. The UN (1989) estimated the 1990 population to be 10.47 million and the projected population in 2000 to be 10.91 million (Table 2.1, Chapter 2). The population of the LA Basin could reach 15.7 million by 2010 (SCAG, 1990).

Climate The Los Angeles basin has a Mediterranean climate. The meteorology of the area is complex. Local topography and local weather patterns of light winds, sea breezes, subsidence inversions and high solar intensity produce ideal conditions for atmospheric stagnation conducive to pollutant reaction and buildup. The LA Basin is bordered to the east by mountains with a high desert area beyond and it has a very low annual rainfall. Mean annual precipitation is 367 mm with a monthly maximum of around 80 mm falling in February. There is sunshine all year round and the winters are very mild, with a monthly minimum of 12°C in January. In the summer, a warm air cap often forms over the moist cool marine air layer which inhibits vertical mixing. A monthly maximum of 21°C occurs in August. The dominant daily weather pattern is a sea breeze beginning in the morning after sunrise and a land breeze at night. During periods of stagnating high pressure this circulation pattern, which takes pollutants out to sea at night and returns them to land during the day, allows air pollutants to build up in the airshed until the passage of a new weather front.

Industry There is little primary heavy industry remaining in the LA Basin because steel, tyre and car assembly plants have left the area. A refinery area in Long Beach, an iron and steel plant and secondary lead smelters inland have been strictly controlled by the Los Angeles Air Pollution Control District (LAAPCD)/South Coast Air Quality Management District (SCAQMD) since the 1970s. Industry in the area is mainly service-oriented, with some aerospace and electronics manufacturing. Electric power is mainly provided by hydroelectric plants, but there are some power plants using natural gas.

Transport The LA Basin developed with almost no public transport network and consequently the residents must rely on the motor vehicle for almost all transportation. The LA Basin has the worst air quality in the US because of the presence of large numbers of motor vehicles and intensive consumer-oriented service activities. The eight million vehicles in an urban area of 12 million people represent possibly the greatest number of vehicles per person (0.67) in the world. It is estimated that the population will increase by five million residents by the year 2010, which will mean a 68 per cent increase in vehicle-miles travelled, and a 40 per cent increase in trips taken (SCAG, 1989).

16.2 Monitoring

Los Angeles air quality has been a major problem and concern since the early 1940s when the post-war boom brought rapid population growth and industrial expansion. Since 1947 the LAAPCD the precursor of the SCAQMD, began enforcing air pollutant emission controls. Hourly ozone (O_3) concentrations exceeding 0.6 ppm (1,200 µg m^{-3}) were reported and in the 1960s O_3 frequently exceeded

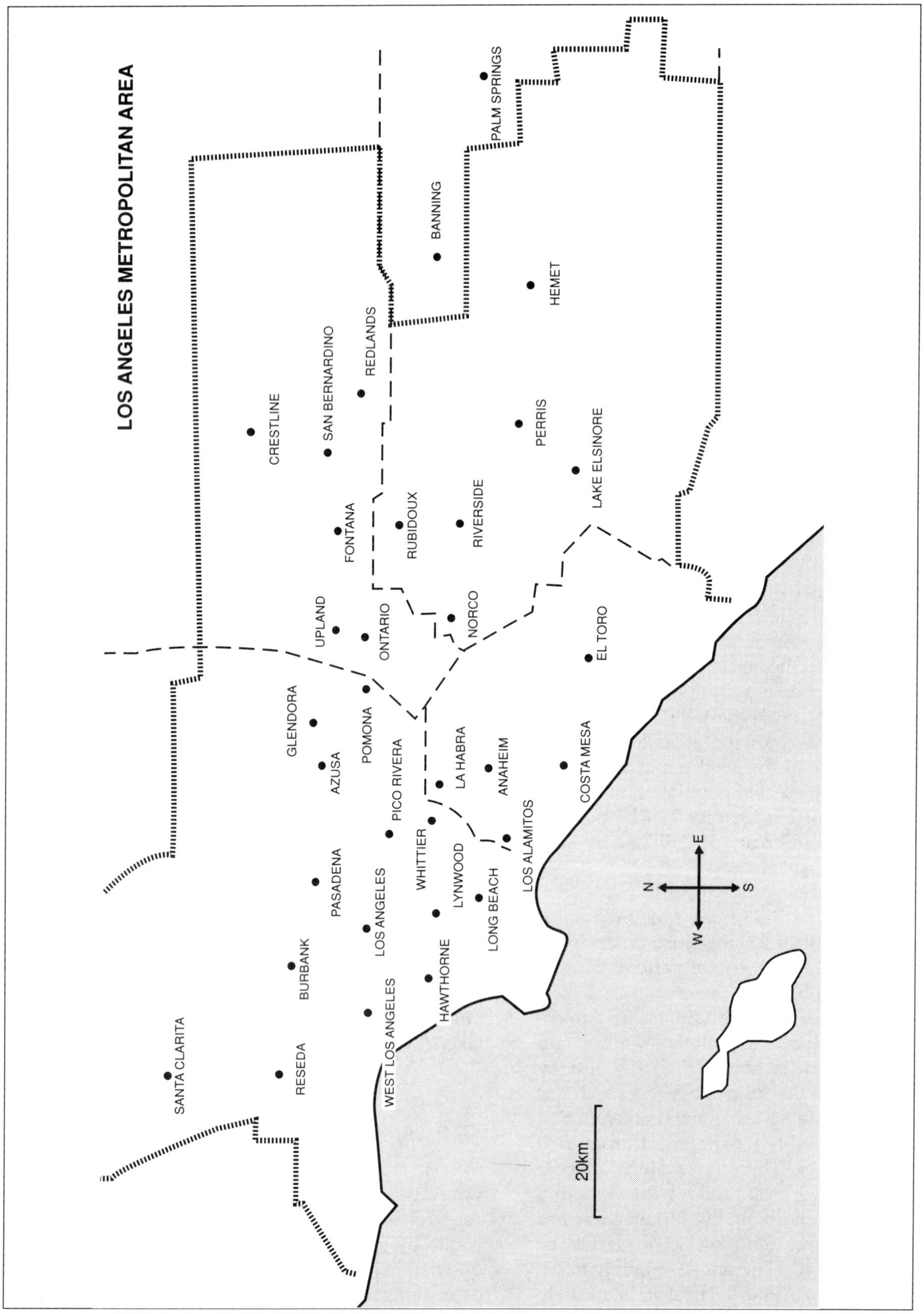

Figure 16.1 *Sketch map of Los Angeles Metropolitan Area*

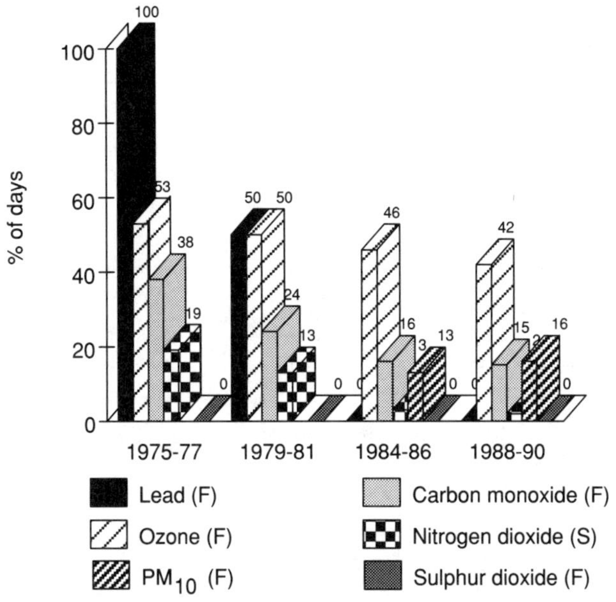

Figure 16.2 *Percentage of days exceeding levels of Federal (F) or State (S) standards ,1975–1990*

After SCAG, 1991

0.5 ppm (1,000 μg m^{-3}). In the 1970s the State of California began instituting vehicle emission standards which were stricter than those promulgated by the US Environmental Protection Agency (USEPA) for the nation at large. Despite their strict controls, the maximum O_3 during the period 1986-1991 only decreased to 0.35 ppm (700 μg m^{-3}) because of an 81 per cent population increase from 1960 to 1990 and allied increases in industrial activity and vehicular traffic. This enormous population increase resulted in many motorists commuting 95–130 km each way to secure affordable single-family housing outside the central area.

The SCAQMD was formed in 1976 when it was recognized that the problem extended well beyond the borders of Los Angeles into neighbouring counties. It now monitors air quality at 37 stations distributed widely over the LA Basin (Figure 16.1). The trends in percentage of days exceeding USEPA and California standards from 1975 to 1989 are shown in Figure 16.2 (SCAG, 1991). Although sulphur dioxide (SO_2), lead (Pb) and nitrogen dioxide (NO_2) problems have been significantly reduced, particulate matter <10 μm (PM_{10}), carbon monoxide (CO) and O_3 are still significant problems for the LA Basin (USEPA, 1991).

16.3 Air Quality Situation

Sulphur dioxide

Emissions Total anthropogenic oxides of sulphur (SO_x) emissions were estimated at approximately 50,000 tonnes per annum in 1987, with mobile sources accounting for 62 per cent (Figure 16.3). It is interesting to note that "off-road" mobile sources are responsible for 39 per cent of total SO_x emissions. Sulphur dioxide

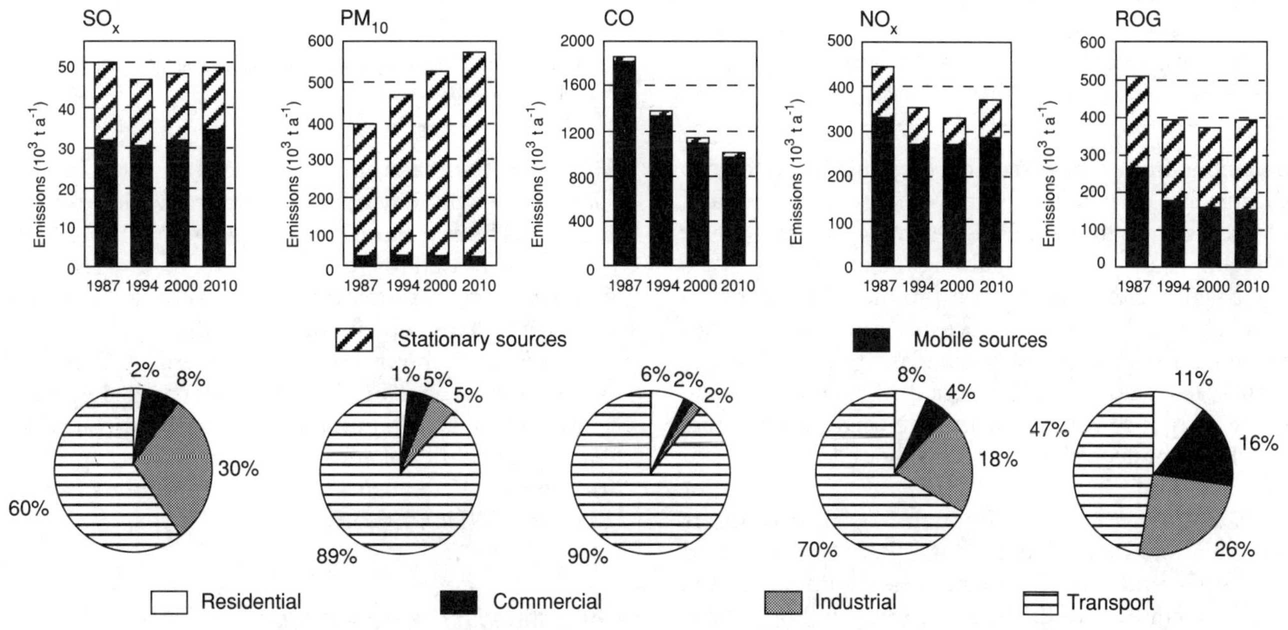

Figure 16.3 *Anthropogenic emissions estimates and projections 1987–2010 and percentage contribution of emissions sources in 1987*

Source: SCAG, 1991

Table.16.1 *Summary of annual mean emission reductions*

Sources		Pollutant (10^3 t a^{-1})				
		SO_x	PM_{10}	CO	NO_x	ROG
Year 2010 Baseline		43	583	1,001	323	418
Tier I Emission Reductions[1]						
Stationary	Area	0	193	4	10	88
	Point	8	7	0	25	17
Mobile	On-road	4	7	402	106	81
	Off-road	15	3	75	40	17
Tier I Total		26	209	481	181	203
Tier II Emission Reductions						
Stationary	Area	1	46	6	7	56
	Point	1	0	0	4	23
Mobile	On-road	0	0	0	0	0
	Off-road	0	0	9	0	2
Tier II Total		2	47	15	10	81
Tier III Emission Reductions						
Stationary	Area	0	10	3	1	44
	Point	1	0	0	0	19
Mobile	On-road	0	0	0	0	0
	Off-road	0	0	19	7	5
Tier III Total		1	10	22	8	68
Year 2010 Remaining Emissions		13	317	482	124	66

[1] Includes emission reductions from regulations adopted between 1 July 1990 and 30 November 1990.

Note: Owing to rounding, totals may not equal sums of columns.
After SCAG, 1990

emissions from refineries and power plants have been drastically curtailed by emission regulations and switching fuels to natural gas. Along with diesel motor vehicles, these are the remaining major sources in the LA Basin. Acid sulphates are produced mainly from photochemical oxidation processes and California has promulgated an air quality standard to control them. Figure 16.3 shows that projected SO_x emissions are likely to reduce further during the early 1990s before stabilizing or even slightly increasing up to 2010.

Ambient Concentrations Although SO_2 already meets the National Ambient Air Quality Standard (NAAQS) and WHO annual mean guidelines, the emission controls of the three-tier plan will bring SO_2 concentrations down even lower (Table 16.1). In 1990 annual arithmetic mean concentrations ranged from 0.2–10 $\mu g\ m^{-3}$, well below the WHO annual guideline. However, the maximum hourly SO_2 as measured at Hawthorne was 886 $\mu g\ m^{-3}$ (0.31 ppm) which exceeds the WHO hourly guideline of 350 $\mu g\ m^{-3}$ (Table 16.2). This was caused by a breakdown at a nearby oil refinery and was not treated as a value requiring changes in state implementation plans.

The State 24-hour sulphate standard of 25 $\mu g\ m^{-3}$ was exceeded once in 1990 at Pasadena (28.4 $\mu g\ m^{-3}$). There is no WHO guideline for sulphuric acid or equivalent acidity of aerosol, but concern is expressed for repeated exposures to concentrations above 10 $\mu g\ m^{-3}$ (WHO, 1987).

Suspended particulate matter

Emissions Total PM_{10} emissions in 1987 were estimated to be around 400,000 tonnes. Emissions are

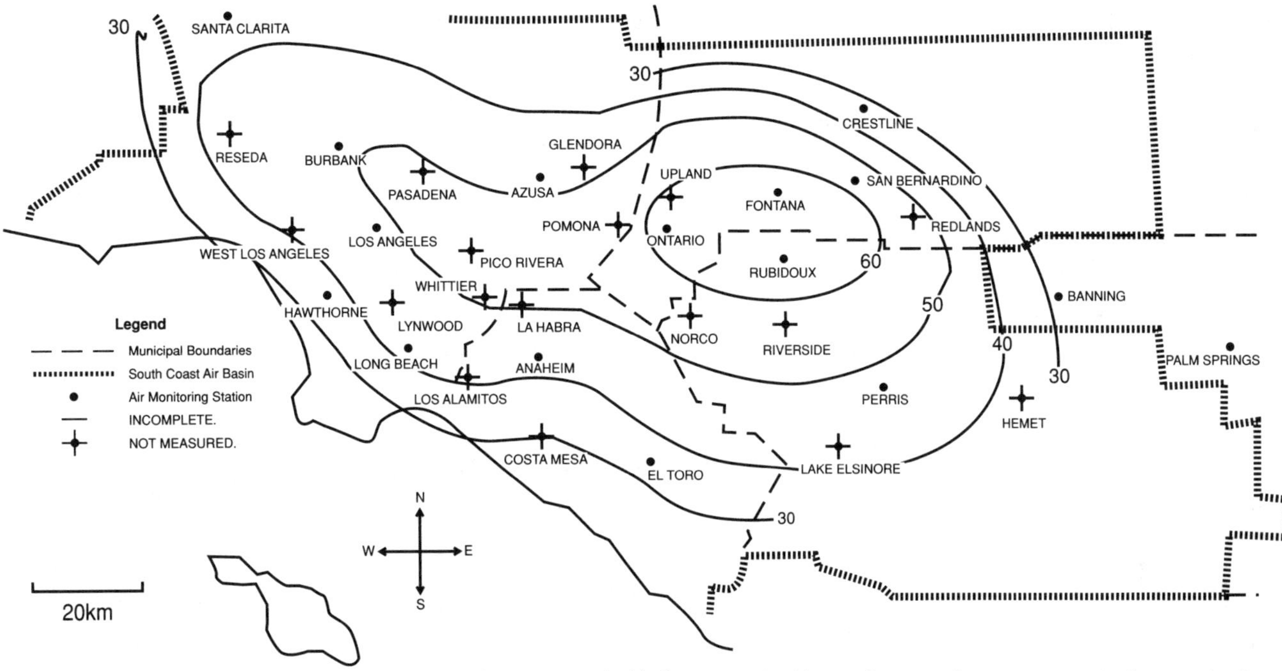

Figure 16.4 *1990 annual mean PM_{10} concentration compared with State standard (annual geometric mean concentration greater than 30 μg m^{-3})*

After SCAG, 1991

primarily from motor vehicle traffic (5 per cent), industry (3 per cent) and other miscellaneous sources (91 per cent) which are mostly related to commercial/domestic activities and, in particular, road dust. Paved road dust stirred by vehicle movement is included in miscellaneous sources (SCAG, 1991) and is much greater in mass than the particles emitted from the motor vehicle exhaust pipe. Domestic burning of household waste has been banned for many years. The three-tier emission reduction programme (Table 16.1) is designed to meet both State and Federal Air Quality Standards by 2010.

Ambient Concentrations Total suspended particulate matter (TSP) and PM_{10} in the LA Basin exceed the State/Federal air quality standards and the WHO guidelines respectively. The highest PM_{10} annual mean in the US (94 μg m^{-3}) has been measured in LA and although the Phoenix, Arizona, area may have more violations of the daily NAAQS because of wind-blown dust, the higher proportion of particulates from anthropogenic sources in the LA basin results in more severe human health impacts. Annual arithmetic mean TSP concentrations in 1990 ranged from 46 μg m^{-3} at Crestline to 115 μg m^{-3} at Fontana; the daily maximum (1,770 μg m^{-3}) was also observed at Fontana.

Daily maxima PM_{10} concentrations range from 88 μg m^{-3} at Crestline to 520 μg m^{-3} at Indio (Table 16.2). Figure 16.4 shows the PM_{10} air quality standards violation distribution in the LA Basin. The westerly winds lead to PM_{10} maxima of photochemical aerosols in the Rubidoux and Indio areas.

Lead

Emissions Petrol sold in the LA Basin is primarily Pb-free, and the little leaded petrol sold only contains 0.026 g l^{-1}. Secondary Pb smelters, although controlled, are now the major sources in the area.

Ambient Concentrations Until 1991, Pb was apparently no longer a problem in ambient air owing to the mandated introduction and use of lead-free petrol (Figure 16.2), although automotive Pb from previously contaminated soil near major roads may be a problem for children who ingest significant quantities. In 1991, several violations of the Pb NAAQS occurred in the vicinity of secondary Pb smelters, leading to a call for a new Pb control plan.

Carbon monoxide

Emissions Man-made CO emissions were estimated to total approximately 1.8 million tonnes per annum in 1987 (Figure 16.3). "On-road" mobile sources

Table 16.2 *Air Quality in the LA Basin in 1990*

	Sulphur dioxide ($\mu g\ m^{-3}$)			Suspended particulate matter: TSP ($\mu g\ m^{-3}$)		Suspended particulate matter: PM_{10} ($\mu g\ m^{-3}$)			Lead ($\mu g\ m^{-3}$)	Carbon monoxide ($mg\ m^{-3}$)		Nitrogen dioxide ($\mu g\ m^{-3}$)		Ozone ($\mu g\ m^{-3}$)
Location	AAM	1-hr Max.	24-hr Max.	AGM	24-hr Max.	AAM	AGM	24-hr Max.	Qtr. mean	1-hr Max.	8-hr Max.	AAM	1-hr Max.	1-hr Max.
Los Angeles	5	57	37	99	211	53	48	152	0.09	14.9	11.3	88	**526**	400
West Los Angeles	6	57	26	62	163	nm	nm	nm	nm	17.2	9.2	61	376	320
Hawthorne	10	**887**	**100**	74	186	41	38	127	0.08	21.8	14.5	64	432	200
Long Beach	9	143	37	82	188	44	41	119	0.09	12.6	10.4	74	508	240
Whittier	5	114	26	nm	nm	nm	nm	nm	nm	13.7	10.3	80	432	380
Resedea	4	57	29	nm	nm	nm	nm	nm	nm	21.8	17.1	64	357	380
Burbank	5	57	31	89	191	52	48	161	0.08	18.3	14.9	90	432	400
Pasadena	4	57	23	70	142	nm	nm	nm	nm	18.3	11.5	89	432	520
Azusa	3	86	23	104	228	55	48	127	nm	8.0	5.8	77	395	460
Glendora	nm	nm	nm	nm	nm	nm	nm	nm	nm	nm	nm	71	357	580
Pomona	nm	nm	nm	nm	nm	nm	nm	nm	nm	14.9	8.6	**104**	395	480
Pico Rivera	**12**	114	40	93	195	nm	nm	nm	0.13	14.9	10.8	94	508	380
Lynwood	9	114	34	102	233	nm	nm	nm	**0.14**	**27.5**	**19.2**	77	489	300
San Clarita	3	29	11	nm	nm	43	39	93	nm	12.6	5.3	59	282	460
Lancaster	nm	nm	nm	79	217	53	44	342	nm	12.6	9.5	38	169	300
La Habra	3	86	20	nm	nm	nm	nm	nm	nm	21.8	11.0	84	414	420
Anaheim	5	57	26	91	422	49	43	158	0.10	19.5	13.4	88	395	360
Los Alamitos	5	86	26	103	834	nm	nm	nm	nm	nm	nm	nm	nm	340
Costa Mesa	2	57	23	nm	nm	nm	nm	nm	nm	14.9	12.3	51	414	300
El Toro	nm	nm	nm	78	132	43	40	88	nm	10.3	6.4	nm	nm	380
Norco	nm	nm	nm	nm	nm	nm	nm	nm	nm	nm	nm	nm	nm	340
Rubidoux	1	86	14	110	274	78	**67**	207	0.08	11.5	7.2	63	301	580
Riverside	nm	nm	nm	96	223	nm	nm	nm	0.08	17.2	8.4	nm	nm	nm
Perris	nm	nm	nm	72	232	59	50	250	nm	nm	nm	53	207	380
Lake Elsinore	nm	nm	nm	nm	nm	nm	nm	nm	nm	nm	nm	nm	nm	380

Table continued

Table 16.2 *Air Quality in the LA Basin in 1990*

	Sulphur dioxide			Suspended particulate matter					Lead	Carbon monoxide		Nitrogen dioxide		Ozone
	(μg m^{-3})			TSP ((μg m^{-3})		PM_{10} (μg m^{-3})			(μg m^{-3})	(mg m^{-3})		(μg m^{-3})		(μg m^{-3})
Location	AAM	1-hr Max.	24-hr Max.	AGM	24-hr Max.	AAM	AGM	24-hr Max.	Qtr. mean	1-hr Max.	8-hr Max.	AAM	1-hr Max.	1-hr Max.
Hemet	nm	nm	nm	nm	nm	nm	nm	nm	nm	nm	nm	nm	nm	440
Banning	nm	nm	nm	60	167	35	29	89	nm	nm	nm	nm	nm	440
Palm Springs	nm	nm	nm	57	170	35	31	83	nm	5.7	2.6	39	169	340
Indio	nm	nm	nm	**131**	1,485	**79**	65	520	nm	nm	nm	nm	nm	320
Upland	3	29	17	93	289	nm	nm	nm	0.07	10.3	7.6	77	357	580
Ontario	nm	nm	nm	91	243	72	61	185	nm	nm	nm	nm	nm	nm
Fontana	0	29	9	116	**1,770**	78	63	**475**	nm	6.9	5.6	74	376	540
San Bernardino	0	29	3	101	289	65	55	235	0.07	10.3	6.9	64	376	580
Redlands	nm	nm	nm	nm	nm	nm	nm	nm	nm	nm	nm	nm	nm	600
Crestine	nm	nm	nm	47	124	37	31	88	nm	nm	nm	nm	nm	**660**

Notes: AAM = Annual Arithmetic Mean AGM = Annual Geometric Mean nm = Pollutant not monitored Bold figures denote highest values at any site
After SCAG, 1991

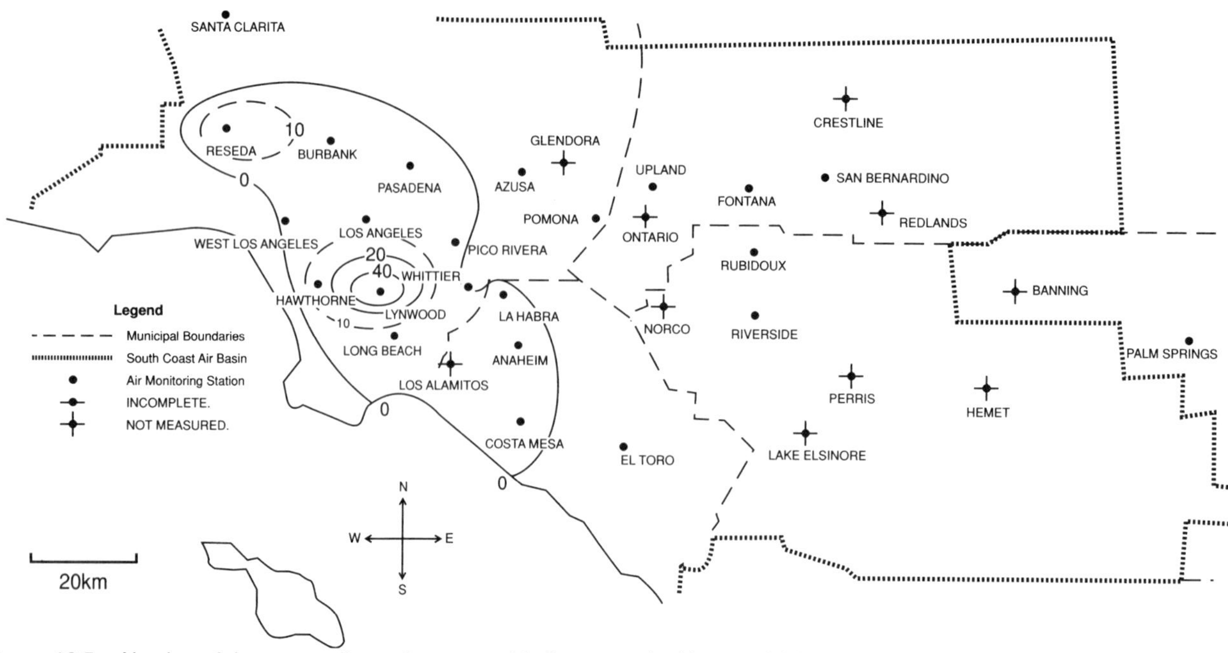

Figure 16.5 *Number of days exceeding carbon monoxide State standard in 1990 (eight-hour mean greater than 9.0 ppm)*

After SCAG, 1991

account for 87 per cent of this total, "off-road" mobile sources a further 11 per cent. Area sources (domestic) and point sources (industry) were estimated to contribute 1 per cent each. The SCAQMD has placed strict controls on industrial emissions of CO and the State of California has instituted strict controls on CO emissions from motor vehicles. These emission reductions through California controls in vehicles have produced a striking reduction of in-vehicle exposures to CO as measured in San Francisco (Ott et al., 1991). As San Francisco and LA have identical motor vehicle emission standards, a similar decrease in Los Angeles, although unmeasured, probably occurred. Emissions of CO are forecast to decrease further to below 1 million tonnes per annum by 2010 (Figure 16.3).

Ambient Concentrations The distribution of CO pollution in the LA basin is shown in Figure 16.5. The extent of CO pollution decreased significantly from the early 1960s to the mid-1980s when the pollution levels and number of exceedences of the NAAQS began to rise. In spite of the strictest CO emission controls in the nation, these increases are thought to arise because of increased traffic congestion in the LA Basin. In 1990 the maximum one-hour CO mean of 27.5 mg m^{-3} and eight-hour average of 19.2 mg m^{-3} measured at Lynwood (Table 16.2) indicate the continuing potential for significant exposures that can bring carboxyhaemoglobin (COHb) in non-smokers up to the 3–4 per cent range.

Oxides of nitrogen

Emissions Estimated total oxides of nitrogen (NO_x) emissions in 1987 were over 440,000 tonnes per annum. Seventy-six per cent of this total are attributable to mobile sources (55 per cent to "on-road" mobile sources). The NO_x emission reductions proposed for O_3 control shown in Table 16.1 were designed to reduce NO_2 concentrations and it is estimated that the air quality standards will be met by the year 2000. However, a National Research Council report (National Research Council, 1991) has criticized the reliance on hydrocarbon controls for reducing O_3; it has concluded that it may be more critical to reduce NO_x emissions than hydrocarbons, as NO_x appears to be the limiting reactant in the system when hydrocarbon emissions have been significantly underestimated. It is thus anticipated that the proposed emission controls in Table 16.1 will have to be tightened in the future. Projected estimates for the LA Basin show that emissions should decrease until the year 2000 but, after that, they will stabilize or even increase depending on the level of population growth.

Ambient Concentrations Nitrogen dioxide is the reddish-brown gas that produces the distinctive colour of the "smog" haze over the LA Basin. The reduced visibility often prevents the viewing of the surrounding mountains from the coastal area which is a major complaint of residents of the area. According to the USEPA (1991), Los Angeles is the only city

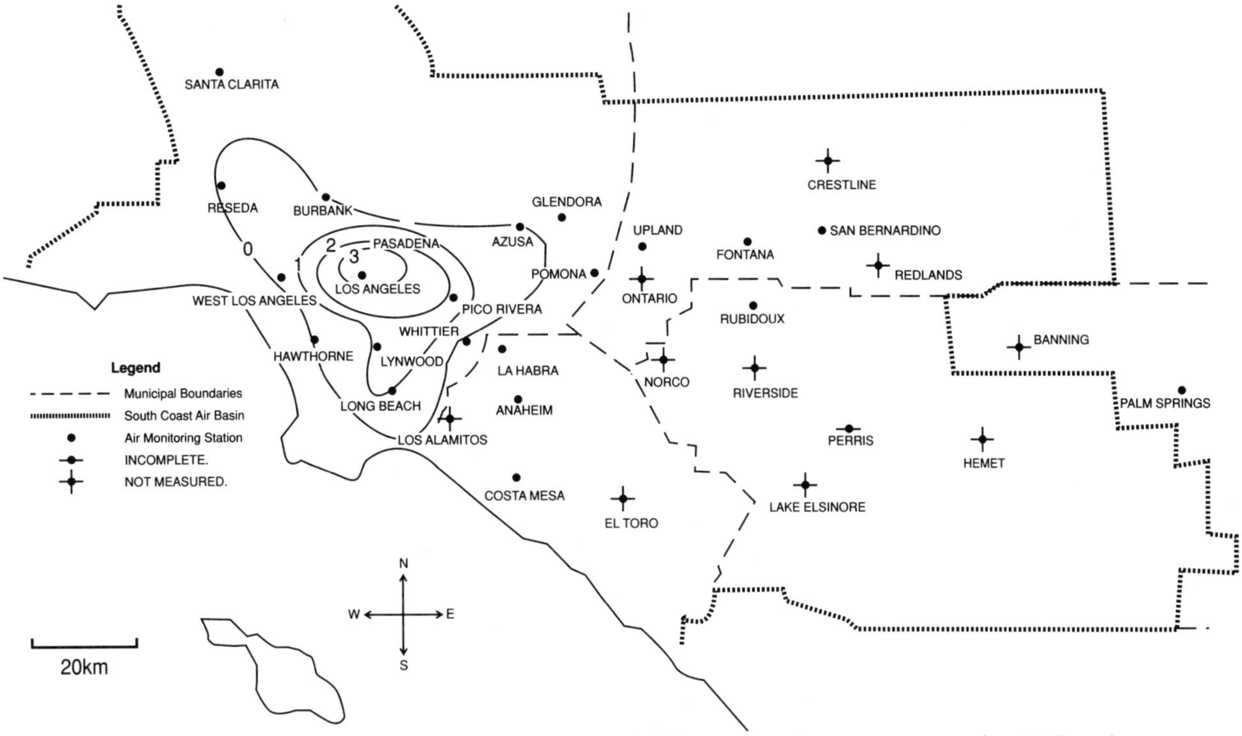

Figure 16.6 *Number of days exceeding State nitrogen dioxide standard in 1990 (one-hour mean greater than 0.25 ppm)*

After: SCAG, 1991

in the United States that has failed to attain the Federal NO_2 NAAQS. As shown in Figures 16.2 and 16.6, NO_2 provides a serious potential for high exposures. The WHO one-hour guideline of 400 μg m^{-3}, corresponding to 0.22 ppm NO_2, was exceeded in 1990 at 8 of the 24 stations reporting NO_2 data. The highest reported hourly value was 526 μg m^{-3} (0.28ppm) which occurred at the Los Angeles station. Annual mean concentrations in 1990 ranged from 39 μg m^{-3} at the Palm Springs station to 104 μg m^{-3} at the Pomona station.

Since NO_2 is so highly reactive the indoor environment provides some protection, but gas appliances used for cooking can over-compensate for this and increase personal exposures to higher levels than the ambient. A study of personal exposures to NO_2 in the LA Basin in a WHO/UNEP GEMS/HEAL study (Matsushita and Tanabe, 1991) showed that nine randomly chosen people tested for one year had weekly mean NO_2 exposures ranging from 15 ppb (28 μg m^{-3}) to 30 ppb (56 μg m^{-3}) NO_2 which fall below under the NAAQS.

Ozone

Emissions Ozone is a secondary pollutant that is formed in the lower atmosphere by complex photochemical reactions between NO_x and reactive organic compounds (ROG) in the presence of intense solar radiation. Owing to this complicated genesis, increasingly strict emission controls on NO_x and hydrocarbons have been enforced (SCAG, 1990 and Lloyd et al., 1989). In the past emissions of reactive hydrocarbons (over 500,000 tonnes per annum in 1987) and NO_x have been controlled in the LA Basin by strict industrial emissions standards and motor vehicle standards primarily for the purpose of controlling O_3. The SCAQMD has recently instituted a three-tier plan to achieve emission reductions for ROG and NO_x in the amounts shown in Table 16.1. They are designed to meet the California and Federal air quality standards by the year 2010. These emissions controls are to be applied to industry, motor vehicles, vehicle fuel and consumer product formulations, and consumer lifestyle activities. Some amendments may be necessary to comply with new specific Federal requirements of the Federal Clean Air Act of 1990. The Emission Controls of Tiers one, two, and three represent the most severe air quality management requirements ever proposed for any city. The LA Basin is also seriously pursuing an economic incentive programme, using marketable permits to supplement or substitute for some of the more stringent control requirements.

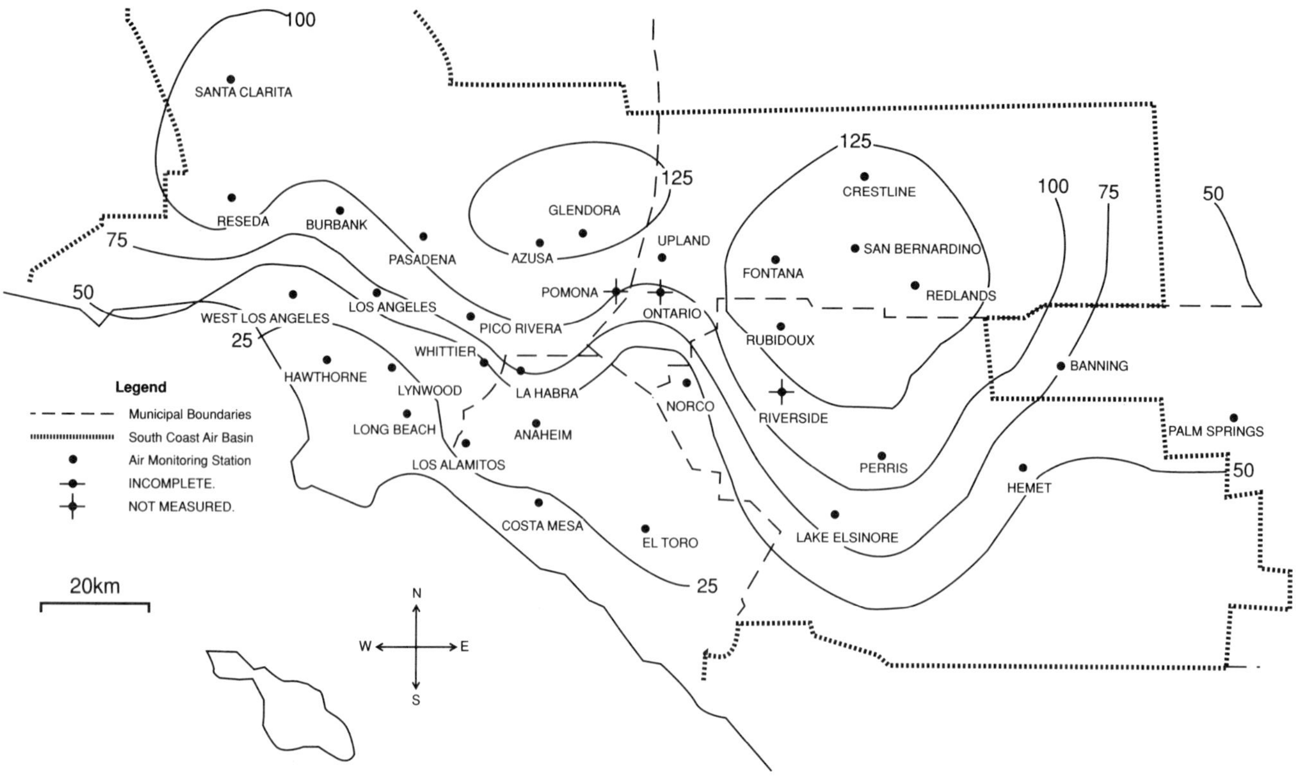

***Figure 16.7** Number of days exceeding State ozone standard in 1990 (one-hour mean greater than 0.09 ppm)*

After SCAG, 1991

Ambient Concentrations Los Angeles has the most serious O_3 problem in the USA (USEPA, 1991). In 1990 the maximum hourly O_3 concentration was 660 μg m^{-3} (0.33 ppm) at the Crestline site; here the US NAAQS was exceeded on 103 days (and at Glendora) and the State standard on 144 days. Figure 16.7 shows the spatial pattern produced by the complicated photochemical reactions that occur as the primary emissions of NO_x and ROG emitted in the downtown area react as the air mass drifts westward. The State standard was exceeded at all stations in 1990.

The indoor environment is protective as O_3 reacts rapidly. Consequently, outdoor exposures during peak hours are to be avoided by sensitive populations such as asthmatics; young school children are kept indoors at school on such days.

16.4 Conclusions

Air Pollution Situation The overall air pollution situation in the LA Basin is the worst in the USA (USEPA, 1991). Of the megacities in this book, only Mexico City has higher O_3 levels. As discussed below, the stringent emission control actions taken by the SCAQMD are necessary to prevent the continuation of the deleterious health effects of high air pollution concentrations that have been documented in several studies. These control activities will continue to place great constraints on the living patterns of LA Basin residents, and the even stricter controls planned for the coming decade will affect everyone living there.

The predominant temporal pattern is for primary emissions to be generated in the downtown LA area in the early morning, followed by an onshore sea breeze that leads to maxima of secondary pollutants such as O_3, PM_{10} (including photochemical aerosols) and visibility-reducing particles further inland.

Control Measures California has taken advantage of the provisions of the Clean Air Act of 1970 which allows a State to promulgate air quality standards more stringently than the Federal EPA standards. In addition, the State of California has its own Clean Air Act that places additional incentives for strict regulations to attain the standards.

The three-tier plan to reduce emissions and meet air quality standards is highly commendable and will take a great deal of effort. The first tier will apply currently available technologies through a combined effort of Federal, State, Regional and Local authorites. The second tier will require research and development of existing technologies to make improvements, and the third tier will require technological breakthroughs in the next decades to achieve the goals. These stringent control actions taken so far and planned for the future are very costly, but are highly justified because they have undoubtedly prevented very serious air pollution episodes from occurring.

Health Effects Owing to the high air pollution in the LA Basin many epidemiological studies have been conducted there. A study of Chronic Obstructive Respiratory Disease (CORD) (Detels et al., 1981) was summarized as follows:

"Questionnaires, spirometry, and the single-breath nitrogen test were administered to 3,192 participants (25–39 years of age) in an area exposed to low concentrations of all pollutants (Lancaster, California) and to 2,369 similar participants living in an area exposed to high concentrations of photochemical oxidants, NO_2, and sulphates (Glendora, California)... ". These results suggest that long-term exposure to high concentrations of photochemical oxidants, NO_2, and sulphates at place of residence may result in measurable impairment in both current smokers and never smokers.

A study of Chronic Obstructive Pulmonary Disease (COPD) in confirmed non-smokers (Seventh-day Adventists) also found a significant effect of photochemical smog (Hodgkin et al., 1989):

"The prevalence of respiratory symptoms, as ascertained by questionnaire, was evaluated in non-smokers who had lived for at least 11 years in either a high photochemical pollution area or a low photochemical pollution area. The risk estimate for "definite" COPD, as defined in this study, was 15 per cent higher in the high pollution area ($p = 0.03$), after adjusting for sex, age, race, education, occupation exposures, and past smoking history. Past smokers had a risk estimate 22 per cent higher than never smokers ($p = 0.01$)."

Another Seventh-day Adventist population group was studied (Euler et al., 1988) and the results were summarized as follows:

"A total of 7,445 Seventh-day Adventist (SDA) non-smokers who were 25 years of age or older and had resided at least 11 years in areas of California with high to low photochemical air pollution were included in this study. A significant association with chronic symptoms was seen for total oxidants above 10 pphm (196 $\mu g\ m^{-3}$). Chronic respiratory disease symptoms were not associated with relatively low NO_2 exposure levels in this population. When these pollutant exposures were studied in exposures to total suspended particulates (TSP) and SO_2, only TSP exposure above 200 mg m^{-3} showed statistical significance ($p < 0.01$). Exposure to TSP is either more strongly associated with symptoms of chronic obstructive pulmonary disease than the other measured exposures or is the best single surrogate representing the mix of pollutants present."

A third study of Seventh-day Adventists (Abbey et al., 1990) found elevated risks of chronic disease with exposure to TSP and O_3:

"Cancer incidence and mortality in a cohort of 6,000 non-smoking California Seventh-day Adventists was monitored for a six-year period and relationships with long-term cumulative ambient air pollution were observed. For all malignant neoplasms among females, risk increased with increasing exceedence frequencies of all thresholds of TSP except the lowest one, and those increased risks were highly statistically significant. Multivariate analyses which adjusted for past and passive smoking, and occupational exposures, indicated statistically significant ($p < 0.05$) elevated relative risks ranging up to 1.5 for incidence of asthma, definite symptoms of airway obstructive disease (AOD), and chronic bronchitis with TSP in excess of all thresholds except the lowest one but not for any thresholds of O_3. Mean concentration of O_3, however, was significantly associated with asthma incidence."

A recent assessment (Hall et al., 1992) suggests that all the residents of the LA Basin experience ozone-related symptoms on an average of up to 17 days each year and face an increased risk of death in any year of 1/10,000 as a result of elevated PM_{10} exposure. It was also estimated that attaining the state air pollution standards could save 1,600 lives a year in the region (at a cost of US$10 billion).

In summary, epidemiology studies consistently show that there has been a significant impact upon the health of the population of Los Angeles, resulting in increased health costs for the population and possible reduced life expectancy for the most seriously affected individuals.

16.5 References

Abbey, D. E., Mills, P. K., Petersen, F. and Beeson, W. L. 1990 *Long Term Ambient Concentrations of Total Suspended Particulates and Oxidants as Related to Incidence of Chronic Disease in California Seventh-day Adventists*, Abstract, Second Annual Meeting of the International Society for Environmental Epidemiology (ISEE), Berkeley, CA, 13–15 August 1990.

Detels, R., Sayre, J. W., Coulson, A. H., Rokaw, S. N., Massey, F. J. Jr, Tashkin, D. P. and Wu, K-W. 1981 The UCLA Population Studies of Chronic Obstructive Respiratory Disease, *American Review of Respiratory Disease* **124**, 673–680.

Euler, G. L., Abbey, D. E., Hodgkin, J. E. and Magie, A. R. 1988 Chronic obstructive pulmonary disease symptom effects of long-term cumulative exposure to ambient levels of total oxidants and nitrogen dioxide in California Seventh-day Adventist residents, *Archives of Environmental Health* **43**, 279–285.

Hall, J. V., Winer, A. M., Kleinman, M. T., Lurman, F. W., Brajer, V. M. and Colome, S. D. 1992 Valuing the health benefits of clean air, *Science* **255**, 812–816.

Hodgkin, J. E., Abbey, D. E., Euler, G. E. and Magie, A. R. 1989 COPD prevalence. In: *Regional Mobility Plan,* Southern California Association of Governments, Los Angeles.

Lloyd, A. C., Lents, J. M., Green, C. and Nemeth, P. 1989 Air quality management in Los Angeles: perspectives on past and future emission control, *Journal of Air Pollution Control Association* **39**, 696–703.

Matsushita, H. and Tanabe, K. 1991 *Exposure Monitoring of Nitrogen Dioxide: an international pilot study within the WHO/UNEP Human Exposure Assessment Location (HEAL) Programme,* WHO/UNEP, Nairobi.

National Research Council 1991 *Rethinking the Ozone Problem,* National Academy of Sciences, Washington D.C.

Ott, W. R., Switzer, P. and Willits, N. 1991 *Trends of in-vehicle CO exposures on a California Arterial Highway over one decade,* Paper presented at "Measuring, Understanding and Predicting Exposures in the 21st Century", 18–21 November 1991, Atlanta.

SCAG 1989 *Regional Mobility Plan,* Southern California Association of Governments, Los Angeles.

SCAG 1990 *Air Quality Management Plan - South Coast Air Basin,* South Coast Air Quality Management District, Southern California Association of Governments, Los Angeles. (Unpublished paper.)

SCAG 1991 *Air Quality Management Plan - South Coast Air Basin,* South Coast Air Quality Management District, Southern California Association of Governments, Los Angeles.

UN 1989 *Prospects of World Urbanization 1988,* Population Studies No. 112, United Nations, New York.

USEPA 1991 *National Air Quality and Emissions Trends Report,* Office of Air Quality Planning and Standards, US Environmental Protection Agency Research Triangle Park.

WHO 1987 *Air Quality Guidelines for Europe,* WHO Regional Publications, European Series No. 23, World Health Organization, Regional Office for Europe, Copenhagen.

17

Manila

17.1 General Information

Geography Metropolitan Manila, the capital of the Philippines, is situated at Latitude 14°36′N and Longitude 120°59′E along the shore of Manila Bay on the west side of Luzon island. The Metropolitan Manila region (or the National Capital Region, NCR) consists of four cities and 13 municipalities (Figure 17.1). Industrial development in the Philippines is focused on Metropolitan Manila which is the country's centre of commerce. Manila city is essentially at sea level (5 m) and occupies approximately 636 km^2 on the deltaic plain of the Pasig River which flows from Laguna de Bay, a lake to the south-east of the city, into Manila Bay (Figure 17.1).

Demography The rapidly expanding population of Metropolitan Manila was 5.9 million in 1980 and 7.0 million in 1985. 1990 estimates were 8.40 million, and the projected population by the year 2000 is 11.48 million. In 1980, approximately 64 per cent of the 48 million people lived in rural areas, while Metropolitan Manila was the country's chief urban population centre with 12 per cent of the country's population. The population density of Manila is approximately 12,600 persons per km^2.

Climate Manila has a tropical rain forest climate with a wet season (June–November) and a dry season (December–May). Mean temperatures are high and do not vary significantly. The mean temperature in January is 25°C and in July 27.8°C. The annual mean temperature is 27°C and the mean relative humidity is 80 per cent. Manila has one monsoon per annum and its annual mean rainfall is 2,159 mm. The mean windspeed at ground level in Manila is 3.0 m s^{-1}, and winds from east to south-east make up nearly half the total directional frequency distribution.

Industry Metropolitan Manila has grown rapidly since the 1950s. The per capita GNP of the Philippines was US$630 in 1988. Since this figure includes rural populations, the per capita GNP within Metropolitan Manila would be higher. In 1986 there were 2,861 industrial facilities operating in Metropolitan Manila, with chemical, food manufacturing, and footwear industries dominating in terms of numbers (Table 17.1). The international port at Manila is a major trading centre of East Asia.

In 1988 Metropolitan Manila had three power stations, all burning Bunker-C oil containing 2–4 per cent sulphur. The total generating capacity of these power stations was 1,210 MW. The largest, the Sucat Power Station, generates 850 MW.

Transport Vehicle registrations in 1988 for petrol- and diesel-driven vehicles are given in Table 17.2. There are about 240,000 cars, 190,000 utility vehicles (such as taxis and jeepneys), and 33,000 trucks registered in Metropolitan Manila. The number of motorcycles and tricycles is about 44,000. Most private motor vehicles are petrol-fuelled, but taxis, jeepneys and buses, which provide the major public transportation, are primarily fuelled by diesel.

Railways connect Manila to other parts of Luzon. The rapid public transport system, the Light Rail Transit, transports 350,000 commuters daily on 30 km of rail system, but the metropolis is still often congested by motor vehicle traffic.

Table 17.1 *Industries in metropolitan Manila*

Industry	Number of units
Chemical and related	372
Food manufacturing	352
Footwear	351
Textile	223
Metal products	216
Basic metal	108
Non-metallic mineral	91
Paper and paper products	73
Rubber products	59

After Manins, 1991

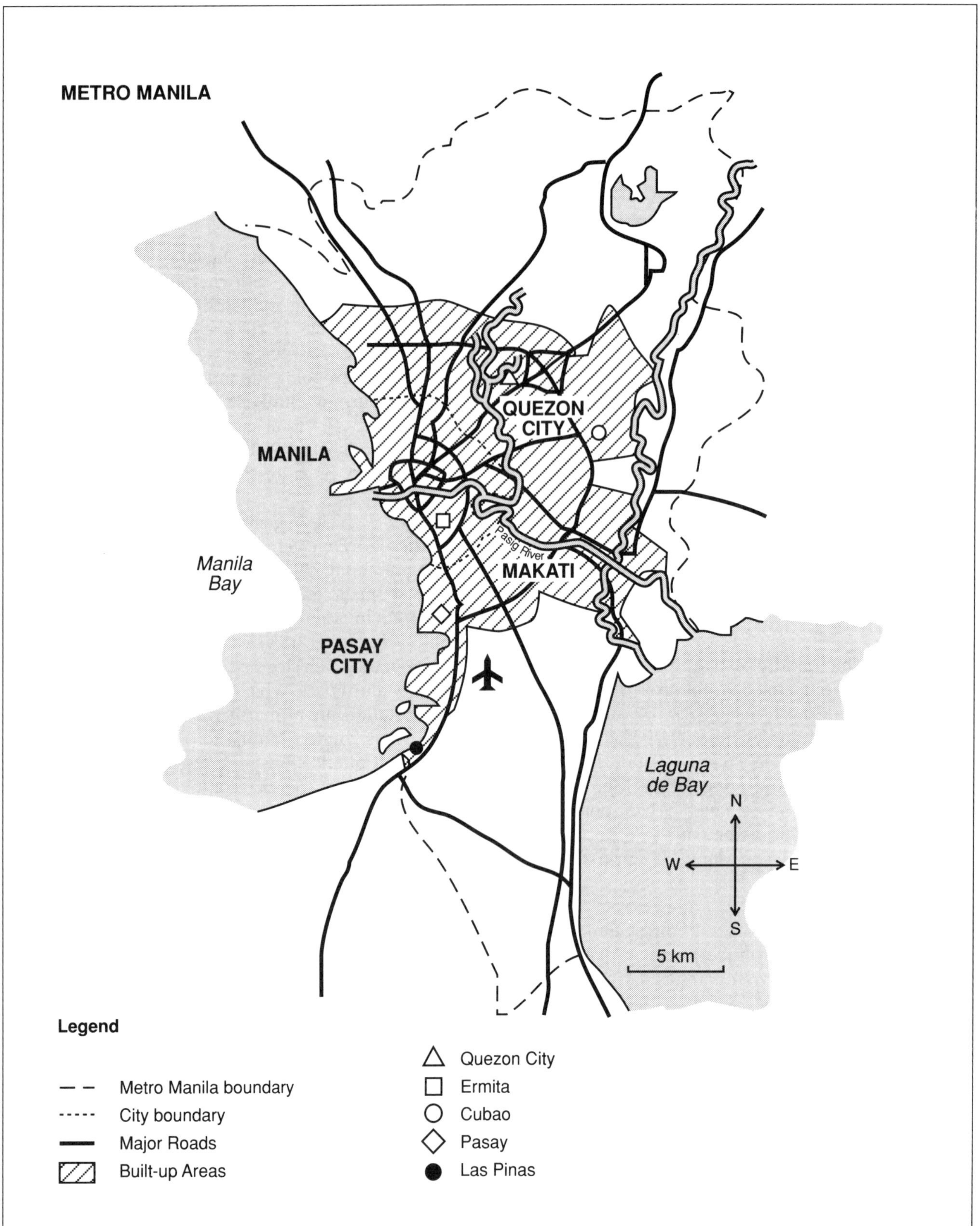

Figure 17.1 *Sketch map of Metropolitan Manila*

Table 17.2 Metropolitan Manila motor vehicle statistics 1988

	Registered number	Fuel economy ($km\ l^{-1}$)	Kilometres per vehicle per annum
Petrol vehicles			
Cars	231,821	12	27,541
Utility	103,138	11	17,973
Truck	4,362	6	50,576
Motorcycles/tricycles	43,677	30	23,692
Diesel vehicles			
Cars	9,200	16	35,365
Utility	86,266	8	23,014
Truck	28,631	3	64,759
Buses	4,060	4	190,296

After Manins, 1991

17.2 Monitoring

Manila has participated in the GEMS/Air programme since 1974. Air quality monitoring has been undertaken continuously at four fixed sites and with two mobile laboratories. Since 1981, when the monitoring equipment was discontinued by its manufacturers, the number of analysers in use gradually declined until 1985 when continuous monitoring ceased. Since 1986, manual sampling for sulphur dioxide (SO_2) and suspended particulate matter (SPM) has been carried out at some sites. Some nitrogen dioxide (NO_2) data were also acquired after 1986.

A WHO site visit report in 1989 stated that monitoring instruments were not being calibrated, and hence monitoring results might be unreliable (Mage, 1989). In 1989, monitoring was being carried out every sixth day for total suspended particulates (TSP) and SO_2 at six to eight monitoring stations, depending on the condition of analytical instruments.

17.3 Air Quality Situation

Sulphur dioxide

Emissions In 1990 a detailed emissions inventory for SO_2, SPM, carbon monoxide (CO), oxides of nitrogen (NO_x), and volatile organic compounds (VOCs) was published by the Philippine Environmental Management Bureau (EMB). Emissions of mobile sources refer to the base year 1988, whereas emissions of stationary sources and aircraft refer to the base year 1987 (Ayala, 1988). However, it was pointed out by Manins (1991) that the EMB inventory does not include all relevant sources and some of the emissions factors used may not be adequate, especially with respect to mobile sources.

EMB emissions data for Metropolitan Manila are summarized in Table 17.3, except for the SO_2 emissions in the industry/commerce sector, for which Manins (1991) showed that the original EMB estimate

Table 17.3 Emissions estimates ($t\ a^{-1}$)

	SO_2	SPM	CO	NO_x
Power stations	83,000	6,000	1,000	9,000
Industry/commerce	31,000[a]	6,000	2,000	1,000
Domestic combustion	n.a.[b]	n.a.	n.a.	n.a.
Petrol vehicles	3,000	3,000	514,000[c]	29,000[d]
Diesel vehicles	31,000[e]	54,000[f]	27,000[g]	76,000[h]
Aircraft	400	100	13,000	4,000
Open refuse burning	n.a.	n.a.	n.a.	n.a.
Total	148,400	69,100	557,000	119,000

[a] Data taken from Manins, who also considered unaccounted industrial fuel use. Estimate by EMB was 6,000 $t\ a^{-1}$
[b] Estimate by Manins is 150 $t\ a^{-1}$
[c] Estimate by Manins is 315,000 $t\ a^{-1}$
[d] Estimate by Manins is 24,000 $t\ a^{-1}$
[e] Estimate by Manins is 11,000 $t\ a^{-1}$
[f] Estimate by Manins is 3,500 $t\ a^{-1}$
[g] Estimate by Manins is 46,000 $t\ a^{-1}$
[h] Estimate by Manins is 36,000 $t\ a^{-1}$
n.a. = No data available

Sources: EMB, 1990 , Manins, 1991

was far too low. Thus the emissions inventories by EMB (1990) and Manins (1991) are very different for several sources (e.g., industry and commerce, and mobile sources).

Table 17.3 shows that total SO_2 emissions are about 148,000 tonnes per annum. The major emissions sources are the three power stations which emit 83,000 tonnes per annum or 56 per cent of the total. Industrial and commercial sources contribute 31,000 tonnes per annum or 21 per cent of the total. The reason for the high SO_2 emissions from combustion sources is that the average sulphur content of the industrial fuel oil is around 3 per cent.

Mobile sources contribute 34,000 tonnes per annum SO_2 (22 per cent of total emissions), with more than 90 per cent of it coming from diesel vehicles. This EMB figure is much higher than the estimate from Manins (1991), who reported mobile sources SO_2 of around 7,000 tonnes per annum. Aircraft (mainly at Manila International Airport) were estimated to emit only relatively low amounts of SO_2. Domestic SO_2 emissions have not been considered in the EMB inventory, but Manins (1991) calculated that the 98 million litres of kerosene consumed in 1988 in Manila for domestic cooking produced approximately 150 tonnes of SO_2.

Ambient Concentrations Sulphur dioxide was monitored in Manila from 1977–1983 at four sites, and since 1986 at three sites. Yet, the two monitoring series are not directly comparable because the 1977–83 data were obtained using automatic hourly samplers and the data since 1986 were from manual samplers run for 24-hour periods. In addition, the methodology was changed from an acidimetric to a colorimetric one when sampling resumed in 1986. Thus, the data since 1986 are considered much more reliable.

As shown in Figure 17.2, annual mean SO_2 concentrations remained relatively constant between 1977 and 1983. Annual mean values were about 50–100 $\mu g\ m^{-3}$ in the four sites. The upward trend during 1981–1983 is likely to be an artifact caused by equipment problems. Compared with the old time series, the more recent data are much lower. In 1989, the annual mean concentration was 20–50 $\mu g\ m^{-3}$ at three sites. Those values are at or below the WHO annual mean guidelines of 40–60 $\mu g\ m^{-3}$. Average daily mean concentrations in 1988 were about 13 $\mu g\ m^{-3}$ at the Pasig site and up to 35 $\mu g\ m^{-3}$ at the Pasay site (see Table 17.4). Maximum daily mean values were reported to be as low as 19 $\mu g\ m^{-3}$ at the Pasig site and as high as 246 $\mu g\ m^{-3}$ at the Ermita site. In a recent modelling study, Manins (1991) predicted peak SO_2 values of 80 $\mu g\ m^{-3}$ (= 0.03 ppm) near the Manila Power Station. The available data show that short-term SO_2 concentrations are generally below the Philippine 24-hour average air quality standard of 0.14 ppm (374 $\mu g\ m^{-3}$) and WHO 98 percentile guideline (not to be exceeded on more than 2 per cent of days) of 150 $\mu g\ m^{-3}$ (= 0.052 ppm).

Table 17.4 *Daily mean SO_2 monitoring data ($\mu g\ m^{-3}$) in 1988*

Station name	Minimum 24-hour mean	Average 24-hour mean	Maximum 24-hour mean
Malabon	5	32	144
Quezon City	3	29	77
Pasig	3	13	19
Ermita	3	27	246
Las Pinas	3	16	128
Pasay	3	35	11

After Manins, 1988

Suspended particulate matter

Emissions As mentioned above, an emissions inventory for anthropogenic particulates and other pollutants was published by the EMB in 1990, but it has been reported that the EMB inventory does not include all relevant sources and some of the

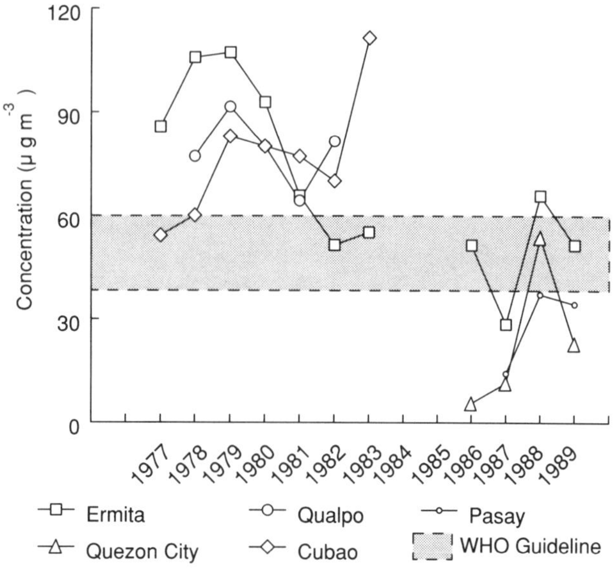

Figure 17.2 *Annual mean sulphur dioxide concentrations in Metropolitan Manila*

After EMB, 1990

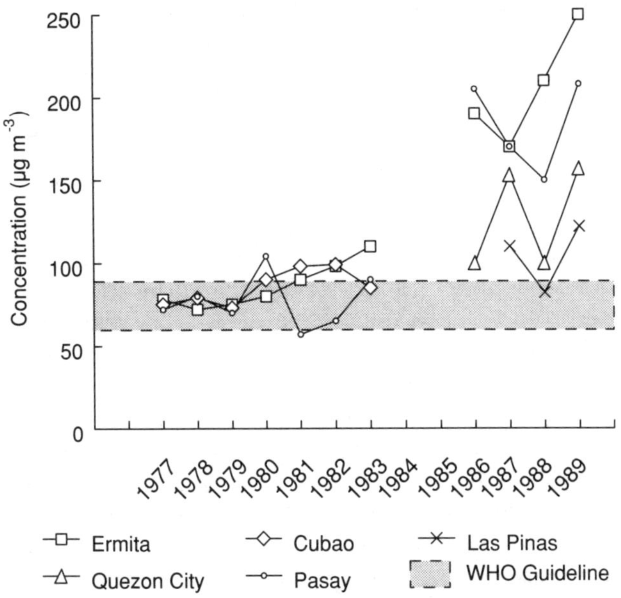

Figure 17.3 *Annual mean suspended particulate matter concentrations in Metropolitan Manila*

After EMB, 1990

emissions factors used may be incorrect, especially with respect to mobile sources (Manins, 1991).

The EMB emissions estimates for particulates are summarized in Table 17.3. Total anthropogenic SPM emissions amount to 69,000 tonnes per annum, with diesel vehicles contributing 54,000 tonnes per annum - nearly 80 per cent. However, this estimate is questioned by Manins (1991) who estimates diesel vehicle SPM emissions to be only 3,500 tonnes per annum. These differences are due to the lack of actual emissions data for Manila motor vehicles. As long as the emissions factors applied to the inventory calculations are taken from other countries' experiences, the errors in the calculations may be very high.

Table 17.5 *Daily mean suspended particulate matter (TSP) (µg m^{-3}) monitoring data in 1988*

Station name	Minimum 24-hour mean	Average 24-hour mean	Maximum 24-hour mean
Malabon	41	227	437
Quezon City	13	166	310
Pasig	29	136	253
Ermita	91	207	316
Las Pinas	26	90	226
Pasay	33	154	251

After Manins, 1988

Compared with diesel vehicles, other SPM sources are minor. Power plants and industry/commerce emit 6,000 tonnes per annum each. Domestic emissions sources have not been calculated in the inventories, but with the predominant use of relatively clean kerosene fuel the contribution of domestic combustion to total SPM emissions is probably small. Aircraft SPM emissions (mainly at Manila International Airport) have been estimated to be very small. An additional source of SPM, which has not been considered in the emissions inventories, is the open burning of waste. This source is of major importance in terms of human exposure as it occurs at ground level. Uncontrolled waste burning generates and releases toxic and carcinogenic compounds often attached to smoke particles.

Ambient Concentrations Total suspended particulates have been monitored in Manila from 1977 to 1983 at four sites, and again since 1986 at six sites. The two monitoring series are not directly comparable because the 1977–83 data were obtained using automatic hourly samplers and the data since 1986 were from manual samplers run for 24-hour periods.

As shown in Figure 17.3, annual mean SPM (TSP) concentrations were relatively constant from 1977 to 1983. Annual mean values were about 50–100 µg m^{-3} at the four sites. The more recent data, however, show a much higher annual mean. In 1989, the annual mean concentration was 120–250 µg m^{-3}. Those values are far above the WHO annual mean guidelines of 60–90 µg m^{-3}. Annual mean TSP concentrations show an upward trend since 1986.

Average daily mean concentrations in 1988 were between 90 µg m^{-3} (at the Las Pinas site) and 227 µg m^{-3} (at the Malabon site) (see Table 17.5). Maximum daily mean values were reported to be as high as 437 µg m^{-3} at the Malabon site. The maximum daily values in all stations exceed the national Philippine 24-hour mean air quality standard of 180 µg m^{-3} and the maximum WHO daily guideline (not to be exceeded on more than 2 per cent of days) of 230 µg m^{-3}.

Lead

Emissions Lead is not included either in the EMB (1990) or in the Manins (1991) emissions inventory. From the traffic statistics given in Table 17.2, it can be estimated that emissions of lead will be in the range of 400–600 tonnes per annum, if a lead content of 0.5–0.8 g l^{-1} in petrol is assumed.

Table 17.6 Lead concentrations ($\mu g\ m^{-3}$) in major thoroughfares in metropolitan Manila, 1987

Station name	Annual arithmetic mean
Malabon	3.04
Quezon City	0.85
Pasig	0.26
Ermita	0.63
Las Pinas	0.31
Pasay	0.57
Paranaque	0.25
Valenzuela	4.35

After EMB, 1990

Ambient Concentrations Data are available for lead concentrations in major thoroughfares in Metropolitan Manila in 1988 (EMB, 1990, see Table 17.6). The annual mean concentrations vary between 0.25–4.35 $\mu g\ m^{-3}$. The upper limit of WHO annual mean guideline range of 0.5–1.0 $\mu g\ m^{-3}$ is far exceeded at two of the eight monitoring sites (Valenzuela: 4.35 $\mu g\ m^{-3}$; Malabon: 3.04 $\mu g\ m^{-3}$). However, lead concentrations were below the WHO guideline range in only three of the eight monitoring sites.

Carbon monoxide

Emissions The EMB emissions estimates for CO are summarized in Table 17.3. According to those figures, total CO emissions amount to 557,000 tonnes per annum, with petrol vehicles contributing 514,000 tonnes per annum or more than 90 per cent. Diesel vehicles were estimated to be the second largest source with emissions of 27,000 tonnes per annum. The EMB data were questioned by Manins (1991), who estimated CO emissions of petrol vehicles to be considerably lower (315,000 tonnes per annum) and emissions of diesel vehicles to be considerably higher (46,000 tonnes per annum) than the EMB figures. However, in both inventories, motor vehicle traffic contributes 96–97 per cent of total CO emissions and thus is by far the dominant emission source. Carbon monoxide emissions of power plants and industry are very small, as normally efficient combustion processes are applied in those sectors.

Aircraft were estimated to contribute 13,000 tonnes per annum, which is high and might be significant for CO ambient air levels in the surroundings of the Manila International Airport. Domestic emissions sources are not included in the available emissions inventories. Owing to low combustion efficiencies of domestic stoves, CO emissions in this sector could be significant.

Open burning of waste, which is a very incomplete combustion process, is an additional important source of CO and has not been considered in the available emissions inventories.

Ambient Concentrations Carbon monoxide has been monitored in Manila from 1977 to 1983 at four sites. However, there are no recent monitoring data. Annual mean concentrations were between 1.5–10 $mg\ m^{-3}$ in 1983. The highest concentrations were recorded at the Ermita site which showed an increasing trend from 1980 to 1983. The extent of CO air pollution is shown clearly when compared with existing guidelines. If annual mean values at Ermita reached the WHO eight-hour mean guideline value of 10 $mg\ m^{-3}$, then it must be expected that short-term concentrations frequently exceed the WHO guideline. Carbon monoxide concentrations are expected to be much higher in traffic congested areas. Since 1983 traffic has greatly increased so that it may be assumed that CO concentrations have also increased.

Oxides of nitrogen

Emissions Oxides of nitrogen include both nitric oxide (NO) and NO_2. Although NO is an order of magnitude less toxic than NO_2, both substances are normally calculated in emissions inventories as NO_2 because NO is ultimately converted to NO_2. The main sources of NO_x are high-temperature combustion processes and the chemical industry (nitric acid and fertilizer production).

An emissions inventory for NO_x was published by the EMB in 1990, but – as mentioned above – the validity of the EMB data has been questioned with respect to completeness of sources and with respect to accuracy (Manins, 1991).

According to the EMB estimates in Table 17.3, total NO_x emissions amount to 119,000 tonnes per annum, with motor vehicles contributing nearly 90 per cent (105,000 tonnes per annum). Of the motor vehicles, diesel vehicles emit considerably more NO_x than petrol vehicles. Manins (1991), estimated a similar amount of NO_x from petrol vehicles, but considerably lower emissions from diesel vehicles (36,000 tonnes per annum) than the EMB.

Power stations were estimated to contribute 9,000 tonnes per annum of NO_x or 7 per cent of the total. This number would seem to be rather low as power stations normally have a very high emission rate of NO_x because of high temperature combustion. Similarly, emissions estimates for industry and commerce (1,000 tonnes per annum) are probably too low. Aircraft contribute 4,000 tonnes per annum or 3 per cent of total emissions.

Domestic emissions have not been estimated, but it can be assumed that NO_x emissions by household stoves are relatively low but may have a high impact on indoor air pollution.

Ambient Concentrations No data are available for ambient NO_x pollution levels.

Ozone

Emissions As shown above, emissions of NO_x are about 119,000 tonnes per annum. There is an EMB (1990) emissions inventory for VOC (150,000 tonnes per annum 'total organic gases', TOG), but this figure is certainly too low as several emissions sources, such as industrial processes and petrol and solvent evaporation, have not been sufficiently considered. There are also natural sources (e.g., vegetation) which should be taken into account in VOC inventories.

Ambient Concentrations No data are available for ozone (O_3) concentrations in ambient air in Manila. Temperatures and insolation are high in the dry season and relatively large amounts of precursor substances (NO_2 and VOC) are emitted, so O_3 concentrations may be very high during daytime in downwind areas of Metropolitan Manila.

17.4 Conclusions

Air Pollution Situation The main air pollution problem in Manila results from the very high SPM, CO, and Pb levels. Short-term and long-term means frequently exceed the WHO guidelines in large parts of the urban area.

Airborne particulate matter is a cause for concern. The major source of these particles is motor vehicles, with a possible locally significant contribution from local industry and waste burning. The highly visible black smoke emitted from diesel vehicles is a very substantial problem in traffic areas as passengers and pedestrians are often exposed to the undiluted exhaust. It has been noted that smoke levels on roads can be so bad that "it is now a common sight to see people riding jeepneys holding handkerchiefs over their mouth and nose to try to reduce their exposure to smoke" (Mage, 1989). A relatively high portion of SPM may be of small physical size and inhalable. However, data are not available to determine the amount of inhalable SPM.

Sulphur dioxide levels in ambient air are low. The Philippines is a semi-tropical country with low emissions of SO_2 from domestic and industrial heating, and there are no strong surface inversions which are typical of the winter conditions of the temperate zones. However, persons directly exposed to diesel traffic exhaust, such as jeepney drivers, may be subjected to high SO_2 concentrations.

No monitoring data are available for NO_x or O_3, but conditions are favourable for ozone formation.

Main Problems Monitoring data suggest that the main pollution problem in Manila stems from the large number of automobiles with no emissions control devices. The main source of SPM, CO and Pb is motor vehicle traffic.

The available emissions data suggest that industrial activities at present make only a relatively small contribution to the pollutant load in the Manila atmosphere. Similarly, domestic combustions seem to have only a small overall impact on emissions.

Other important sources of pollution, which so far have not yet been sufficiently quantified, are frequent open burnings of waste. Although their contribution to total emissions is probably small, emissions from waste burning have a high impact on local ambient air pollution in the affected areas.

Control Measures Emissions inventories are the basis for control actions. The existing emissions inventories for Metropolitan Manila, although preliminary, are far more complete than for many other cities. Yet, they need to be continuously improved, particularly with respect to a better quantification of mobile sources, a more detailed identification of emissions from industrial processes, and a spatial differentiation of emissions. Furthermore, the location and amount of open burning of waste need to be traced.

Monitoring efforts require improvement, including upgrading and calibration of equipment, and the reinstitution of CO measurements owing to current traffic conditions.

Control actions should focus primarily on controlling motor vehicle emissions. At present, vehicle maintenance is generally inadequate and petrol-powered vehicles have no emissions controls. The jeepney diesel engines are often imported from Japan after their economically useful life is over and rebuilt or reconditioned for use in the Philippines. Thus emissions are much higher than those of comparable new engines. The problem of motor vehicle emissions is greatly exacerbated by the rapidly increasing numbers of vehicles. As new vehicles are introduced they replace some of the older "dirtier" engine vehicles, but this process will be slow and the emission reduction might be offset by an additional growth in traffic. The costs of modern emission control, as currently enforced in the USA, Japan and Europe, may be beyond the capacity of the Philippine economy. Gradual introduction of more stringent standards and reliance on inspection and maintenance could be used to mitigate the expected emission increase (Elston, 1991).

17.5 References

Ayala, P. M. 1988 *Metro Manila Air Pollution Emission Inventory for 1987*, Environmental Management Bureau, Department of Environment and Natural Resources, Manila.

Elston, J. C. 1991 *Motor vehicle emission control*, WHO Mission Report, Manila. (Unpublished report).

EMB 1990 *The Philippine Environment in the Eighties*, Environmental Management Bureau, Department of Environment and Natural Resources, Manila.

Mage, D. T. 1989 Assignment Report on Collaboration on Air Quality Management, WHO Western Pacific Regional Centre for the Promotion of Environmental Planning and Applied Studies (PEPAS), Kuala Lumpur. (Unpublished report).

Manins, P. 1988 Personal communication CSIRO, Canberra.

Manins, P. 1991 *Model for Air Pollution Planning and Computerization of Air Quality Data*, WHO Mission Report, Manila. (Unpublished report).

18

Mexico City

18.1 General Information

Geography The Metropolitan Area of Mexico City (MAMC) is situated in the Mexican Basin at Latitude 19°26'N, Longitude 99°07'W and at a mean altitude of 2,240 metres. It covers 2,500 km^2, surrounded by mountains, two of which are over 5,000 metres high. Two valley channels, located to the north-east and north-west, funnel the air to the centre and to the south-west of the city. The MAMC includes the 16 boroughs of the Federal District and 17 conurbated municipalities of the State of Mexico (Figure 18.1).

Demography The population of the MAMC was estimated to be 19.37 million in 1990, more than one-fifth of Mexico's total population (UN, 1989). However, the results of the 1990 National census showed a total population of just above 15 million for the MAMC (INEGI, 1991). Population densities in MAMC range from almost 7,000 persons per km^2 in the centre to 500 persons per km^2 in the least populated zones. Fifty-five per cent reside in the Federal District and 45 per cent in the conurbated municipalities of the State of Mexico. Projections indicate that the population will grow at a rate of 1.4 per cent per annum to 24.44 million by 2000.

Climate The local climate is marine west coast. Temperatures remain relatively constant through the year, with a monthly mean of 15°C (12°C in January, 17.4°C in May). Precipitation occurs mainly during summer, from June to September, and totals 725 mm per annum. This annual precipitation pattern has a strong effect on concentrations of pollutants such as lead (Pb) and suspended particulate matter (SPM).

Due to the particular geographic characteristics and the light winds, ventilation is poor with a high frequency of surface as well as upper air temperature inversions. During winter (November–May) inversions occur up to 25 days per month. The expansion of the urban area and consumption of energy have markedly modified the valley's microclimate. The heat island effect is pronounced in Mexico City and hot spots can be up to 12°C above the temperature in suburban and rural areas (DDF, 1991).

Industry The MAMC is the political, administrative and economic centre of Mexico, accounting for one-third of the country's Gross Domestic Product. There are more than 30,000 industries – of all types and sizes – and 12,000 service facilities in the MAMC. Among the industries in the valley, 250 handle hazardous wastes and 4,000 use combustion or transformation processes generating major atmospheric emissions. MAMC's total energy consumption

Table 18.1 *Percentage energy balance of MAMC, 1986*

	Traffic (%)	Electricity (%)	Industry (%)	Commerce (%)	Others (%)	Total (%)
Petrol	36.0					36.0
Diesel	8.1		5.2	0.4		13.7
Fuel Oil		8.1	1.6	0.3		10.0
Liquid Petroleum Gas				0.8	9.1	9.9
Natural Gas		5.6	13.9		0.5	20.0
Electricity	0.5		4.8	1.7	3.4	10.4
Total	44.6	13.7	25.5	3.2	13.0	100

After DDF, 1989

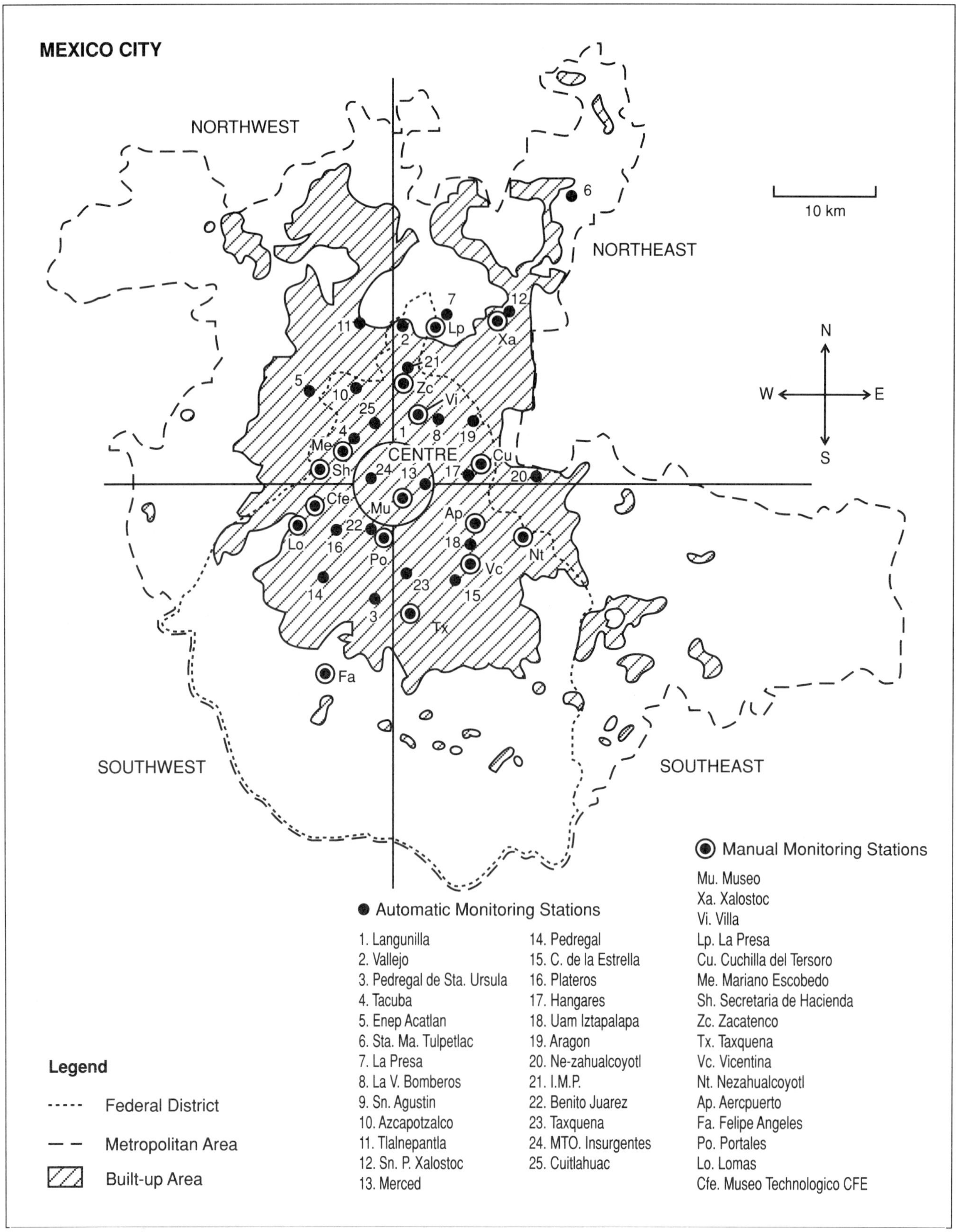

Figure 18.1 *Sketch map of the Metropolitan Area of Mexico City*

Table 18.2 *Trip distribution by mode in the MAMC, 1989*

	Bus	Taxi	Collective Taxi	Metro	Trolley/Tram	Private
Number of vehicles	10,500	56,502	69,561	2,269	450	2,372,180
Million Person trips per day	9.7	1.02	9.0	4.8	0.5	4.4
Percentage of person trips per day	33%	3%	31%	16%	2%	15%

After DDF/EDOMEX, 1989

reached 512 Petajoules in 1986. The corresponding 1986 energy balance by sector (in per cent) is summarized in Table 18.1.

Transport Some 2.5 million motor vehicles (buses, minibuses, taxis, lorries, vans and private cars), are responsible for 44 per cent of the total energy consumption in the MAMC. Motor vehicles are by far the main pollution source in the area, burning 40 thousand barrels of diesel and 1 million barrels of leaded petrol each day. In 1989, there were almost 30 million person trips in the region on an average day, of which about 25 million were made by public transport. Details are given in Table 18.2.

The public transport system comprises a mixture of government-owned and concessionary vehicles. There are 10,500 urban ("Resta100") and suburban buses, which account for one-third of the total trips. There are more than 69,000 "collectivos" or collective taxis (vans). These collective taxis follow a fixed route and are of two main types: a 22-seater (average) minibus, which may carry standing passengers, and a nine-seater van or "combi" type. The latter are being gradually replaced by minibuses. Every day 9 million person trips are made in this way.

The MAMC also has a modern metro network. In 1989 there were eight subway lines with 125 stations and 2,205 carriages in operation. All of them are located in the Federal District. The metro network covers 141 km and accounts for almost 5 million trips per day (Table 18.2). The metro is the fastest transportation in the city, averaging 34 km h^{-1}. Additionally, a recent study found that commuters using the metro are exposed to lower levels of carbon monoxide (CO) than the users of any other transport system (Fernandez-Bremauntz, 1991). Although other electric vehicles exist, such as trolleybuses and a one-line light train, their coverage is very limited and they play a modest role in public transport with only 500,000 trips per day.

During the 0600–0900 hours peak period, over 70 per cent of all person trips are made for going to work or to school. However, this "windowtime" can be split in two phases. Public transport commuters gather by thousands at the (feeding) termini points of the inner transportation network, from 0600–0730 hours. It is only from 0730 hours that private car numbers start to build up. Then car congestion becomes a major problem, as it is during the late afternoon peak, between 1800–2000 hours. There is another increase in traffic around 1400 hours, but it is smaller than the other two peaks.

In 1982, the road system covered 126 km^2 which accounted for more than 8 per cent of the total urban area (JICA, 1988). The system includes multi-lane roads with controlled access – the Peripheral, the Inner Circuit, Tlalpan Avenue and the M. Aleman Viaduct. The road system also has a primary network of Axis of Circulation ("ejes viales"), organized in a perpendicular pattern crossing the city in all directions. The routes of buses, trolleybuses and collective taxis are largely organized following these axes of circulation.

Private car numbers have grown sharply. In the Federal District alone, the number of cars rose from 0.5 million in 1968 to 1.5 million in 1978. It is known that the current number of private cars for the MAMC is 2.4 million (Table 18.2). The economic situation of the country means that the car turnover rate is very low; the average motor vehicle in MAMC is 10 years old.

Although private cars produce up to 63 per cent of total emissions from transport and up to 99 per cent of the CO emitted, they represent only 4.4 million person trips per day (Table 18.2), less than 15 per cent of trips made by all modes of transport.

18.2 Monitoring

Air pollution monitoring in Mexico City started in the 1950s and became more systematic in the late 1960s (OPS, 1982). In 1966, with the support of the Pan American Health Organization, 14 monitoring stations were installed to measure smoke, SPM and sulphur dioxide (SO_2). In the early 1970s, the Mexican authorities and United Nations Environment Programme (UNEP) developed a programme to improve environmental quality in several cities. One

Table 18.3 *Emissions inventory for MAMC 1989 (10^3 t a^{-1})*

Sector	SO_2	SPM	CO	NO_x	NMHC
Energy					
PEMEX*	14.7	1.1	52.6	3.2	31.7
Power Plants**	58.2	3.5	0.5	6.6	0.1
Industry					
Industry	65.7	10.2	15.8	28.8	39.9
Services	22.0	2.4	0.4	3.9	0.1
Transport					
Private cars	3.5	4.4	1,328.1	41.9	141.0
Taxis	0.8	1.0	301.1	9.5	31.9
Combis & minibuses	0.8	1.0	404.4	10.0	42.7
Urban bus	5.2	0.2	6.2	8.0	2.4
Suburban bus	13.0	0.6	12.6	18.2	5.3
Petrol lorries	0.9	1.1	779.5	16.9	67.8
Diesel lorries	20.0	0.9	16.5	26.1	7.2
Other	0.2	0.1	5.0	2.7	1.6
Environmental degradation					
Erosion	0.0	419.4	0.0	0.0	0.0
Forest fires, etc.	0.1	4.2	27.3	0.9	199.7
*Total****	205.7	450.6	2,950.6	177.3	572.1

* closed in 1991
** switched to natural gas in 1991 (100 % by the end of the year)
*** emission values are rounded to the nearest 100 tonnes, therefore the sum of columns may not be the same as the total

After Gobierno de la Republica, 1990

component of the programme was to install a manual network of 22 stations for SO_2 and SPM which was completed by mid-1976 (Marquez, 1977). Finally, with technical assistance from the US Environmental Protection Agency (USEPA) an automatic monitoring network became operational in 1985. The network, known as the Red Automatica or the RAMA, covers the greater part of the MAMC (Figure 18.1), and its stations measure SO_2, CO, ozone (O_3), oxides of nitrogen (NO_x) and non-methane hydrocarbons (HCNM). However, it should be noted that only five stations - one in each zone in addition to the centre - are equipped with a complete set of monitors to measure the five pollutants mentioned above (Figure 18.1). A manual network of 16 stations for SO_2 and SPM is still fully operational as well. The positioning of the sites in the monitoring network is also given in Figure 18.1.

The configuration of the RAMA is listed in Figure 18.1. Air quality measurements are reported daily to the public in the form of an index value or "IMECA". The IMECA calculation makes the criterion value for each pollutant equal to 100 points (Appendix I). A contingency programme exists to cope with prolonged episodes of high pollution levels, when the IMECA passes values of 300 points. During an episode, a series of actions are taken to try to reduce the emission of pollutants. These measures include reduction of activity of highly polluting industry and vehicle circulation restrictions. In extreme cases, primary schools are temporarily closed to prevent potential damage to children's health.

In spite of limited economic resources, the RAMA is operated efficiently by its technicians. Performance of the equipment is checked by personnel from the USEPA, making audits to the RAMA twice a year (Saint, 1991).

Besides the air pollution monitoring networks described above, the National (UNAM) and the Metropolitan (UAM) Universities also have some equipment to carry out monitoring activities. Scientists from both institutions have worked in the field of air quality evaluation for many years (DDF, 1982; Marquez, 1977).

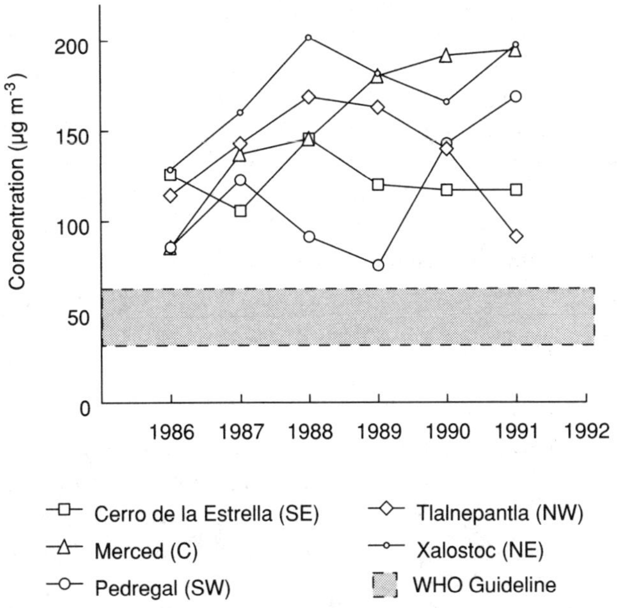

Figure 18.2 *Annual mean SO2 concentrations*

Note: 1991 data to 31 September only
After Gonzalez-Garcia, 1991

18.3 Air Quality Situation

Sulphur dioxide

Emissions Table 18.3 shows that industry and power plants were the main sources of SO_2 in the MAMC in 1989. In addition, the PEMEX refinery produces an estimated 15,000 tonnes of SO_2 per annum. However, in 1991 the two power plants in the MAMC switched to natural gas and the PEMEX refinery was closed by the government on environmental grounds. Diesel-engined buses and lorries are the only other significant sources of SO_2 in the MAMC.

Although fuels in the MAMC have progressively been improved in terms of sulphur content, both fuel oil and diesel still have relatively high amounts of sulphur - 3.5 per cent and 1.2 per cent respectively (DDF, 1987). In order to reduce emissions, natural gas has gradually replaced (heavy) fuel oil for electricity generation in the power plants. Additionally, a number of minibuses, buses, commercial lorries and vans have been retrofitted to enable them to use gas (Gobierno de la Republica, 1990). However, 43 per cent of all SO_2 emissions are produced by industrial sources, a sector that has seen little improvement in terms of pollution control.

Ambient Concentrations In the late 1960s and early 1970s, annual mean SO_2 levels were in the range 40–190 µg m^{-3}, with daily maxima occasionally as high as 500-900 µg m^{-3} (OPS, 1982). Figure 18.2 shows SO_2 annual (arithmetic) means and daily maxima from 1986 to 1991 for five specific sites of the Red Automatica. Here, mean SO_2 levels ranged from 50-160 µg m^{-3}, well above the WHO guideline values. For this period, the north-east zone (Xalostoc station) registered the highest levels for the MAMC, which is not surprising if one considers the high density of industrial units in the zone. This is confirmed by the 1989 isopleths for the mean SO_2 levels (Figure 18.3) although the highest levels, 170 µg m^{-3} (60 ppb), have moved now to the north-west.

In summary the SO_2 situation is:

- mean levels range from 80 to 200 µg m^{-3};
- daily maxima are between 200 and 550 µg m^{-3};
- no clearcut trends have been noted over the five years in the 1986–1991 period;
- the 1986–1991 SO_2 levels are still comparable to those of the past 20 years; thus
- implying that WHO guidelines are not met at a significant number of locations and that the national air quality standard is exceeded occasionally in some specific sites.

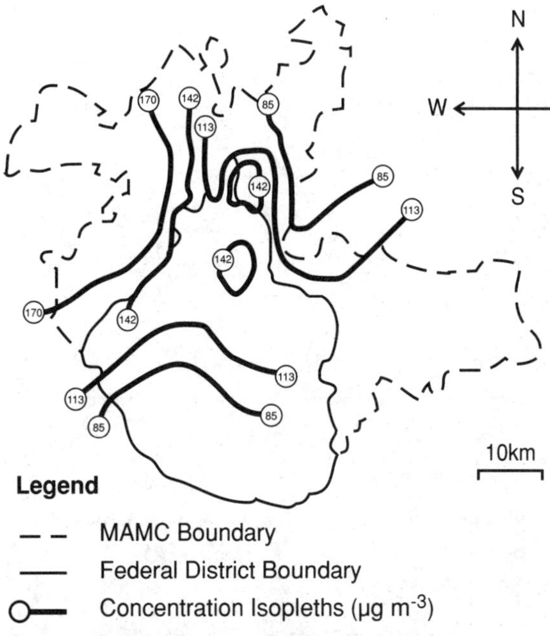

Figure 18.3 *Annual mean sulphur dioxide distribution in MAMC, 1989 (µg m^{-3})*

After Gobierno de la Republica, 1990

Suspended particulate matter

Emissions In the MAMC emissions of SPM include natural particulates (such as soil, biological processes, rubbish) and anthropogenic particulates emitted by motor vehicles, industrial and/or combustion processes or formed in the atmosphere as a result of transformations of other contaminants (e.g., nitrates and sulphates).

Emission estimates for the most important sources of SPM in the MAMC are given in Table 18.3. Environmental degradation, including phenomena such as erosion, forest fires and others, is obviously the overruling origin of SPM in MAMC's ambient air.

Ambient Concentrations Historical data for the SPM levels in Mexico City during the late 1960s and early 1970s reveal that mean levels at 12 different, but unidentified, locations were generally in the 60–150 μg m^{-3} range, with the exception of one site where mean levels reached 240 μg m^{-3} (Fuentes-Gea and Garcia-Gutierrez, 1990). The corresponding maximum daily values varied between 200 and 1,000 μg m^{-3}. Up-to-date information given in Figure 18.4 for a limited number of specific sites for the 1976-1991 period illustrates that average SPM levels certainly did not decrease as a function of time. The current concentration range is between 100 μg m^{-3} and 500 μg m^{-3} (Nezahualcoyotl and Xalostoc, both in the north-eastern quadrant).

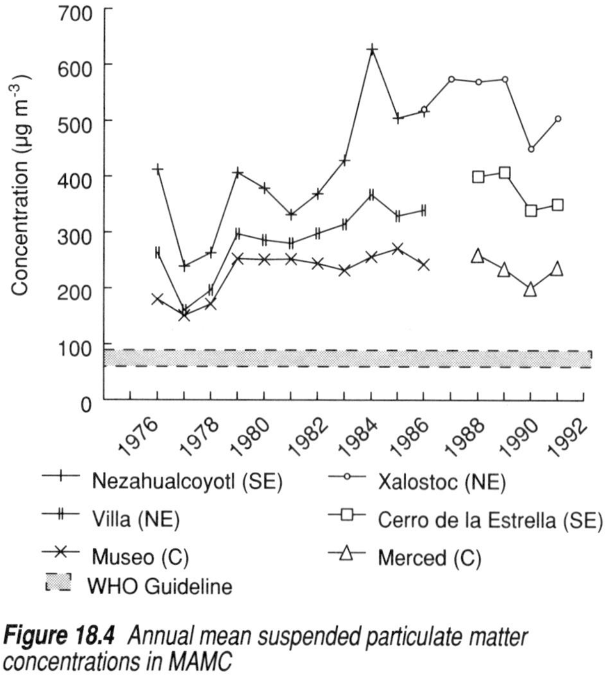

Figure 18.4 *Annual mean suspended particulate matter concentrations in MAMC*

Note: 1991 data to 31 September only
Source: Gonzalez-Garcia, 1991

The national air quality standards and the WHO guidelines are frequently exceeded at more than one place in the MAMC. In this context it is useful to remember that the particulate matter < 10 μm (PM_{10}), are the particulates with the most pronounced impact on health as well as on visibility. It has been suggested that in the MAMC the PM_{10} may represent between 40 and 60 per cent of the total SPM reported (from measurements based on high-volume sampling).

Lead

Emissions Lead emissions are not quantified in Table 18.3; however, it can be assumed that petrol motor vehicles are undoubtedly the major source of ambient Pb in Mexico City. The lowering of the maximum Pb content of petrol from 3 ml tetraethyl-lead per gallon to less than 1 ml per gallon has undoubtedly played an important role in decreasing Pb emissions in recent years.

Ambient Concentrations Annual mean levels at different locations are given as a function of time for the 1977-1991 period in Figure 18.5. Of note is the continuous decrease from 1982 to 1986 in four of the five reporting sites, followed by a sharp increase over the next two years and a pronounced decrease to levels under the national (quarterly) Pb evaluation standard of 1.5 μg m^{-3} by 1991 (for which incomplete data are available).

Carbon monoxide

Emissions Transport emissions amount to 2.85 million tonnes per annum, corresponding to 97 per cent of the CO total (Table 18.3). Emission standards for petrol-powered vehicles were, until 1990, very permissive. From the 1993 car model onwards, the maximum allowed CO emission will be lowered to 2.11 g km^{-1} (Table 18.4). However, recent assessments consider that the CO emitted by an average car in the MAMC is presently as high as 60 g km^{-1} (DDF, 1989).

Ambient Concentrations Carbon monoxide is currently measured at 15 RAMA stations, all of which use non-dispersive infrared absorption monitors. Levels of CO show large variations through the metropolitan area. Variation depends on the source, meteorological conditions (notably wind speed) and most importantly the particular micro-environment where the monitoring station is located. The RAMA provides information mainly on "background" or ambient

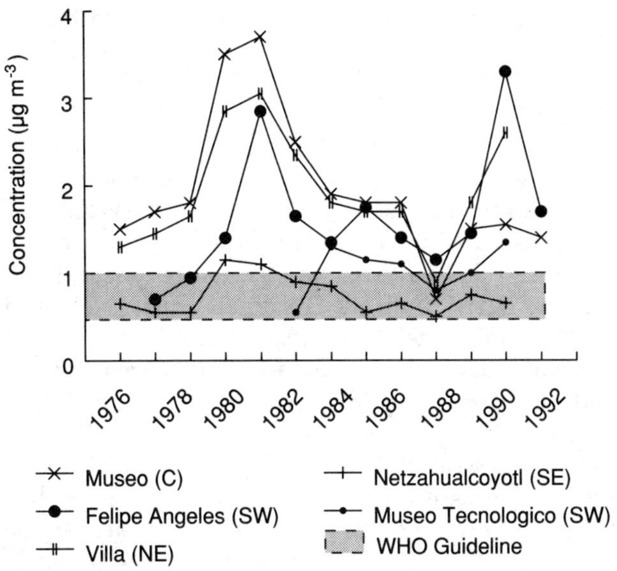

Figure 18.5 *Annual mean lead concentrations in MAMC*

Note: 1991 data to 31 September 1991 only
After Gonzalez-Garcia, 1991

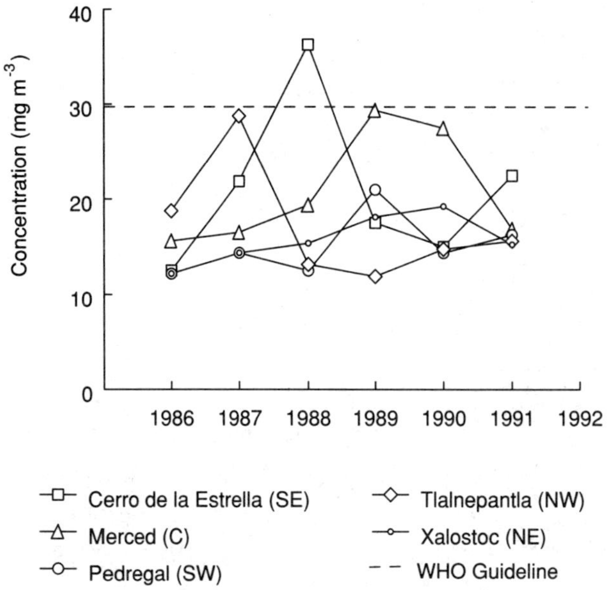

Figure 18.6 *Maximum eight-hour mean carbon monoxide concentrations in MAMC*

Note: 1991 data to 31 September only
After Gonzalez-Garcia, 1991

levels. Therefore, additional measurements are necessary at street level to evaluate exposures of people spending time near the vehicles and roads.

A north-west station (Cuitlahuac) usually registers the highest CO levels in the city, although there is no evidence to suggest that the north-west zone has the worst CO problem. However, it is well known that the station receives a strong influence from the road vehicles stopping at the traffic lights a few metres away from the station. More evidence from some other site-year combinations during the 1986–1991 period illustrates the present CO situation in the MAMC (Figure 18.6).

The information given in Figure 18.6 indicates that exceedences of WHO guidelines and national air quality standards (NAQS) for CO occur at several locations. A recent exposure assessment study (Fernandez-Bremauntz, 1991) confirmed that CO levels measured at fixed site stations even underestimate actual exposures of pedestrians, street sellers and commuters.

As expected, CO levels peak between 0700 and 0900 hours, when low temperatures, atmospheric stability (inversions) and heavy vehicular traffic occur simultaneously. There is another peak in the evening around 2000 hours, but it is usually lower than the morning one. Carbon monoxide pollution levels during the week show a slight increase from Monday to Friday, and decrease significantly on weekends, mainly on Sunday.

The altitude of the city, and consequently the relatively low content of oxygen of its atmosphere, plays a double role in the CO problem. First, it causes CO emissions to increase, due to an incomplete combustion process. Second, it exacerbates the health effects attributed to CO, especially among highly susceptible population groups, such as children, pregnant women and asthmatics.

Table 18.4 *Petrol vehicles emission standards*

Year of model	Maximum allowed emission (g km^{-1})		
	Hydrocarbons	Carbon monoxide	Nitrogen oxides
1985	2.00	22.00	2.30
1990	1.80	18.00	2.00
1991–92	0.70	7.00	1.40
1993 onwards	0.25	2.11	0.62

Source: Espinosa and Medina, 1990

Oxides of nitrogen

Emissions The relative importance of the different categories of local sources of NO_x are given in Table 18.3. Motor vehicle emissions account for more than

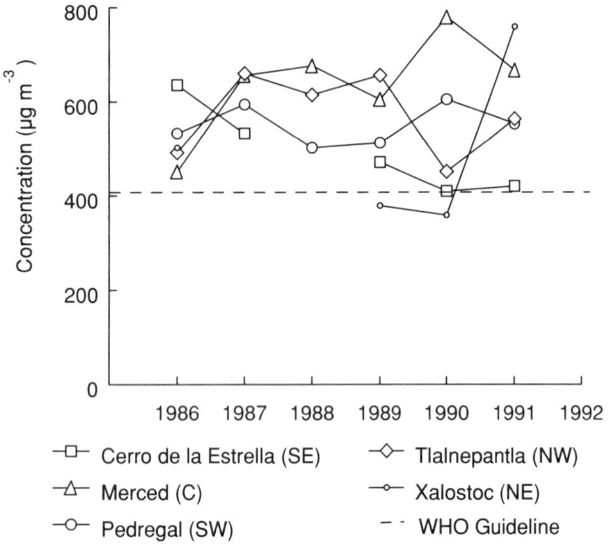

***Figure 18.7** Maximum hourly nitrogen dioxide concentrations in MAMC*

Note: 1991 data to 31 September only
After Gonzalez-Garcia, 1991

three-quarters of the total emissions. As a consequence traffic emissions will predominantly determine the ambient levels measured in the metropolitan area of Mexico City.

Ambient Concentrations The maximum hourly nitrogen dioxide (NO_2) levels at five different locations are represented for the years 1986-1991 in Figure 18.7. For each of the sites the overall situation seems to be stationary with less or more pronounced variations from one year to another, and no clearcut trend for either annual mean levels or for maximum hourly levels. Mean NO_2 levels are in the 60-110 ppb range (113-207 $\mu g\ m^{-3}$), while hourly extremes go from 160-380 ppb (301-714 $\mu g\ m^{-3}$). The latter are often above the WHO guideline (400 $\mu g\ m^{-3}$), as well as above the national air quality standard of 210 ppb (395 $\mu g\ m^{-3}$) for a number of site-year combinations as shown in Figure 18.7. The frequency with which safe levels of NO_2 were exceeded during the period 1986-1991 nevertheless did not exceed 5 per cent of the days in one year in the whole city, not even in areas of high industrial activity or intense motor vehicle traffic.

Besides the possible health effects, NO_x have other implications from the point of view of air quality. They are the basic precursors of O_3, the major air pollution problem in the MAMC. Furthermore, they are ultimately transformed in the atmosphere into acids and nitrates which, like sulphates, are particles which decrease visibility and contribute to acid deposition.

Ozone

Emissions Ozone is a pollutant that is not emitted directly, but is formed as a product of very complex reactions of NO_x and hydrocarbons in the presence of sunlight. Taking into account the local topographical, meteorological and the relatively high NO_x and HCNM emissions (Table 18.3), Mexico City and its surroundings is almost the ideal place to generate O_3 in the local boundary layer.

Ambient Concentrations Ozone levels in Mexico City are exceptionally high. As is shown in Figures 18.8 and 18.9, O_3 represents an air quality problem in the whole city, but there are areas that are more seriously affected. During the period 1986–1989, the greatest frequency of excess O_3 levels occurred in the south-western zone (254 days) (Figure 18.8). In that zone during the same period, the levels were exceeded by an average of at least four hours daily which increased the risks associated with this pollutant.

Annual mean O_3, as reported by five sites over six years, fluctuated around 200 $\mu g\ m^{-3}$ with lows around 100–150 $\mu g\ m^{-3}$ and highs between 300–400 $\mu g\ m^{-3}$. As already mentioned, the worst situation occurs in the south-west sector with the monitoring site "Pedregal"

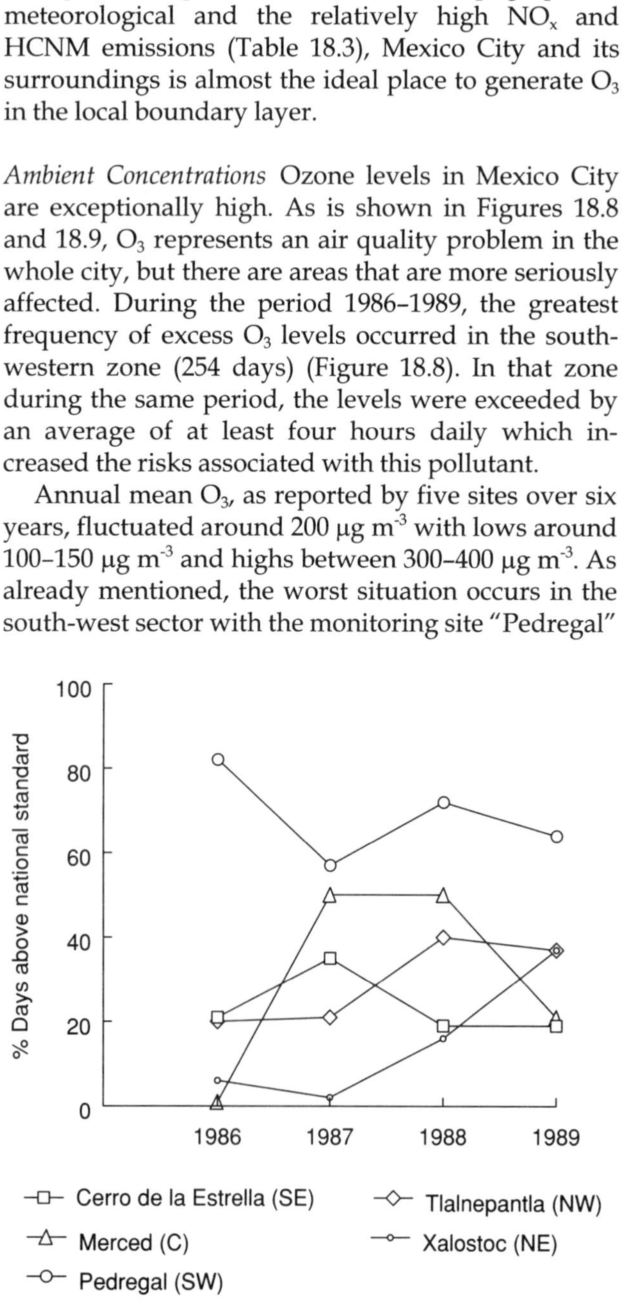

***Figure 18.8** Percentage of days above national ozone standard, 1986–1989*

After DDF, 1989

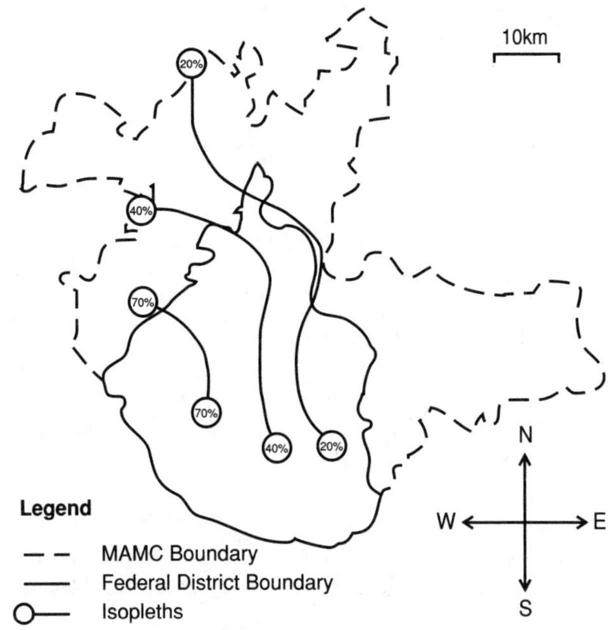

Figure 18.9 *Percentage of days with at least one exceedence of the national ozone standard in MAMC, 1987–1989*

After Gobierno de la Republica, 1990

as a typical example. Hourly O_3 levels often reach 600 $\mu g\ m^{-3}$ there, with extreme values up to 850–900 $\mu g\ m^{-3}$. The number of hours during which the national air quality standard of 110 ppb is exceeded is consequently high; 80 to 100 hours per month is not unusual.

18.4 Conclusions

Air Pollution Situation The MAMC has very serious air pollution which has increased with the growth of the city, population, employment in industry, and traffic. The main cause of air pollution in the city is energy consumption. It is therefore necessary to take account of the city's economic development and future trends when considering the functional relationship between well-being, energy consumption and pollution.

Air pollution in the MAMC, owing to suspended dust and other particles, is an old problem. However, pollution as we know it today began about 50 years ago with the growth of industry, transportation and population.

Control Measures The deterioration in MAMC's air quality has been so rapid that the city must act urgently on two fronts: first, transforming the economic foundation of the city with non-polluting activities to replace the old industries following a comprehensive strategy that halts pollution growth through the development of better technologies and, secondly, the use of better fuels, emissions controls and protection of wooded areas.

In his inauguration speech in December 1988, President Carlos Salinas de Gortari said: "I am giving precise, urgent and imperative instructions to the Mayor of the Federal District to act immediately and efficiently to promote community participation in the fight against pollution." In response to that instruction, a Comprehensive Pollution Control Programme for the Mexico City Metropolitan Zone, as described in DDF (1991), was developed. The ranked priority areas for action in this strategic plan are:

1. **The Oil Industry** - in its refinery activities, distribution and reformulation of fuels;

2. **Transportation** - in its cargo, collective and individual modes, both in urban and environmental efficiency and in technological advances, rational use of energy and pollution emission control;

3. **Private Industry and Service Facilities** - in technological and productive modernization, as well as in energy efficiency and pollution emission control;

4. **Thermoelectric Power Plants** - as the largest consumers of fuels in the city, in the continuous use of clean fuels;

5. **Reforestation and Ecological Restoration** of the deforested soils, areas without sewerage, creation of ecological reserves, and open air garbage dumps transformation;

6. **Research, Ecological Education and Social Communication** by the agencies responsible for air-quality monitoring, research and social communication.

These actions will not enable the city to recover the air quality it had half a century ago, since in the Valley of Mexico instead of 1.5 million inhabitants there are now over 15 million and instead of being an agricultural and mining country, the population's employment and current income are sustained by industry, transportation and services.

The Comprehensive Programme will require a great effort from the Government and society to handle the problem. It is a medium-term programme: one for the 1990s. This programme requires a strategic common effort that must be sustained for decades.

18.5 References

DDF 1982 *Registro de vehiculos automotores en el Distrito Federal*, D. G. Autotransporte Urbano, Departamento del Distrito Federal, Mexico DF.

DDF 1987 *La contaminacion atmosferica en la Zona Metropolitana de la Ciudad de Mexico y las medidas aplicadas para su control*, Departamento del Distrito Federal, Internal Report. (Unpublished report.)

DDF 1989 *Programa Integral Contra la Contaminacion Atmosferica*, Programa de Emergencia, Proyectos, Departamento del Distrito Federal, Mexico DF.

DDF 1991 *Comprehensive Pollution Control Programme for the Mexico City Metropolitan Zone*, Departamento del Distrito Federal, Mexico DF.

DDF/EDOMEX 1989 *Programa Integral del Transporte*, Departamento del Distrito Fedral y Gobiemo del Estado de Mexico, Mexico DF.

Espinosa, M. E. and Medina, L. 1990 *Air pollution from mobile sources in the MAMC*, 83rd Annual Meeting of the Air and Waste Management Association, Pittsburg.

Fernandez-Bremauntz, X. 1991 *Commuters' Exposure to Carbon Monoxide in the MAMC*, PhD Thesis, Imperial College, University of London, London.

Fuentes-Gea, V. and Garcia-Gutierrez, A. 1990 *Air Pollution Trends in Mexico City, TSP (1976–1984)*, 83rd Annual Meeting of the Air and Waste Management Association, Pittsburg.

Gobierno de la Republica 1990 *Programa integral contra la contaminacion atmosferica*, Gobierno de la Republica, Mexico DF.

Gonzalez-Garcia, R. 1991 Personal communication, Secretaria de Desanrollo Urbano y Ecologia, Mexico DF.

INEGI 1991 *Resultados preliminares del censo de poblacion 1990*, Instituto Nacional de Estadistica, Geografia e Informatica, Mexico DF.

JICA 1988 *The study on air pollution control plan in the Federal District, Final Report*, Japan International Co-operation Agency, Mexico DF.

Marquez, M. E. 1977 *Informe sobre la calidad del aire en algunas ciudades del pais*, Proyecto Mexico PNUD, Subsecretaria de Mejoramiento del Ambiente, Mexico DF.

OPS 1982 *Red Pan Americana de Muestro de la Contaminacion del Aire, Informe Final 1967-1980*, Publication CEPIS 23, Organizacion Pan Americana de la Salud, Lima.

Saint, C. 1991 Personal communication, US Environmental Protection Agency, Washington DC.

UN 1989 *Prospects of World Urbanization 1988*, Population Studies No. 112, United Nations, New York.

19

Moscow

19.1 General Information

Geography Moscow is the capital and administrative centre of the Russian Federation. It is the most populous city in the country. The entire city covers an area of 994 km^2; the area within the Moscow ring-road is 880 km^2 (Figure 19.1).

The city lies in the broad, shallow valley of the Moskva River, a tributary of the Oka and thus of the Volga, at the centre of the European Russian plain (Latitute 55°45'N Longitude 37°34'E), at an altitude of 156 m. Moscow is surrounded by mixed forests. The city is crossed from the north-west to the south-east by the Moskva River; the Yauza River meets the Moskva River in the centre of the city. Industrial premises and public utilities occupy about 25 per cent of the urban territory; parks and public open space occupy a further 20 per cent.

Demography Moscow had a population of 9.39 million in 1990, 1 million more than in 1980 (UN, 1989) (Figure 1.1). UN estimates suggest that Moscow will have a population of 10.11 million by the year 2000. The main influences on population dynamics are employment and trade. The population density within the Moscow highway circle is 10,000 people km^{-2}.

Climate Moscow has a continental cool summer climate. The annual mean temperature is 4.4°C. Maximum mean monthly air temperature in July reaches 20°C; the minimum of -10°C is observed in January. For the period 1980–90 the absolute maximum and minimum air temperatures recorded were 36°C and -38°C. Moscow has a pronounced urban heat island effect; the air temperature in the centre of the city is 1–2°C higher than in the periphery. This produces the possibility of atmospheric pollution in the centre of the city owing to the advection of polluted air from peripheral industrial zones.

Prevailing winds are north-easterly, easterly and south-westerly. The mean wind speed in the centre is 1.8–2 m s^{-1} and increases to 5–7 m s^{-1} in south-eastern and southern districts. Calms (wind speed 0-1 m s^{-1}) occur for 44 per cent of the year in the city centre. These calms, which often coincide with surface temperature inversions (26 per cent of the year), provide favourable conditions for pollutant accumulation. Ground-based inversions are most frequently observed at night and in the early morning in April–June and August–October. Elevated temperature inversions also occur between December and March.

Mean precipitation is moderate, 575 mm a year, and peaks in July and August with brief downpours and thunderstorms; 60–65 per cent of all days of the year have precipitation. A considerable part of this precipitation falls as snow, beginning in October. Permanent snow cover is usually established by mid-November, lasting until mid-April.

Industry Moscow is the centre of an area of intense industrial development, the Central Industrial Region. Moscow's industry was once dominated by textiles (worsted, cotton, linen and silk); however, today engineering and metalworking are the principal industries. Motor vehicle manufacturing is the most important industry in Moscow; the Likhachyov Auto Works is one of the largest factories in the city. Machine tools and precision engineering and electrical, electronic and radio-engineered goods are major specialities. The Hammer and Sickle Steel Works is the only major steel works in Moscow providing specialized, high-grade steel for the engineering industries. Most steel, however, is transported from the Donetzk Basin and the Urals. Moscow has a large chemical industry and many of the industry's products are derived from Moscow's oil refinery which processes oil piped from the Volga-Urals oil field. Timber processing industries are also found in the area, ranging from furniture making to pulp and paper, with associated printing and publishing works.

Moscow's 1,600 industrial enterprises and industrial zones are distributed evenly throughout the city (see Figure 19.1). Strict planning restrictions have been placed on new industries wanting to develop in the city. Relocation of heavy industry out of the city centre has been slow.

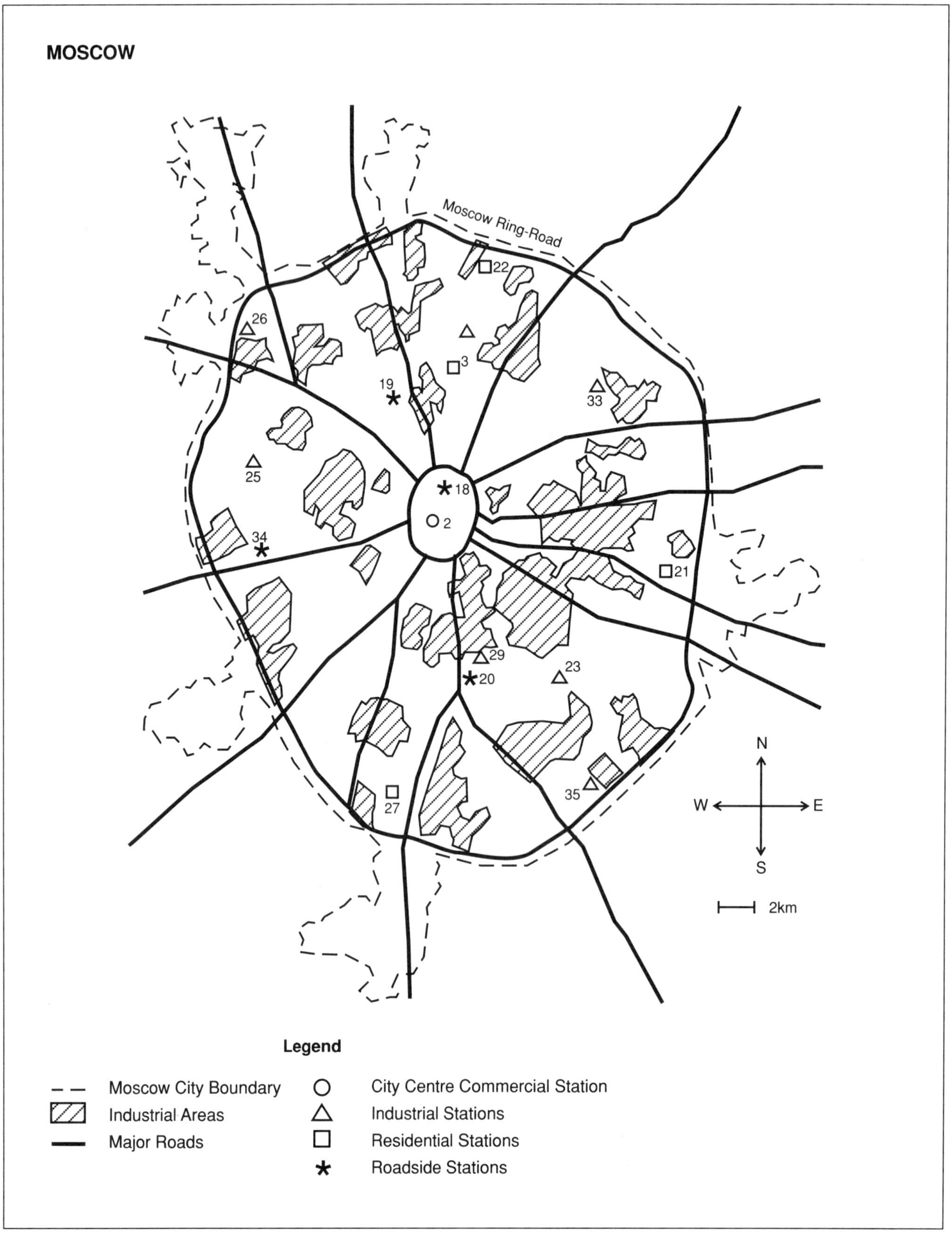

Figure 19.1 *Sketch map of Moscow*

Transport Moscow has a highly developed system of motorways which generally follow a spoked wheel (orbital and radial) pattern. The construction of the Moscow orbital highway considerably diminished the flow of motor transport through the city and has contributed to the overall decrease in air pollution attributable to motor vehicles in the central area. Currently up to 90 per cent of transit freight traffic passes the city by the orbital highway.

Moscow's motor vehicle population in 1985 consisted of 135,000 freight vehicles (including 25,000 buses) and 500,000 passenger cars and rose to a total of 665,000 in 1989. It is projected that the total number of motor vehicles will double between 1995 and 2000.

A considerable amount of passenger traffic in Moscow is carried by surface public transport (buses, trolley buses and trams). In total, surface public transport carried 3,935 million passengers in 1989 (60 per cent by bus), compared with 2,741 million on the underground railway (Finansyi i Statistika, 1990). An increase in the number of bus passengers was observed in the period 1980–87. The total distance of bus routes is much longer than that of the other types of passenger transport added together.

19.2 Monitoring

Moscow is not included in the GEMS/Air programme at present. The State Service for Observations and Control of Environmental Pollution Levels (OGSNK) is responsible for information on pollutant levels in the environment and is supported by local authorities and Republican Committees for Hydrometeorology. Information on environmental quality is transferred to the organizations of the state for legislative and planning purposes and also to the media. The observational network of monitoring stations in Moscow is a constituent of the OGSNK. The network is supervised by the Moscow Centre for Hydrometeorology and Control of the Environment.

Air pollution control is the responsibility of the Moscow City Committee for Nature Protection and the Moscow City Council of Deputies. The main legislative acts concerning air quality are: the Atmospheric Protection Act, adopted by the Supreme Soviet in 1980; resolution of the USSR Council of Ministers on the Reconstruction of Nature Protection Measures (1988); and relevant resolutions of Moscow City Council.

The Institute of Global Climate and Ecology (IGCE) Institute of Applied Geophysics and the Main Geophysical Observatory (MGO) participate in assessments of air pollution in Moscow. These institutions are responsible for the development and improvement of methods for air pollution monitoring, the assessment of pollutant emissions, and for determining pollutant pathways and chemical transformation in the atmosphere.

The air pollution monitoring network in Moscow consists of 19 observational stations. Air is sampled on filters or sorbents with subsequent analysis for four priority pollutants - carbon monoxide (CO), nitrogen dioxide (NO_2), suspended particulate matter (SPM) and sulphur dioxide (SO_2). Observations are carried out three or four times a day at 0700 hours, 1300 hours, 1900 hours and 0100 hours. Air samples are collected in sorption tubes; filters are exposed for 20–30 minutes. No comparisons have been made between these methods and more conventional 24-hour based or continuous monitoring methods employed by GEMS/Air and national monitoring authorities. Urban measurements are made throughout the Russian Federation using similar methods to those used in Moscow. A detailed description of the methods used is given in the *Manual for atmospheric pollution observations* (Gidrometeoizdat, 1979) and *Temporal methodics for chemical analysis of the atmosphere samples with sampling on solid film sorbents* (Gidrometeoizdat, 1989).

The observational programme is given in Table 19.1. Together with the compulsory basic programme of systematic monitoring, measurements of particular pollutants important in specific districts are also performed. These include phenols, sulphate, formaldehyde, ammonia, carbon black, hydrogen sulphide and heavy metals (Osipov, 1989). Besides these systematic observations, surveys of air pollution are undertaken with the help of mobile laboratories and special investigations are made following the complaints of residents.

19.3 Air Quality Situation

Sulphur dioxide

Emissions Total SO_2 emissions were estimated to be approximately 130,000 tonnes per annum in 1990. An overall decrease in SO_2 emissions and ambient concentrations was observed between 1987 and 1990 (Figure 19.2). The most probable reason for this was the reduction in coal and heavy fuel oil use for power generation purposes. For the period 1987–90 the use of coal was reduced by over 70 per cent and heavy fuel oil by approximately one-third. The share of industrial sources in the total emission of SO_2 was minor (10 per cent in 1990) in comparison with power generation (Rovinsky et al., 1991).

Table 19.1 *The observational programme of air pollution monitoring stations in Moscow*

Station number	SO_2	NO_2	NO	CO	SPM	Pb
1	+	+		+		
2	+	+		+	+	
3		+		+		
18	+	+	+	+	+	
19		+		+		+
20				+	+	
21	+	+		+	+	
22	+	+	+	+	+	+
23	+	+		+	+	
24		+		+	+	
25	+	+		+	+	+
26		+		+	+	
27	+	+	+	+	+	+
28	+	+		+	+	
29	+	+	+	+	+	+
33		+		+	+	
34		+		+	+	
35		+		+	+	+
36		+		+	+	

After Rovinsky et al., 1991

Ambient Concentrations Routine monitoring of SO_2 has been conducted since 1980. However, because of changes in the methods used in 1987, the data on SO_2 are only given for the period 1987–90. Sulphur dioxide concentrations are measured using the pararosaniline-formaldehyde method with air sampling on thin-film sorbent. Annual mean SO_2 concentrations varied in the vicinities of monitoring stations from 1–9 μg m^{-3}. These values are extremely low considering emissions estimates and must call into question the accuracy and sensitivity of the monitoring method. It is not clear whether the problem stems from the extrapolation of 20-minute exposure data to annual means, the representivity of the network, the analysis of samples, or data handling. Comparison of the SO_2 method employed in Moscow with ISO approved methods should be seen as a major priority for the authorities. The highest maximum (20-minute exposure) concentrations of up to 240 μg m^{-3} were registered in industrial districts located in north-west and south-east sectors of the city (MCGNPS, 1987, 1988, 1989, 1990; MCGNPS, 1991). Once again, even maximum values would appear to be very low when compared with cities with similar estimated emissions.

Sulphur dioxide levels increase in winter, linked with the increase in fuel consumption, and the highest SO_2 concentrations are observed in the industrial zones (Table 19.2). The mean levels of SO_2 presented do not generally exceed the WHO guidelines (40–60 μg m^{-3}) or the national standard (50 μg m^{-3}) (Rovinsky et al., 1991).

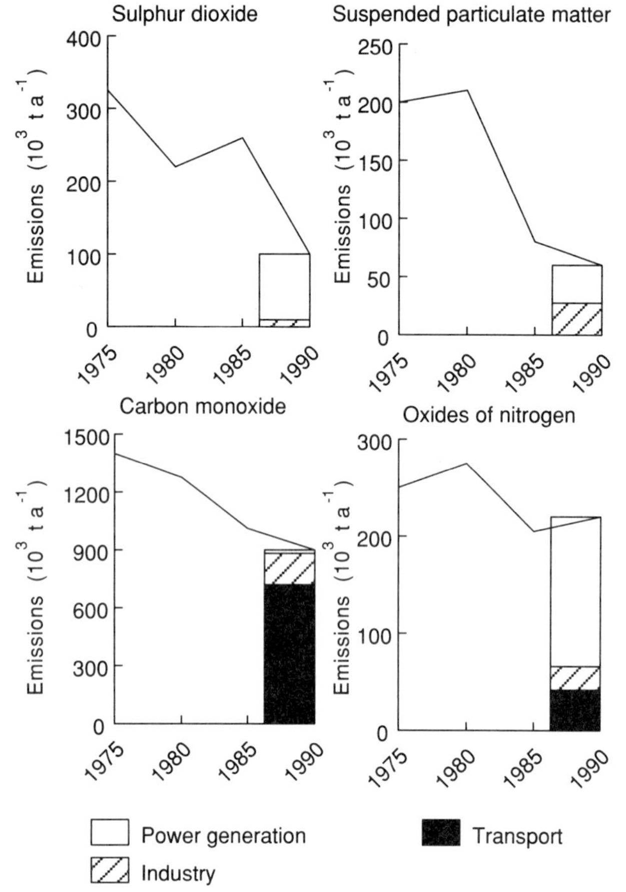

Figure 19.2 *Estimated total emissions 1975–1990 and emissions by source*

Source: Rovinsky et al., 1991

Suspended particulate matter

Emissions The most important sources of SPM in Moscow are fuel combustion by-products from energy production. This source was responsible for 54 per cent of total anthropogenic SPM emissions in 1990 (Figure 19.2). The remaining 46 per cent was attributed to industry (Rovinsky et al., 1991).

Ambient Concentrations Suspended particulate matter is measured gravimetrically after air sampling on filters. Concentrations of SPM for the period 1980–88 were rather stable; mean annual concentrations of 150–250 μg m^{-3} were observed (MCGNPS, 1991). A

Table 19.2 *Annual mean pollutant concentrations in air in different zones of Moscow*

Zone	Station numbers	SPM ($\mu g\ m^{-3}$)	SO_2 ($\mu g\ m^{-3}$)	CO ($mg\ m^{-3}$)	NO_2 ($\mu g\ m^{-3}$)
Central	2	100	0.1	2	100
Industrial	23, 24, 25,26, 29, 33,35, 38	100	3.0	2	100
Motorways	18, 19, 20,34	100	–	3	70
Residential	3, 21, 22,27, 28	100	1.0	2	80
Green	37	–	–	2	60

After Rovinsky et al., 1991

considerable decrease in SPM levels in air is noted between 1988 and 1990 (Figure 19.3).

Reductions in the use of liquid and solid fuel by power stations probably account for the decrease in SPM concentrations in recent years. Mean annual levels varied noticeably in different parts of the city in 1980–86 from 100–300 $\mu g\ m^{-3}$ (MCGNPS, 1991). However, between 1988 and 1990 these differences were practically absent, and homogeneous distribution of SPM pollution over the city was observed. Maximum 20-minute SPM concentrations often reach 1000–1500 $\mu g\ m^{-3}$ in single episodes and are registered most frequently in the north-western (station 26) and south-eastern (stations 23, 29) industrial zones of the city (CVGMO, 1981, 1982, 1983, 1984, 1985, 1986; MCGNPS, 1987, 1988, 1989, 1990; MCGNPS, 1991). These levels greatly exceed the national standard (150 $\mu g\ m^{-3}$) and WHO guideline (230 $\mu g\ m^{-3}$) for mean daily concentrations.

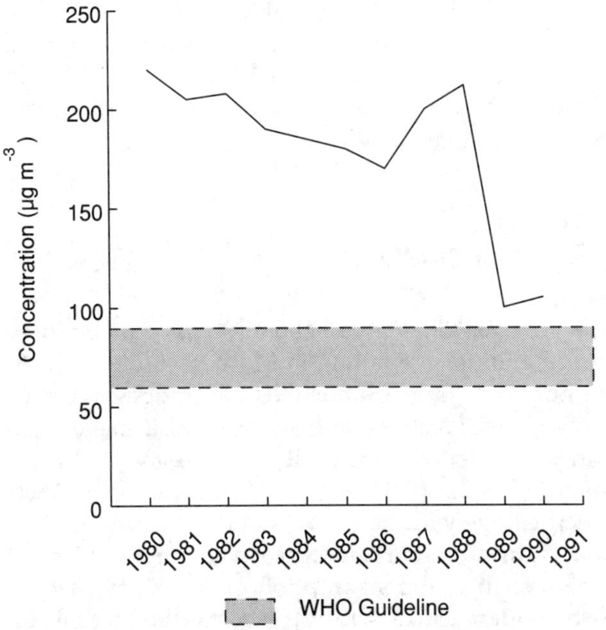

Figure 19.3 *Annual mean suspended particulate matter concentrations at all sites in Moscow*

After Rovinsky et al., 1991

The seasonal variation of SPM concentrations is not explicit, but a slight increase is observed in April–July and October–December.

Lead

Emissions Although not quantified, motor vehicles are believed to be the major source of lead (Pb) emissions. Control measures now forbid the sale of leaded petrol in Moscow. A reduction in motor vehicle traffic has also helped to reduce Pb levels over the last three years (Rovinsky et al., 1991).

Ambient Concentrations Lead concentrations are measured by atomic-absorption spectrometric analysis. The Pb content in air in 1985-90 varied from 0.01–0.04 $\mu g\ m^{-3}$ (mean annual concentrations) (MCGNPS, 1991). The highest Pb in air is registered in the south-east sector of the city (station 29), where concentrations have reached 0.25 $\mu g\ m^{-3}$ (maximum) and 0.09 $\mu g\ m^{-3}$ (mean annual values) (CVGMO, 1981, 1982, 1983, 1984, 1985, 1986; MCGNPS, 1987, 1988, 1989, 1990; MCGNPS, 1991). The levels detected do not exceed the national standard for air quality (0.3 $\mu g\ m^{-3}$) or the WHO recommended guideline (1 $\mu g\ m^{-3}$).

Carbon monoxide

Emissions Anthropogenic emissions of CO were approximately 900,000 tonnes in 1990 (Figure 19.2). It is estimated that in 1990 motor vehicles produced 80 per cent of CO emissions and that industry produced 18 per cent (Rovinsky et al., 1991).

Ambient concentrations Carbon monoxide concentrations are determined using infrared absorption techniques. Annual mean CO concentrations ranged from 2–6 mg m^{-3} in 1980–90 (MCGNPS, 1991).

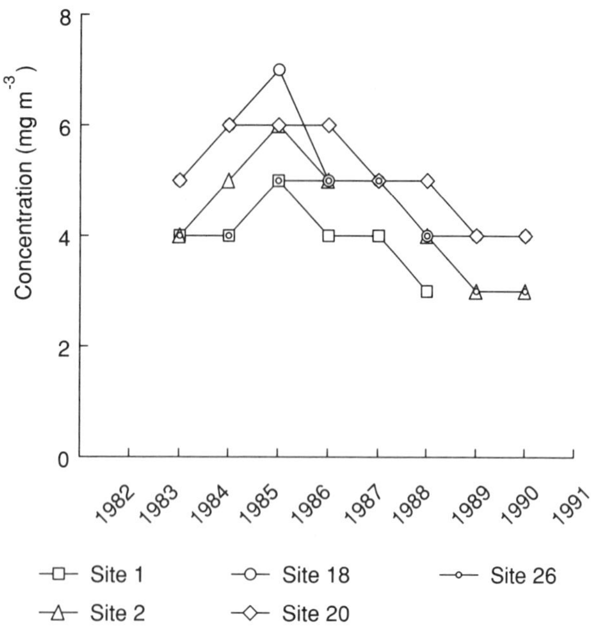

***Figure 19.4** Annual mean carbon monoxide concentrations at selected sites*

After Rovinsky et al., 1991

Concentrations were 20–30 per cent higher in the central (station 2) and southern (station 20) sectors. The maximum concentrations (20-minute exposure) of 13–44 mg m^{-3} were observed in these sectors of the city, considerably above the national air quality standard of 5 mg m^{-3} (CVGMO, 1981, 1982, 1983, 1984, 1985, 1986; MCGNPS, 1987, 1988, 1989 1990; MCGNPS, 1991). Maximum concentrations occur in both winter (February) and summer (June and July). Surface temperature inversions and low wind speed favour high levels of CO and are most frequent in the summer, at night and early in the morning. The reduction of CO emissions for 1980–90 (Figure 19.2), and the corresponding decrease in ambient CO concentrations, is greatest over the latter five years (Figure 19.4).

Oxides of nitrogen

Emissions The major source of oxides of nitrogen (NO_x) in Moscow is electricity generation which accounts for 70 per cent of the total anthropogenic NO_x emissions of 210,000 tonnes per annum. The contribution of motor vehicles to total anthropogenic NO_x did not exceed 19 per cent in 1990. Emissions of NO_x from electricity generation and motor vehicles increased by 8 and 3 per cent respectively between 1980 and 1990, but emissions from industrial sources diminished by 12 per cent (Bekker and Boikova, 1990; Rovinsky et al., 1991).

Ambient Concentrations The measurements of NO_2 are made using the Gries-Illosvay reagent procedure. Air is sampled using a thin-film sorbent. Mean daily NO_2 concentrations exceeded the national air quality standard (40 µg m^{-3}) in the central, north-western, and eastern sectors of the city. Daily mean NO_2 concentrations (mean daily) reached 100-150 µg m^{-3} which is comparable to the air quality guideline recommended by WHO (150 µg m^{-3}) (CVGMO, 1981, 1982, 1983, 1984, 1985, 1986; MCGNPS, 1987, 1988, 1989, 1990; MCGNPS, 1991).

Nitrogen dioxide concentrations increased in all sectors of the city between 1986–90 (Figure 19.5). Figure 19.2 shows the increase in NO_x emissions over this period. The highest NO_2 levels occur between February and May as a result of low wind speed and surface temperature inversions.

The reduction of NO_2 pollution should be a priority for improving the air quality of the city.

Ozone

Ozone (O_3) is not considered a problem in Moscow. Occasional studies show that, even when suitable meteorological conditions occur, levels rarely exceed guidelines. Scavenging of O_3 by NO is thought to be an important factor. The contribution of the "urban plume" to downwind rural concentrations is not known. Total hydrocarbon measurements show that the evaporation of fuel from storage tanks and industry is high in comparison with other urban areas in the Russian Federation.

19.4 Conclusions

Air Pollution Situation Generally, ambient pollutant concentrations are falling in Moscow. Motor vehicles account for a large proportion of anthropogenic emissions. Power stations and industry also make a significant contribution to pollutant emissions in the city (Figure 19.2) (Bekker and Boikova, 1990). The electricity generating stations in the city appear to be the main emission sources of NO_x, SO_2 and SPM. Changes in fuel consumption over the last decade, with wider use of natural gas instead of coal and heavy fuel oil, have had a great influence on emissions. The increase in NO_x emissions since 1985 could be as a result of natural gas use, but may also be linked with other air pollution sources.

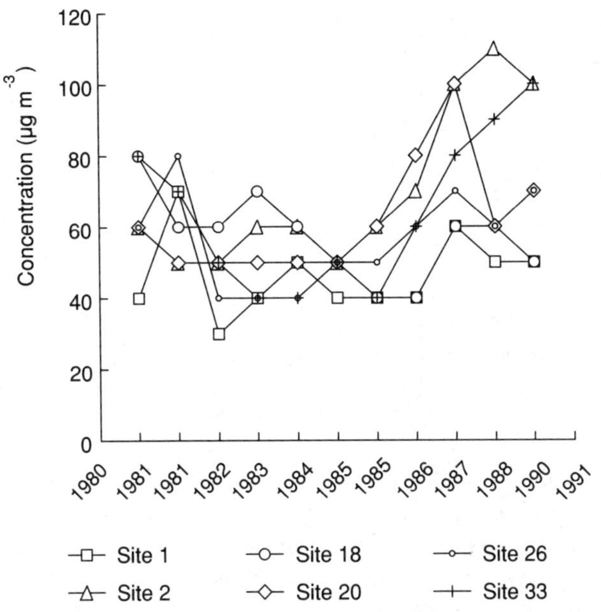

Figure 19.5 *Annual mean nitrogen dioxide concentrations at selected sites*

After Rovinsky et al., 1991

Main Problems Comparisons of ambient air quality data and emissions estimates with cities of similar size and socio-economic structure suggest that monitoring data values are rather low, especially for SO_2. There may be problems in the monitoring methodology. It is not clear whether these problems arise from the chemical analysis of the samples or in the subsequent interpretation of the data. It is suggested that quality assurance (and quality control) audits are performed to test the current monitoring methodology against internationally accepted (ISO) methods. The introduction of an automated monitoring network would greatly improve Moscow's air quality management options.

19.5 References

Bekker, A. A. and Boikova, R. A. 1990 *Basic problems of improving air and aquatic basins of Moscow and Moscow region*, All-Union Institute of Scientific and Technical Information (VINITI) **9**, Moscow (in Russian).

CVGMO 1981 *Review of the atmospheric pollution and harmful substance emissions in Moscow city, towns and settlements of Moscow region for 1980*, Central High-altitude Hydrometeorological Observatory, Moscow (in Russian).

CVGMO 1982 as CVGMO, 1981, for 1981.

CVGMO 1983 as CVGMO, 1981, for 1982.

CVGMO 1984 as CVGMO, 1981, for 1983.

CVGMO 1985 as CVGMO, 1981, for 1984.

CVGMO 1986 as CVGMO, 1981, for 1985.

Finansyi i Statistika 1990 *Moscow in Figures* (Statistics annual), Moscow, Finansyi i Statistika (in Russian).

Gidrometeoizdat 1979 *Manual for atmospheric pollution observations*, Gidrometeoizdat, Leningrad (in Russian).

Gidrometeoizdat 1989 *Temporal methodics for chemical analysis of the atmosphere samples with sampling on solid film sorbents*, Gidrometeoizdat, Leningrad (in Russian).

MCGNPS 1987 *Review of the atmospheric pollution in Moscow city and towns and settlements of the Moscow region for 1986*, Moscow Centre for Hydrometeorology and Environmental Observations, Moscow (in Russian).

MCGNPS 1988 as MCGNPS, 1987, for 1987.

MCGNPS 1989 as MCGNPS, 1987, for 1988.

MCGNPS 1990 as MCGNPS, 1987, for 1989.

MCGNPS 1991 *Yearbook of the atmospheric pollution in Moscow city and towns and settlements of the Moscow region for 1990*, Moscow Centre for Hydrometeorology and Environmental Observations, Moscow (in Russian).

Osipov, Yu. S. 1989 *The state of the environment over the Moscow agglomeration*, All-Union Institute of Scientific and Technical Information (VINITI) **1–2**, Moscow (in Russian).

Rovinsky, F. Y., Egorov, V. I. and Starodubsky, I. A. 1991 *Urban Air Quality in Moscow*, Institute of Global Climate and Ecology, Moscow. (Unpublished report).

UN 1989 *Prospects of World Urbanization 1988*, Population Studies No. 112, United Nations, New York.

20

New York

20.1 General Information

Geography New York City (NYC) is located 10 m above mean sea level in the north-eastern United States (Latitude 40°45'N, Longitude 74°00'W), at the mouth of the Hudson River which flows into the Atlantic Ocean. The city consists of five boroughs – Manhattan, The Bronx, Brooklyn, Queens and Staten Island and covers an area of 787 km^2. Manhattan is an island bounded by the Hudson River and East River; traffic to and from Manhattan is by several bridges and tunnels. Areas of northern New Jersey and of south-west Connecticut are in the airshed and contribute emissions which can affect air quality in NYC. The New York City Metropolitan Area (NYCMA) which contains parts of New York State including Long Island and areas of New Jersey and Connecticut (Figure 20.1) is quite level and there are no mountains that provide barriers to pollutant transport to and from the area. The NYCMA covers an area of 3,585 km^2.

Demography New York has always been one of the largest cities of the USA and its pattern of development has been established for many decades. Recent population growth from immigration has been partially offset by outward migration. The population and projected growth of the New York urban agglomeration is given in Table 2.1 (UN, 1989). This classification is different to that for New York City (1990 population 7.34 million) and NYCMA (1990 population 18.7 million).

Climate New York has a continental warm summer climate with cold winters and hot humid summers. Annual monthly temperatures range from zero to 25°C. Annual mean precipitation is 1,120 mm. There is good ventilation of the area from the ocean, but periods of atmospheric temperature inversion and stagnation can occur in both winter and summer. Extremes of cold weather result from air moving south-east from the Hudson Bay region of Canada. During winter the inversion conditions, when prolonged, led to the high pollution episodes of the 1950s and 1960s. Extremes of hot weather are associated with air masses moving overland from a Bermuda high pressure system. Summer stagnation periods lead to high ozone (O_3) concentrations.

Industry New York City is a major world trade and financial centre. Industries have concentrated in the urban area (as with most megacities) because of the proximity to major markets and easy access to transport. Consumer-related activities and light industry predominate within the city. These include clothing and other finished products; printing and publishing; and food products. Heavy industries, primarily chemicals and petroleum refining, are located across the Hudson River in the Bayonne-Elizabeth section of New Jersey. Other industries include machinery manufacture; textile products; leather products; paper products; motor vehicle and aircraft manufacturing; and shipbuilding. Electrical power generation by the Consolidated Edison Corporation has been moved from within the city to outer regions with limited direct impact on the city.

Transport New York City is a major air, water and land transport centre of the east coast of America. The J. F. Kennedy airport in Queens is one of the busiest in the world and other airports, La Guardia and Newark, also serve the metropolitan area. New York Harbour is one of the world's busiest and shipping emissions, along with the airport emissions, make a noticeable contribution to the emissions of the area. New York City has an extensive rapid transit network with subways and bus lines throughout, and commuter bus and rail lines link it to Long Island, upstate New York, Connecticut and New Jersey.

The use of motor vehicles in the NYCMA is very difficult because of the extreme congestion, especially in Manhattan. The limited availability of on-street parking and high cost of off-street parking makes Manhattan, with a rate of 12 vehicles per 100 people, one of the lowest areas of motor vehicle registration in the USA. Although the population of NYC is growing, the number of registrations of private vehicles declined between 1970 and 1980, but has started to grow again in the last decade. In 1985 motor vehicle registrations in NYC totalled 1.62

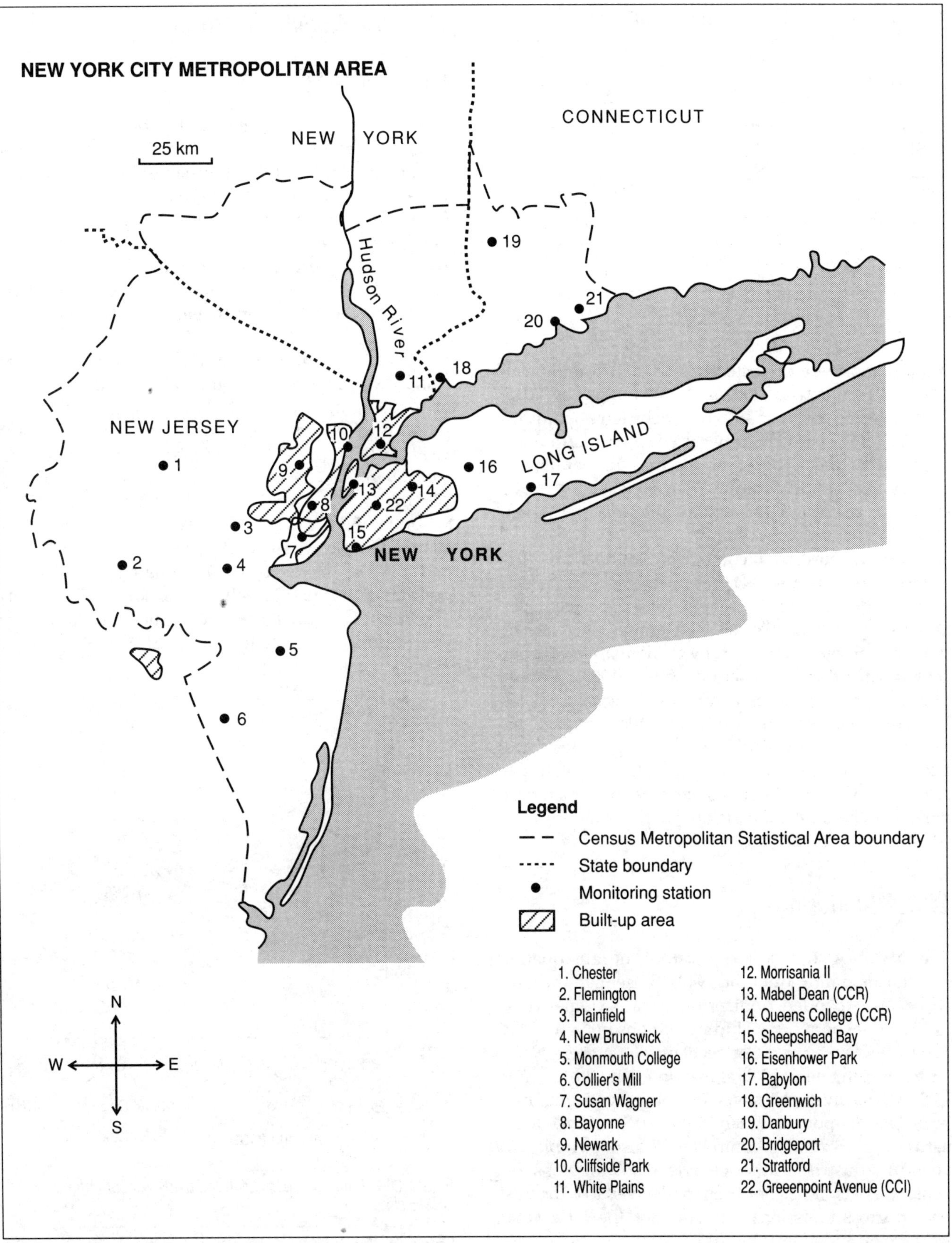
NEW YORK CITY METROPOLITAN AREA
25 km
NEW YORK
CONNECTICUT
Hudson River
NEW JERSEY
LONG ISLAND
NEW YORK
N
S
W
E
Legend
Census Metropolitan Statistical Area boundary
State boundary
Monitoring station
Built-up area
1. Chester
2. Flemington
3. Plainfield
4. New Brunswick
5. Monmouth College
6. Collier's Mill
7. Susan Wagner
8. Bayonne
9. Newark
10. Cliffside Park
11. White Plains
12. Morrisania II
13. Mabel Dean (CCR)
14. Queens College (CCR)
15. Sheepshead Bay
16. Eisenhower Park
17. Babylon
18. Greenwich
19. Danbury
20. Bridgeport
21. Stratford
22. Greeenpoint Avenue (CCI)

Figure 20.1 *Sketch map of New York City Metropolitan Area*

Table 20.1 Gross estimated emissions, 1985 (10^3 t a^{-1})

Pollutant	New York City	New York Metropolitan Area
SO_x	55	349
SPM (PM_{10})	56	289
CO	368	1,752
NO_x	121	513
VOC (ROG)	217	872

After US NAPAP, 1985

million and were projected to rise to 1.78 million in 1990, 1.91 million by 2000 and 2.09 million by 2015. The projected increase in motor vehicle registrations in NYC of 25 per cent between 1985 and 2015 compared with the 4 per cent increase in population during this period indicates that congestion can be expected to expand throughout the area (Bergman, 1988).

The transport problem is compounded by the deterioration of the NYC subway system which shows a decrease in passengers of 1.5 per cent per annum from 1948-1988 and an annual increase of daily motor vehicle traffic flow to Manhattan of 2 per cent during the same period (Allen, 1989). Consequently strict emission controls on motor vehicles will be required to counteract the increase in registered vehicles and lower average speed expected. New York State has recently adopted the stricter California emission standards for new motor vehicles to be sold and registered in the State.

20.2 Monitoring

The NYCMA has a long history of air quality monitoring dating from the 1940s. Since then the network has expanded from primarily a NYC network to a regional network including stations in New York State, New Jersey and Connecticut. The present NYC air monitoring network is shown in Figure 20.1.

An extensive emission control programme has been based upon a switch from coal to fuel oil to natural gas. Emissions controls have also been placed on local industries. Although a detailed emissions inventory exists, it has not been made available for this report; gross emissions estimates for the base year 1985 are provided in Table 20.1.

20.3 Air Quality Situation

Sulphur dioxide

Emissions Anthropogenic oxides of sulphur (SO_x) emissions totalled an estimated 54,000 tonnes per annum in NYC in 1985. Emissions for the NYCMA were far greater at almost 350,000 tonnes per annum (Table 20.1). Sulphur dioxide (SO_2) emissions have decreased significantly over the past 20 years as a result of lowering of the sulphur content of bituminous coal and residual fuel oil (USEPA, 1991b), followed by a shift to natural gas and light oils.

Ambient Concentrations Annual mean SO_2 concentrations have been within the WHO guideline range of 40-60 μg m^{-3} since 1985 (Figure 20.2). A rare high pollution episode owing to stagnation in January 1989 caused daily SO_2 levels to reach 230 μg m^{-3} and 190 μg m^{-3} at Greenpoint Avenue (a maximum of 269 μg m^{-3} was reached at the Mabel Dean High School) (NYDEC, 1990). The SO_2 and total suspended particulates (TSP) problems of the 1950s and 1960s have been eliminated and these pollutants are now largely under control in winter. The highest annual arithmetic mean reported by USEPA in 1989 was 60 μg m^{-3} (USEPA, 1991b).

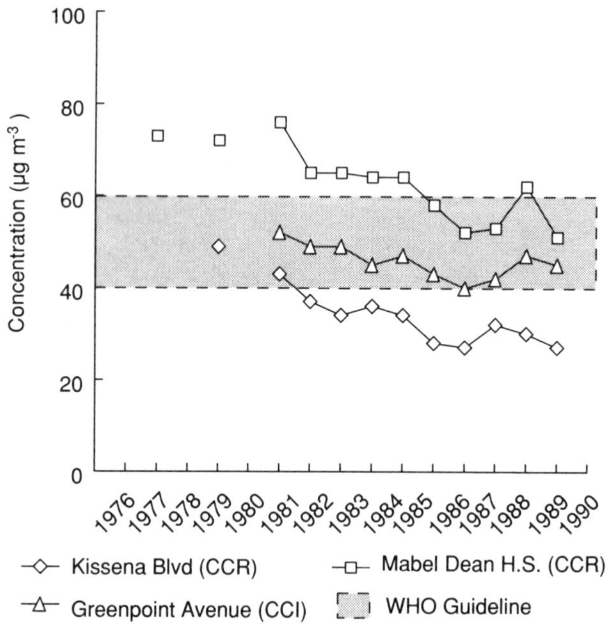

Figure 20.2 Annual mean sulphur dioxide concentrations at GEMS sites

Source: GEMS data

Suspended particulate matter

Emissions In 1985 total estimated emissions of particulate matter <10 μm (PM_{10}) in NYC were approximately 55,000 tonnes per annum and in NYCMA approximately 290,000 tonnes per annum.

Over the past 25 years (1965–1990) a striking reduction in annual TSP has occurred in New York State and the NYCMA via enforcement of both Federal and New York State particulate emission regulations (USEPA, 1991b). These included laws concerning incinerators, industrial processes, fossil fuel combustion and diesel-fuelled motor vehicles. The minimization of allowable fuel sulphur content also caused a shift to cleaner, less particulate-emitting fuels such as natural gas.

Ambient Concentrations Figure 20.3 shows the annual mean TSP monitored at three GEMS/Air stations as decreasing slightly from 1975 to 1989, all well below the upper WHO guideline of 90 μg m^{-3} annual arithmetic mean (but within the guideline range of 60–90 μg m^{-3} annual arithmetic mean). Figure 20.4 shows the daily variation of TSP at Greenpoint Avenue (City Centre Industrial) during 1989. The maximum observed value of 150 μg m^{-3} is well under the WHO 98 percentile guideline and the former pattern of winter maxima for TSP has almost disappeared. The NYCMA now monitors PM_{10} and in 1989 the maximum annual arithmetic mean was 44 μg m^{-3} with a daily maxima of 130 μg m^{-3} at the PS 314 station (NYDEC, 1990). In 1989 the USEPA reported a maximum annual mean in NYC of 66 μg m^{-3} – the National Ambient Air Quality Standards (NAAQS) is 50 μg m^{-3} (USEPA, 1991b).

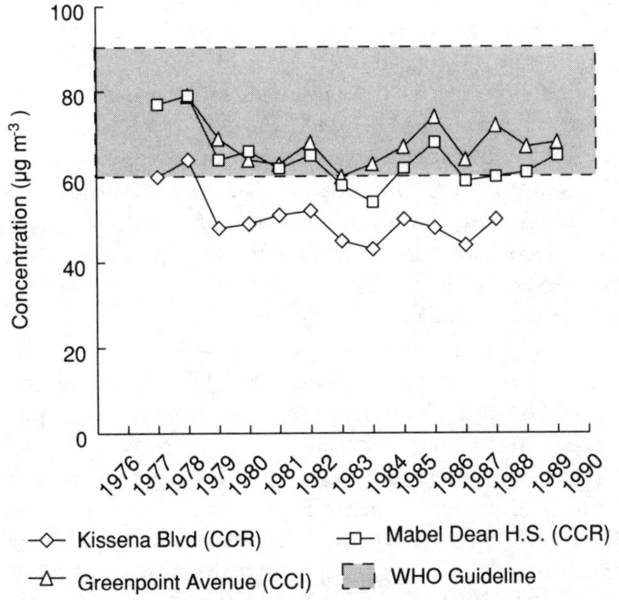

Figure 20.3 *Annual mean suspended particulate matter (TSP) concentrations at GEMS sites*

Source: GEMS data

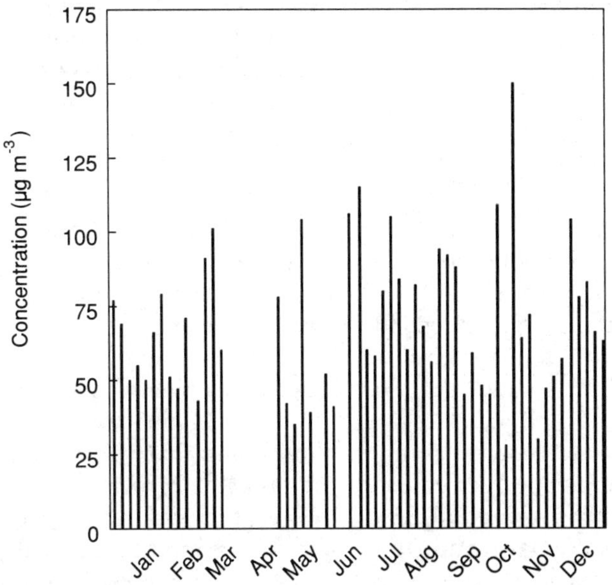

Figure 20.4 *Daily suspended particulate matter (TSP) concentrations at Greenpoint Avenue (CCI) 1989*

Note: 6-day sampling frame. Zero levels indicate days when no measurements were taken.
Source: GEMS data

Lead

Emissions Lead (Pb) emissions are now primarily from motor vehicles using leaded fuel, which has a low lead content of 0.026 g l^{-1}. Emissions estimates for Pb are not available.

Ambient Concentrations The USEPA report that the NAAQS and the WHO guideline for Pb are met in the NYCMA. In 1989 the quarterly maximum Pb concentration in NYC was 0.12 μg m^{-3}. However, in Newark, New Jersey the quarterly maximum reached 0.41 μg m^{-3} (USEPA, 1991b).

Carbon monoxide

Emissions Carbon monoxide (CO) emissions in NYC were estimated to be around 368,000 tonnes per annum in 1985 and over 1.75 million tonnes per annum in the NYCMA. Carbon monoxide has decreased significantly in the NYCMA primarily through the Federal automotive emission controls (NYDEC, 1984). New York State has recently adopted the California emissions standards for all new cars sold which, along with some petrol formulation changes, should help to meet the NAAQS in the future.

Ambient Concentrations The number of days on which either the CO standard of 9 ppm (11.25 mg m^{-3}) eight-hour mean or 35 ppm (43.75 mg m^{-3}) one-hour mean were exceeded dropped from 94 in 1980 to 4 in 1989 at the special trend stations as shown in Figure 20.5. In 1989 the second highest eight-hour mean at the 59th Street-Lexington Avenue, Manhattan station was 12 ppm (15 mg m^{-3}) indicating that vehicle CO emission reductions of the order of 25 per cent are still required (USEPA, 1990). Strict enforcement of traffic laws and controls, such as banning of private cars in certain areas, should encourage the use of public transport and reduce traffic congestion.

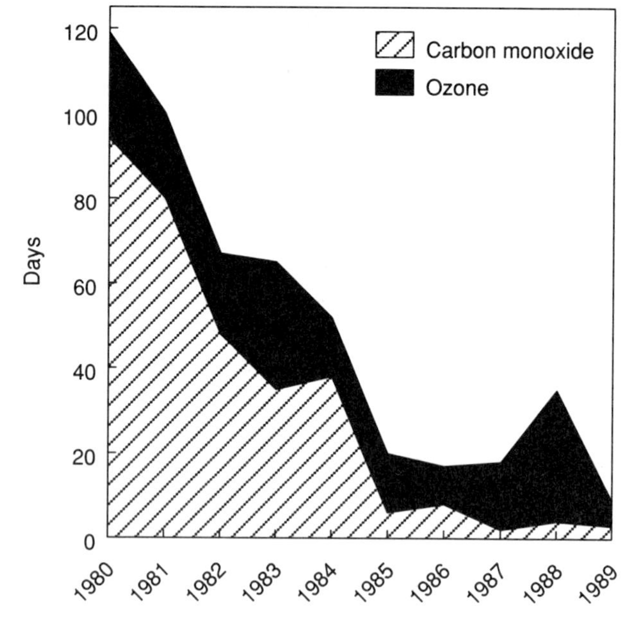

Figure 20.5 *Number of days pollutants standard index was greater than 100 at CO and O_3 trend sites*

After USEPA, 1991b

Oxides of nitrogen

Emissions It is estimated that man-made oxides of nitrogen (NO_x) emissions in NYC totalled over 120,000 tonnes per annum in 1985 and over 513,000 tonnes in the NYCMA. Once again the strict Californian emissions standards for all new cars sold, recently adopted by New York State, should help to reduce NO_x emissions in the future.

Ambient Concentrations Nitrogen dioxide (NO_2) concentrations in the NYCMA have decreased, through motor vehicle emission controls, as part of the national strategy to meet O_3 air quality standards. In 1990, NYC met the US national ambient air quality standard of 94 μg m^{-3} annual arithmetic mean with a value of 87 μg m^{-3}. However, the WHO guidelines for maximum allowable 24-hour mean (150 μg m^{-3}) and one-hour mean (400 μg m^{-3}) were not met during 1990; daily maximum of 160 μg m^{-3} and hourly maximum of 402 μg m^{-3} were recorded.

Ozone

Emissions Ozone is formed in the troposphere by complex photochemical reactions between NO_x and reactive organic compounds (ROCs) in the presence of sunlight. Annual emissions of ROCs in 1985 totalled 216,000 tonnes in NYC and over 870,000 tonnes in NYCMA.

The fall in O_3 maxima discussed below is primarily a result of automotive emission controls on NO_x and ROCs, plus stricter inspection maintenance of motor vehicles and a requirement for "Reasonably Available Control Technology " (RACT) to be applied to stationary hydrocarbon emission sources. The NYCMA, by adopting the stricter California standards for new motor vehicles, will further reduce vehicle emissions but the increase in vehicles and traffic congestion (more time spent at low-speeds or idle) may prevent the standards from being met. A further complication to this situation is the chemical structure of the air mass reaching the NYCMA. Passage over "upwind" industrial and metropolitan areas which extend from Washington DC to Boston MA means that the NYCMA daily emissions of NO_x and hydrocarbons are injected into an air mass already containing these precursors plus an initial content of O_3 remaining from the previous day's reactions.

Ambient concentrations Areas of O_3 maxima in the NYCMA are "downwind" of the city proper depending on wind direction, e.g., in New Jersey or Connecticut. For example, in 1990 O_3 maxima of 0.197 ppm (421 μg m^{-3}) occurred on 4 August in Madison CT (about 80 miles from NYC) and on 11 September at Monmouth College, NJ whilst the NYC and New York State stations were all under 0.12 ppm (257 μg m^{-3}) (USEPA, 1991a). Figure 20.5 shows that the O_3 situation is improving as the number of violations of the NAAQS (0.12 ppm O_3 one-hour mean, the expected value not to be exceeded more than once per year) has decreased from almost 120 to less than 10. The design value for O_3 control has also decreased from 0.272 ppm to 0.197 ppm during this period, representing a required 40 per cent reduction in O_3 to meet the 0.12 ppm standard (NYDEC, 1990).

20.4 Conclusions

Air Pollution Situation Prior to the 1950s coal was the primary fuel for industry, electric power generation and domestic heating. In NYC there was a long series of winter pollution episodes of high TSP and SO_2 concentrations caused by a combination of persistent stagnation and inversion. Only the good atmospheric dispersion of the area prevented air pollution disasters from reaching the magnitude of the 1952 London episode when thousands of people were seriously affected. Major episodes took place in November 1953 and January/February 1963, culminating in an episode during the 1966 Thanksgiving weekend (last weekend of November) which resulted in significant excess mortality and morbidity for NYCMA residents. In the late 1960s when a change to fuel oil was under way, many apartment houses were still heated by coal and the power plants which supplied electricity for the NYCMA were partially coal-fired. Since then strict pollution controls have been enforced resulting in the phasing out of coal and the introduction of natural gas and low-sulphur fuel oil for domestic heating and power production.

By 1975, the excessive air pollution had been controlled and air quality management programmes were instituted that met the NAAQS for TSP. By 1984 WHO guidelines for SO_2 were met (USEPA, 1991b). The major problems in the NYCMA today are associated with the primary automotive pollutants CO and the secondary pollutant O_3 produced from the automotive emissions of ROCs and NO_x.

Main Problems The NYCMA has solved the air quality problems of the 1950s and 1960s by significantly reducing fossil fuel combustion emissions of TSP and SO_2. The remaining air quality problems of the 1980s and 1990s are related to emissions from motor vehicles. Although implementation of plans to reduce air pollution from motor vehicles has produced marked reductions in CO and O_3, further significant reductions in emissions will be required to meet the NAAQS. In summary, extensive emissions controls are still required in the NYCMA, as well as in surrounding areas, to protect the "down-wind" areas from excessive O_3 exposures.

Health Effects Although the winter air pollution episodes of the 1950s and 1960s were studied in great detail by epidemiologists, there are few new data on health effects of more recent air pollution. Modern problems may relate more to high O_3 in summer which, with high temperature and humidity, leads to respiratory problems of the elderly and other sensitive individuals.

20.5 References

Allen, J. G. 1989 *Regional Transportation Status 1988*, PT 1244801, New York Metropolitan Transportation Council, New York.

Bergman, J. E. 1988 *Highway Travel Forecasts 1980-2015*, PT 1239801, New York Metropolitan Transportation Council, New York

NYDEC 1984 *Air Quality Implementation Plan, Control of Carbon Monoxide and Hydrocarbons in New York City Metropolitan Area*, AIR P-174 (8/83), New York Department of Environmental Conservation, New York.

NYDEC 1990 *New York State Air Quality Report, Ambient Air Monitoring System Annual 1989*, DAR-90-1, New York Department of Environmental Conservation, New York.

UN 1989 *Prospects of World Urbanization 1988*, Population Studies No. 112, United Nations, New York.

USEPA 1990 *New York City Carbon Monoxide Traffic Site Survey 1981-1989*, US Environmental Protection Agency Region II, New York.

USEPA 1991a *Ozone Air Quality 1990: New Jersey and New York*, US Environmental Protection Agency Region II, New York.

USEPA 1991b *National Air Quality and Emissions Trends Report 1989*, US Environmental Protection Agency, Research Triangle Park.

US NAPAP 1985 *Emissions Inventory*, US National Acid Precipitation Assessment Program, Washington DC.

21

Rio de Janeiro

21.1 General Information

Geography The metropolitan area of Rio de Janeiro is located at Latitude 22°54'S, Longitude 43°10'W on the Atlantic coast of Brazil. It comprises the following municipalities (Figure 21.1): Rio de Janeiro, Niterói, São Gonçalo, São João de Meriti, Duque de Caxias, Nova Iguaçu, Nilópolis, Itaguaí, Itaboraí, Petrópolis, Magé, Maricá, Mangaratiba and Paracambi. Rio de Janeiro is situated on the Bay of Guanabara. The Pão de Açúcar (Sugar Loaf) is situated at the entrance of the bay and the city lies on a plain on the western shore which is interrupted by several rocky mountains, an offshoot of the Serra do Mar mountain range. The urban areas range from sea level to 885 m in altitude.

Demography The total population of the 14 municipalities is 11.12 million, 80 per cent of the total population of the State of Rio de Janeiro and is projected to rise to 13 million by 2000 (UN 1989). This population is unevenly distributed as almost 60 per cent lives in the city of Rio de Janeiro (area 1,171 km^2). The total area covered by the metropolitan area is approximately 6,500 km^2, corresponding to 14 per cent of the State's total surface area.

Climate The local climate is tropical rain forest with monthly mean temperatures varying between 21°C (June–July) and 26°C (January–February). On an annual basis there are some 2,250 hours of sunshine with a low of 150 hours per month in August–September and a high of 200 hours per month in January–March. Space heating is thus not required and plays no role in local air pollution problems. The annual total amount of precipitation fluctuates around 1,000 mm with monthly extremes of 150 mm per month in the January–March quarter and under 50 mm per month in the June–August period. The climate is heavily influenced by an onshore/offshore sea breeze.

Industry Important industries include metallurgy, machinery, clothing and footwear, textiles and mineral products. Rio de Janeiro is also a major financial centre. The city has one of the highest per capita incomes in Brazil and retail trade is substantial.

21.2 Monitoring

The metropolitan area can be subdivided into four sub-regions I to IV (FEEMA, 1989) as shown in Figure 21.1. With 18 monitoring sites, most of the monitoring activity has concentrated in Region III, where the first sulphur dioxide (SO_2) and suspended particulate matter (SPM) measurements were reported in 1967 (OPS, 1982). In contrast, Region I has only two regular monitoring sites, Region II has none and Region IV has only one.

Monitoring started in 1967 in the city of Rio de Janeiro with seven sites where ambient SPM levels were determined as daily averages (high-volume sampling technique). With the support of the Pan American Health Organization (PAHO) (Alvarez, 1978), those seven sites were combined in 1968 into one monitoring network with one sulphur-smoke sampling station (in CENTRO or GEMS1,) and 20 stations for the monthly determination of settled dust and sulphates. From 1980 this network was gradually scaled-up with the addition of a number of sites outside the city of Rio de Janeiro. By 1982 the maximum configuration was reached with 39 sites for settled dust, 23 for sulphates, 23 for SPM, 13 for SO_2 and one for carbon monoxide (CO). Table 21.1 summarizes the characteristics of the SPM and SO_2 sites operational in 1984. Of the 24 sites listed, 16 are located in the city of Rio de Janeiro, while only 8 cover 6 of the neighbouring municipalities. The 3 sites also reporting to GEMS are located in the city of Rio de Janeiro. GEMS1, a Centre City Commercial (CCC) site, corresponds to CENTRO in Table 21.1, GEMS2, a Suburban Industrial (SI) site to Penha and GEMS3, a Suburban Residential (SR) site to Copacabana.

In 1975 Findacao Estadual de Engenharia do Meio Ambiente (FEEMA) received an automatic monitoring station for CO, hydrocarbon (HC), ozone (O_3), SO_2, total sulphur, hydrogen sulphide (H_2S), nitric

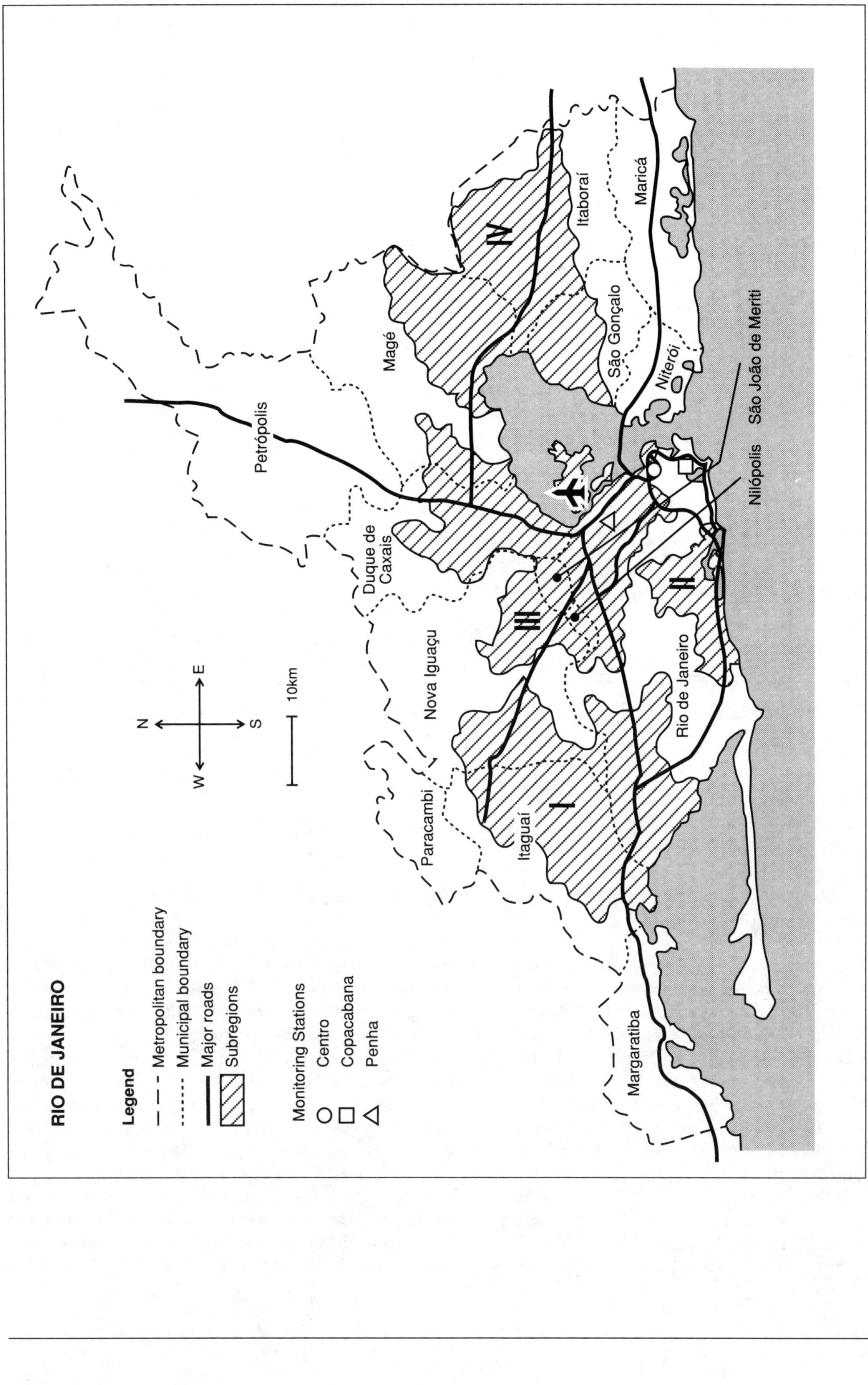

Figure 21.1 Sketch map of metropolitan area of Rio de Janeiro

***Table 21.1** Suspended particulate matter and sulphur dioxide monitoring sites, 1984*

Site	SPM	SO_2	h(m)[1]	d(m)[2]	Influencing Sources
Bangu	x		26	300	Reduced traffic
Benfica	x	x	20	30	Intense traffic, refinery
Bonsucesso	x	x	4	50	Intense traffic, industry
Centro (GEMS1)	x	x	15	20	Intense traffic
Copacabana (GEMS3)	x	x	3	5	Intense traffic
EMPA	x		12	50	Intense traffic
Ilha do Governador	x	x	3	500	Airport, refinery
Iraja	x	x	27	100	Intense traffic
Maracanã	x	x	9	3	Intense traffic
Meier II	x	x	20	10	Regular traffic, industry
Penha (GEMS2)	x		8	50	Regular traffic, industry
Rio Comprido	x	x	40	100	Industry, traffic
Santa Tereza	x		6	6	Little traffic
São Cristóvão	x		5	5	Industry, traffic
Santa Cruz	x	x	12	50	Industry, traffic
Inhauma	x		35	20	Reduced traffic
Itaguaí		x	n.a.	n.a.	n.a.
Niterói	x		26	4	Regular traffic
Nova Iguaçu	x		15	4	Regular traffic
Duque de Caxias 1	x	x	15	50	Reduced traffic
Duque de Caxias 2		x	5	20	Refinery, traffic
Nilópolis	x		20	20	Regular traffic
São Gonçalo	x		15	30	Intense traffic
São João de Meriti	x		20	30	Reduced traffic

[1]h(m): sampling height above ground level
[2]d(m): distance to traffic
x: indicates pollutant monitored
n.a.: no data available

After FEEMA, 1989

oxide (NO) and nitrogen dioxide (NO_2) from the World Health Organization (WHO). However, because of instrumentation problems and through lack of technical back-up and spare parts, only partial results were obtained during the first year of operation.

Within the framework of a Brazilian-Japanese co-operative programme, a second fully automatic monitoring site for the main traffic-related pollutants was installed in the suburb of Gavea (Figure 21.1) in 1986. The lack of spare parts again proved to be a major problem when operating this station continuously in order to generate consistent data sets. In 1987 a co-operative agreement between Brazil and the Federal Republic of Germany resulted in the acquisition of a full-size mobile laboratory equipped with meteorological instrumentation and CO, O_3, SO_2 and SPM monitors. This laboratory is used at various locations to study the local influence of heavy traffic (FEEMA, 1989).

21.3 Air Quality Situation

Sulphur dioxide

Emissions Sulphur dioxide emissions are mainly caused by the industrial use of heavy fuel oil, with a sulphur content of up to 5 per cent, and coal with up to 3 per cent sulphur content. Emissions from mobile sources contributed significantly (55 per cent) to the "area source" sector. Diesel-fuelled motor vehicles, and other small but multiple (area) sources, account for 16 per cent of the total SO_2 emissions (Table 21.2), but can contribute significantly to local ambient SO_2 concentrations due to their street-level emission height. An up-to-date emissions inventory is not available so the downward trend noted in Table 21.3 for the industrial oxides of sulphur (SO_x) and SPM emissions has not been confirmed by more detailed analysis.

Table 21.2 *Emissions Inventory 1978 for the metropolitan area of Rio de Janeiro (10^3 t a^{-1})*

	SPM		SO_x		CO		NO_x		HC	
Municipality	PS	AS	PS	AS	PS	AS	PS	AS	PS	AS
Rio de Janeiro	54	27	62	17	1	487	5	33	8	107
Duque de Caxias	5	6	67	3	1	23	9	3	13	5
Itaboraí	8	1	7	–	–	4	1	–	1	1
Itaguaí	1	2	1	1	–	6	–	1	–	1
Magé	8	1	2	–	–	5	–	–	–	1
Mangaratiba	1	–	–	–	–	1	–	–	–	–
Maricá	1	1	–	–	–	2	–	–	–	–
Nilópolis	–	1	–	–	–	4	–	–	–	1
Niterói	3	2	1	2	–	37	–	3	–	8
Nova Iguaçu	21	7	4	2	–	24	–	3	–	6
Paracambi	–	1	–	1	–	1	–	–	–	–
Petrópolis	1	3	2	1	–	19	–	1	–	4
São Gonçalo	35	2	11	3	1	18	1	2	–	4
São João de Meriti	–	2	–	1	–	8	–	1	–	2
Subtotals	138	56	157	31	3	639	16	47	22	140
Total	194		188		642		63		162	

Note: Emissions smaller than 500 tonnes are not reported
Point sources (PS) – industrial stack releases
Area sources (AS) – the composite of commercial and residential sources, as well as mobile (traffic) and fugitive sources (road dust).

After Lopes et al., 1980

Table 21.3 *Trend in industrial point source emissions (10^3 t a^{-1})*

Pollutant	1978	1981	1983
SO_x	149 (157)	117	98
SPM	122 (138)	71	66
CO	3 (3)	4	2
HC	25 (22)	39	33

Note: Data from Table 21.2 appears in brackets
Sources: Lopes et al., 1982; FEEMA, 1989

Ambient Concentrations Ambient SO_2 levels exceed WHO guidelines and Brazilian air quality standards at many different locations. Until 1978 only the GEMS1, or CCC site at Centro (Table 21.1), generated daily SO_2 concentrations. In 1978 the network shown in Table 21.1 became operational. Owing to many start-up problems, it was three years before representative yearly data sets of daily means were obtained from all sites. From 1981 to 1984 the network was fully operational, but in 1985 all monitoring seems to have ended (FEEMA, 1989). Available information is thus limited to 1967 (one site), 1978 (three

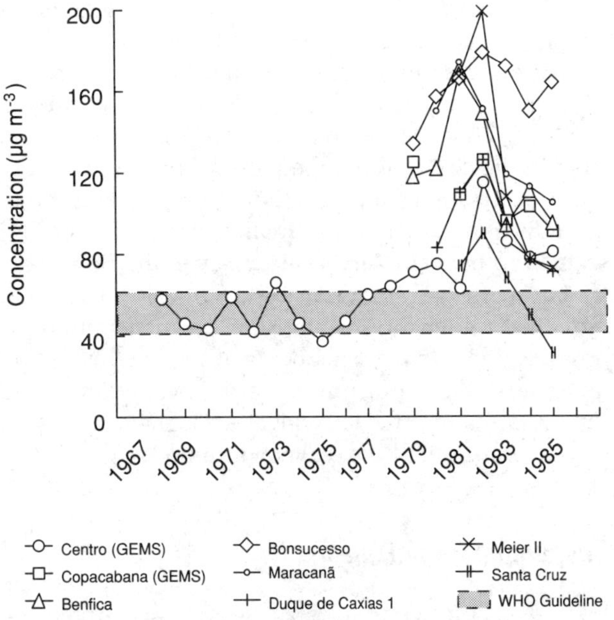

Figure 21.2 *Annual mean sulphur dioxide concentrations at selected sites*

Sources: GEMS data; Justen, 1991; OPS, 1982

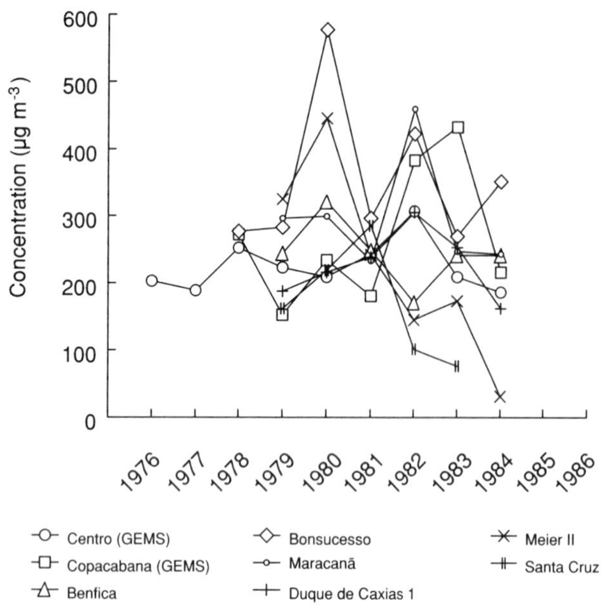

***Figure 21.3** Annual maximum sulphur dioxide concentrations at selected sites*

Sources: GEMS data; Justen, 1991

sites), 1984 (13 sites) (Figure 21.2). Maximum daily levels are also presented in Figure 21.3. Data from the GEMS1 site suggest an upward trend till 1980/1981, followed by a downward trend until the end of monitoring in 1984.

The WHO annual and daily guidelines have been exceeded since 1976 in all sites, with the exception of Santa Cruz and Itaguaí in 1983 and 1984. The national primary air quality standard for the annual mean was not met in Benfica, Bonsucesso, Copacabana (GEMS3), Iraja, Maracana and Rio Comprido, all six sited in the city of Rio de Janeiro. With respect to extreme daily values the situation improved with time. By 1984 the national air quality standard (NAQS) seemed to be met everywhere, as even the Bonsucesso site, with its measured maximum of 351 µg m^{-3}, remained under the specified 365 µg m^{-3} level not to be exceeded more than once a year. However, the data sets are sometimes very incomplete. The recorded maximum daily level at a given site is in fact a lower limit of the real maximum.

Suspended Particulate Matter

Emissions Fugitive (road dust) sources accounted for more than 56,000 tonnes per annum of total SPM emissions in 1978. Total SPM emissions from industrial sources were estimated to be approximately 66,000 tonnes per annum in 1973, compared with 122,000 tonnes per annum in 1978.

Ambient Concentrations Combining data and information from different sources (OPS, 1982; FEEMA, 1989; Justen, 1991), made it possible to reconstruct the time series of annual arithmetic mean levels and daily maxima (Figure 21.4 and Figure 21.5). Most of the annual data sets of daily means are highly incomplete, as the local sampling scheme of one filter every sixth day is seldom realized. Consequently, some annual data sets only have some 10–15 valid daily measurements.

The trend in mean SPM levels (Figure 21.4) over the past 20 years seems to have two components: a stationary or slightly upward trend up to the end of the 1970s, followed by a downward trend leading to a stationary situation by the end of the 1980s. The restricted number of available data years, and the incompleteness of the individual data sets (with n = 10–60 days per annum) mean that the trend in Figure 21.5 of the measured maximum daily levels is not clear. Fluctuating patterns are noted, with large year to year variations in a given site. The lower 150 µg m^{-3} limit of the WHO daily guideline is not met in at least 50 per cent of the sites, while the Brazilian air quality standard of 240 µg m^{-3}, not to be exceeded more than once a year, is certainly exceeded at more than one location.

Despite a decrease in recent years, mean SPM levels are still exceeding the WHO guidelines and the Brazilian air quality standards in a number of sites (Figure 21.6). Bonsucesso, Duque de Caxias, EMPA, Iraja II, Inhauma, São João de Meriti and São Gonçalo are the most polluted sites. Figure 21.4 clearly illustrates that the SPM levels measured in the three GEMS sites are low and that at least half of the non-GEMS sites show higher mean and extreme levels (Figure 21.5). The lowest levels of SPM pollution are recorded in the Santa Tereza site, with means under 60 µg m^{-3} and extremes below 150 µg m^{-3}, followed by Santa Cruz and Bangu.

A more detailed analysis on a monthly basis of the available data (FEEMA, 1989) reveals a seasonal cycle, with increased levels during the winter period from May to September, in all but two sites. This is obviously due to the less favourable atmospheric dispersion conditions during that part of the year.

The only available information with respect to the levels of inhalable particulate matter (IPM) comes from a study by Daisy et al. (1987), measuring simultaneously total suspended particulates (TSP) and IPM during two winter weeks in 1984 in São Cristóvão. Twelve-hourly TSP levels varied between 25 and 220 µg m^{-3}, with an average of 100 µg m^{-3}, while the corresponding IPM levels averaged 70 µg m^{-3} with extremes from 20 to 100 µg m^{-3}. Owing to its limited size, this study only indicates that the levels of IPM in

Rio de Janeiro deserve further investigation. Other specific SPM studies (Trindade et al., 1980; Trindade et al., 1981; Carvalho, 1982) in general confirm the findings summarized in Figures 21.4 and 21.5.

In 1983 atomic absorption spectrometry was introduced on the high-volume glass fibre filters (monthly composite samples) to determine the average ambient zinc (Zn), iron (Fe), manganese (Mn), copper (Cu) and lead (Pb) levels. Referring to a FEEMA report (FEEMA, 1989) the yearly levels during the 1984–1987 period in the monitoring network can be summarized as follows:

Zn From less than 0.1 µg m^{-3} in Copacabana to almost 1 µg m^{-3} in Iraja and Nilopolis.

Fe From a low 0.3 µg m^{-3} in Santa Tereza to almost 3.5 µg m^{-3} in Bonsucesso and 3 µg m^{-3} in Nova Iguaçu.

Mn From below 0.02 µg m^{-3} at various locations up to 0.2 µg m^{-3} in São Cristóvão (where the average Fe levels also reach 2 µg m^{-3}) and 0.1 µg m^{-3} in Nova Iguaçu (with average Fe levels of 3 µg m^{-3}).

Cu from below 0.03 µg m^{-3} (Ilha do Governador, Penha) to 0.3 µg m^{-3} in Niteroi.

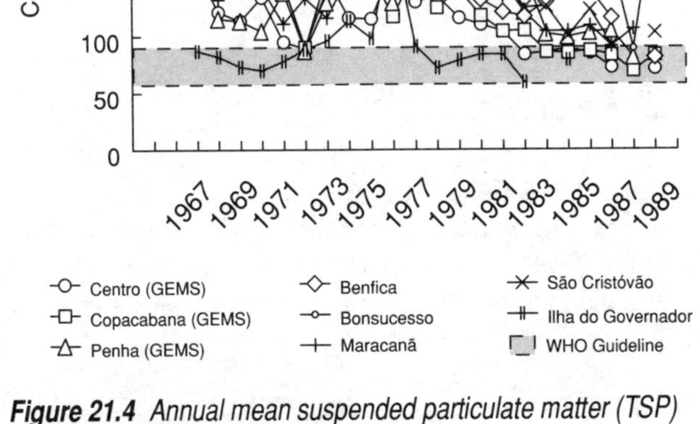

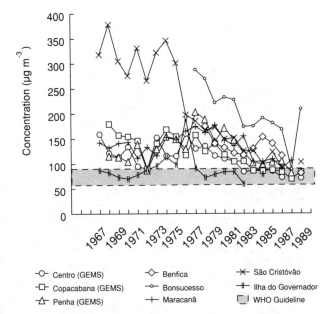

Figure 21.4 *Annual mean suspended particulate matter (TSP) concentrations at selected sites*

Sources: GEMS data; Justen, 1991; OPS, 1982

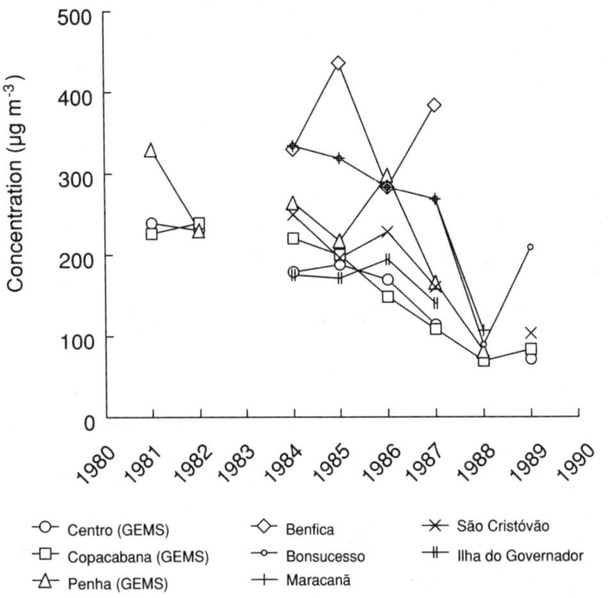

Figure 21.5 *Annual maximum suspended particulate (TSP) concentrations at selected sites*

Sources: GEMS data; Justen, 1991

Lead

Ambient Concentrations Measurements of Pb were made on a monthly basis using the TSP samples. FEEMA (1989) reported that between 1984–1987 Pb concentrations ranged from 0.1 µg m^{-3} in Ilha do Governador up to yearly averages of 0.7–0.8 µg m^{-3} in EMPA, Maracana and Bonsucesso.

Approximately 80 per cent of all measured annual means during the 1984–1987 period were below 0.5 µg m^{-3}; the WHO annual guideline is 1–1.5 µg m^{-3}. The gradual decrease of the tetraethyl lead content of the local (alcohol-blended) petrol, and the switch to alcohol-fuelled vehicles, undoubtedly reduced the ambient Pb levels as a function of time (Trindade et al., 1981; Trindade and Pfeiffer, 1982; Manohar et al., 1982; Daisy et al., 1987).

Carbon monoxide

Emissions In 1978 estimated CO emissions totalled 642,000 tonnes per annum of which 98 per cent was attributed to motor vehicles. Rio de Janeiro has experienced a growth in motor-vehicle traffic since 1978 and therefore emissions of CO are likely to have increased despite improvements in engine efficiency.

Ambient Concentrations During the 1984–1987 period a number of hourly CO concentration measurements were carried out at several locations where intensive traffic was expected to generate elevated CO levels. The findings of this monitoring campaign can be summarized as (FEEMA, 1989):

- Mean hourly concentrations reached 2.5 mg m^{-3} at 0800 hours, 3.75 mg m^{-3} at 2000 hours and 6.25 mg m^{-3} at 2100 hours.
- Maximum hourly levels ranged from 7.5 mg m^{-3} during the night to 18.75 mg m^{-3} between 0900 hours and 1400 hours, and up to 35 mg m^{-3} between 1900 hours and 2100 hours. However, all CO levels above 32 mg m^{-3} were measured in February 1985. In the absence of more detailed information on monitoring conditions, it is impossible to analyse the relevance of the higher one-hour values.
- National Air Quality Standards were never exceeded, while the WHO one-hour guideline was occasionally exceeded.

Oxides of nitrogen

Emissions Ninety-two per cent of oxides of nitrogen (NO_x) emissions in 1978 were attributed to mobile sources. Total anthropogenic emissions were estimated to be in the region of 63,000 tonnes per annum. Increasing motor-vehicle numbers and movements since then mean that current emissions from mobile sources are far greater. However, without specific data on the motor-vehicle fleet, it is impossible to make even rough approximations of the current NO_x emissions load.

Ambient Concentrations No NO_2 data have been reported and therefore no assessment can be made. Oxides of nitrogen emissions in 1978 are consistent with other cities which experience high NO_2 levels and are likely to have increased through the 1980s.

Ozone

Emissions The age and unknown accuracy of the emissions data make it difficult to comment on the O_3-forming potential in Rio de Janerio. However, the climate and topography of the area make the area ideal for photochemical reactions.

Ambient Concentrations No ambient O_3 data have been reported. The local climate and topography and the experiences of São Paulo suggest that O_3 may be, or may become, a very serious problem in Rio de Janeiro. Co-ordinated monitoring of O_3 should be initiated to identify and quantify the scale of photochemical pollution.

21.4 Conclusions

Air Pollution Situation The local climate, the proximity of the ocean and the complex topography (Figure 21.1) often result in unfavourable pollutant dispersion conditions, i.e., strong temperature inversions, high frequency of calms, specific poorly ventilated local areas. The area also has ideal conditions for photochemical reactions. In addition, the rapid and unplanned expansion of the urbanized and industrialized areas with uncontrolled mixtures of multiple pollutant sources, leads to a city with a unique airshed.

The ambient pollution levels reported should not to be interpreted or extrapolated as being representative of the entire metropolitan area of Rio de Janeiro, but only for the areas in the vicinity of the monitoring sites.

Main Problems For a very complex and large area, such as the metropolitan area of Rio de Janeiro, the available emissions and ground-level concentration data are insufficient to analyse the local situation with respect to air quality and its consequences.

Major technical and operational problems have undoubtedly hampered the various and prolonged attempts of the local organizations to evaluate the actual air pollution situation for a number of pollutants. From the available data it is nevertheless obvious that the metropolitan area of Rio de Janeiro still has a SPM problem, and possibly also a SO_2 problem, at a number of locations.

21.5 References

Alvarez H. 1978 Pan American Air Monitoring Network. In: *Studies in Environmental Science 2: Air Pollution Reference Measurement Methods and Systems*, Elsevier Scientific Publishing Company, Amsterdam, 95–97.

Carvalho, N. R. V. 1982 Chlorides, sulphates and nitrates in Rio de Janeiro. In: *Proceedings of the 5th International Clean Air Congress 1980*, Argentina, 293–297.

Daisy, J. M., Miguel, A. H., de Andrade, J. B., Pereira, P. A. P., and Tanner, R. L. 1987 An overview of the Rio de Janeiro aerosol characterization study, *Journal of Air Pollution Control Association* **37**, 15–23.

FEEMA 1989 *Qualidade do ar na Regiao Metropolitana do Rio de Janeiro 1984/1987*, Findacao Estadual de Engenharia do Meio Ambiente, Rio de Janeiro.

Justen, R. 1991 TSP and SO_2 data from 1968 to 1989, Personal Communication.

Lopes, I. M. R., de Souza Araujo, J. R., Abaurrea, J. A., Moreira Lopes, A. F. and Camara Fernandes, J. O. 1982 Emission de contaminantes atmosfericos en la region metropolitana de Rio de Janeiro. In: *Proceedings of the 5th International Clean Air Congress 1980*, Argentina, 283–292.

Manohar, H., Gandhi, G., Pereira, C. A. S. and Silveira, F. L. R. 1982 X-ray fluorescence analysis of aerosols in the metropolitan city of Rio de Janeiro. In: *Proceedings of the 5th International Clean Air Congress 1980*, Argentina, 206–216.

OPS 1982 *Red Pan Americana de Muestro de la Contaminacion del Aire, Informe Final 1967-1980*, Publication CEPIS 23, Organizacion Pan Americana de la Salud, Lima.

Trindade, H. A. and Pfeiffer, W. C. 1982 Relationship between ambient lead concentrations and lead in gasoline in Rio de Janeiro, *Atmospheric Environment* **16**, 2749–2751.

Trindade, H. A., Oliveira, A. E., Pfeiffer, W. C., Londres, H. and Costa-Ribeiro, C. L. 1980 Meteorological parameters and concentration of total suspended particulates in the urban area of Rio de Janeiro, *Atmospheric Environment* **14**, 973–978.

Trindade, H. A., Pfeiffer, W. C., Londres, H. and Costa-Ribeiro, C. L. 1981 Atmospheric concentration of metals and total suspended particulates in Rio de Janeiro, *Environmental Science and Technology* **15**, 84–89.

UN 1989 *Prospects of World Urbanization 1988*, Population Studies No. 112, United Nations, New York.

22

São Paulo

22.1 General Information

Geography São Paulo (Latitude 23°37'S, Longitude 46°39'W) is situated approximately 60 km from the south-east coast of Brazil at a mean altitude of 800 m above mean sea level. The Greater São Paulo Area (GSPA) covers approximately 8,000 km^2 with a complicated topography dominated by hills 650–1,200 m high. The GSPA has 37 municipalities (Figure 22.1), including the city of São Paulo - the state capital. Approximately 5,000 km^2 of the GSPA is urbanized. This urbanized area mainly covers the Paulistano Plateau, 715–900 metres above sea level.

Demography The total population of GSPA was 18.42 million in 1990, more than 10 per cent of Brazil's total population. The rapid increase over the past decades is noteworthy: from almost 5 million in 1960 to 8 million in 1970, 12.5 million in 1980 and heading towards a forecasted total of 23.60 million by the year 2000 (UNEP, 1989; UN, 1989; CETESB, 1990).

Climate The local climate can be summarized as tropical rain forest with a dry winter (May–August), with a minimum monthly mean temperature of 15°C, followed by a wet summer (September–April) with a maximum monthly mean temperature of 22°C. Annual mean precipitation is 1,930 mm. During winter subsidence inversions occur 50 per cent of the time with "unfavourable dispersion conditions" as a consequence (CETESB, 1989, 1990).

Industry The State of São Paulo is responsible for more than 50 per cent of Brazil's GNP of approximately 300 billion US dollars. The economically most important region in São Paulo State - and in the country - is the Regiao Metropolitana de São Paulo (RMSP) or GSPA.

Energy statistics (CESP, 1989a) show the following secondary energy consumption for the GSPA in 1988:

Petrol	1.5 million tonnes per annum
Diesel oil	1.7 million tonnes per annum
Fuel oil	1.6 million tonnes per annum
Hydrated ethanol	1.7 million tonnes per annum
Liquid petroleum gas	0.6 million tonnes per annum
Electricity	33.8 TW

Apart from this, a significant fraction of São Paulo State's coal (600,000 tonnes per annum), wood (1.1 million tonnes per annum), kerosene (500,000 tonnes per annum), gas (1 million tonnes per annum), coke (850,000 tonnes per annum), (anhydrous) ethanol (680,000 tonnes per annum) and sugar cane pulp (5.7 million tonnes per annum) consumption (CESP, 1989b) also takes place in the GSPA. As 95 per cent of electric power generation is hydroelectric in São Paulo State (CESP, 1989b), the air pollution impact of electricity generation is negligible.

Transport There are 4 million vehicles registered in the GSPA, of which over 3 million are registered in São Paulo. The Department of Traffic (DETRAN), gives the following number of vehicles by fuel type in 1990: 1.2 million alcohol-fuelled, 2.2 million petrol-fuelled and 300,000 diesels. Companhia de Tecnologia de Saneamento Ambiental's (CETESB) estimates for the 1988 traffic-related emissions are shown in Figure 22.2 (CETESB, 1990).

22.2 Monitoring

Air pollution monitoring started in 1968 with the systematic measurement of the daily average sulphur dioxide (SO_2) and smoke levels at one site (OPS, 1982). By 1973 some 10 monitoring sites of this type were operational at different locations, and three of these sites were submitting data to GEMS (CCC2, CCM3 and CCM5). This manual network (Rede Manual) expanded to 14 monitoring sites by 1977, but was reduced to six sites in the early 1980s. Those six sites, as represented on the map in Figure 22.1, are still operational. Three of those sites are GEMS sites, namely the ones in Aclimaçao (CCC2), Tatuapé (CCM5) and Pinheiros (CCR6). All six sites are located in the central part of São Paulo city and are not representative for the whole GSPA.

SÃO PAULO

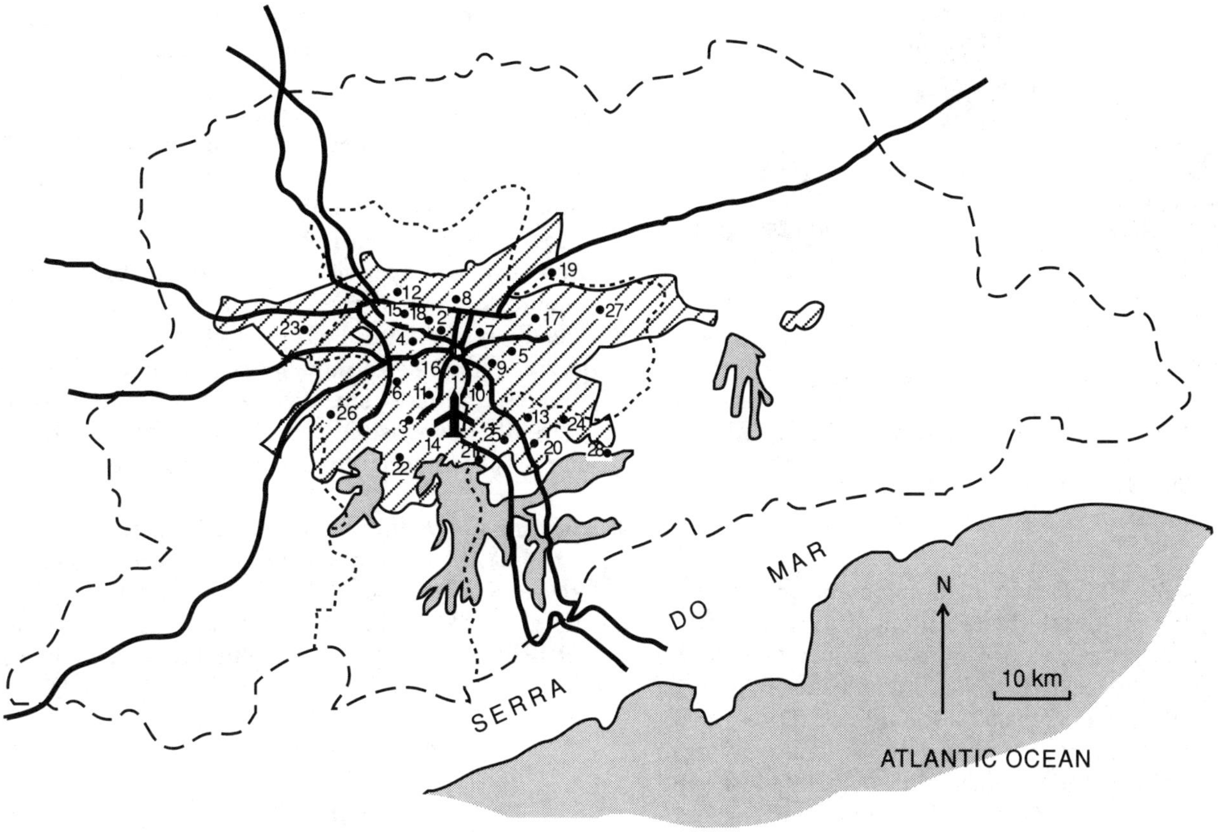

Legend

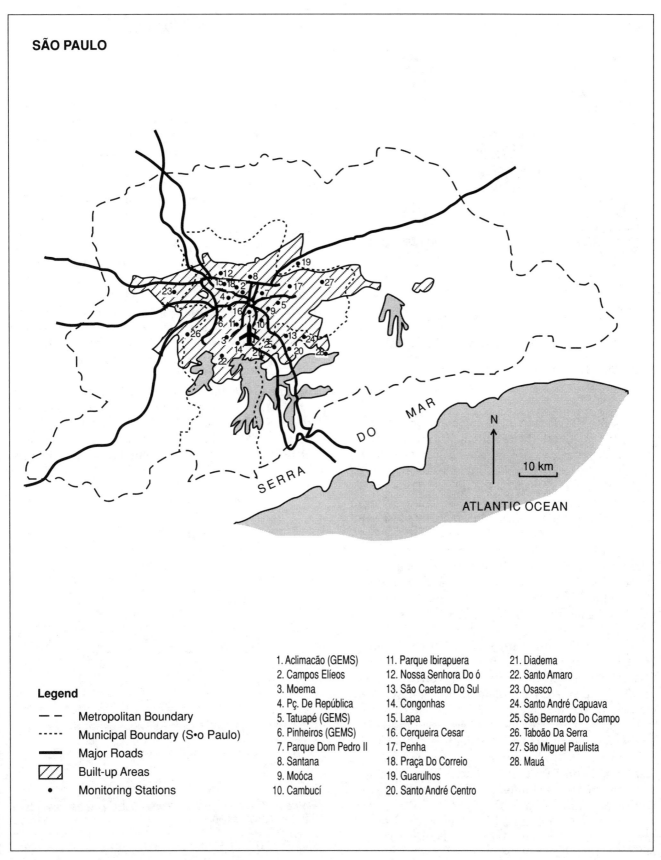

- – – Metropolitan Boundary
- ----- Municipal Boundary (S•o Paulo)
- ▬ Major Roads
- ▨ Built-up Areas
- • Monitoring Stations

1. Aclimacão (GEMS)
2. Campos Elíeos
3. Moema
4. Pç. De República
5. Tatuapé (GEMS)
6. Pinheiros (GEMS)
7. Parque Dom Pedro II
8. Santana
9. Moóca
10. Cambucí
11. Parque Ibirapuera
12. Nossa Senhora Do ó
13. São Caetano Do Sul
14. Congonhas
15. Lapa
16. Cerqueira Cesar
17. Penha
18. Praça Do Correio
19. Guarulhos
20. Santo André Centro
21. Diadema
22. Santo Amaro
23. Osasco
24. Santo André Capuava
25. São Bernardo Do Campo
26. Taboão Da Serra
27. São Miguel Paulista
28. Mauá

Figure 22.1 *Sketch map of the Greater São Paulo Area*

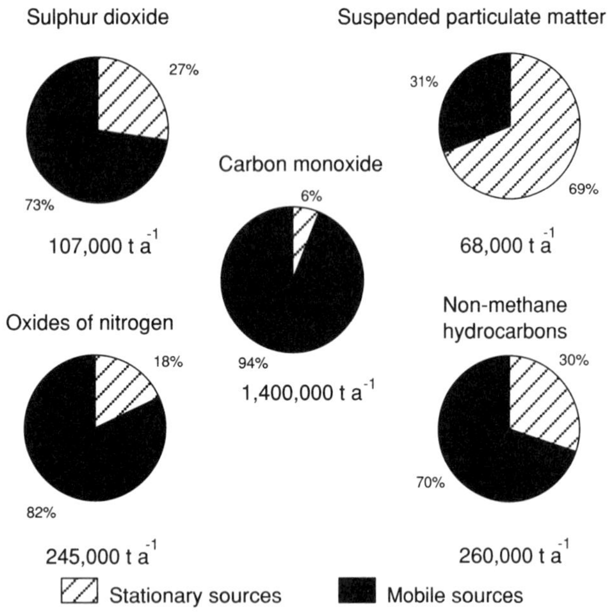

***Figure 22.2** Estimated pollutant emissions in GSPA, 1988*

Source: CETESB, 1990

In 1981 CETESB started an automatic or telemetric monitoring network (Rede Automatica) with 25 fixed monitoring sites and two mobile laboratories. Figure 22.1 gives the location of the different sites within the GSPA (the three Cubatao sites are not represented). All sites are in the central part of São Paulo city or in the neighbouring municipalities (south-east and west). No site is common to the manual and the automatic network. The latter was supplemented in 1983 with (non-automatic) high-volume total suspended particulates (TSP) samplers at 10 of the 25 monitoring sites. At the same time, one high-volume TSP sampler was also installed at Pinherios (CCR6) in the manual network.

At four of the automatic monitoring sites (Parque Dom Pedro II, Ibirapuera, São Caetano do Sul and Osasco) the size distribution and the carbon content of the suspended particulate matter (SPM) were studied in more detail during a special campaign from November 1987 to October 1988. Lead (Pb) levels were monitored in 1983 and in 1987 (a similar campaign in 1978, also run at four sites, had only the São Caetano do Sul site in common). Two other special campaigns, determining aldehyde levels, were run at the Parque Dom Pedro II, Moóca, Congonhas and Praça do Correio sites of the automatic network from July 1980 to June 1981 and January 1985 to February 1986 (CETESB, 1990).

22.3 Air Quality Situation

Sulphur dioxide

Emissions Figure 22.2 shows the estimated anthropogenic emissions in 1988. Nefussi et al. (1977) reported SO_2 emissions of 600,000 tonnes per annum for the GSPA in 1976. This had dropped to 280,000 tonnes per annum by 1981 (CETESB, 1985) and subsequently stabilized at around 110,000 tonnes per annum by 1988/1989 (CETESB, 1989, 1990). This means an overall reduction by a factor of over five. However, mean ambient levels were only reduced by a factor of 2.5. This is because emission reductions were achieved by controlling industrial point sources - switching from high-sulphur fuel oil to low-sulphur fuel, biomass, natural gas or electricity. The basic SO_2 load made up by the smaller but numerous low-level sources, and more specifically by traffic emissions, is presently responsible for 73 per cent of total SO_2 emissions and has remained almost unchanged.

Ambient Concentrations Figure 22.3 summarizes the SO_2 data for the manual network. After a slight increase of SO_2 in the 1970s, levels fell significantly in the early 1980s and seem to have stabilized since 1984. National air quality standards (NAQS) were

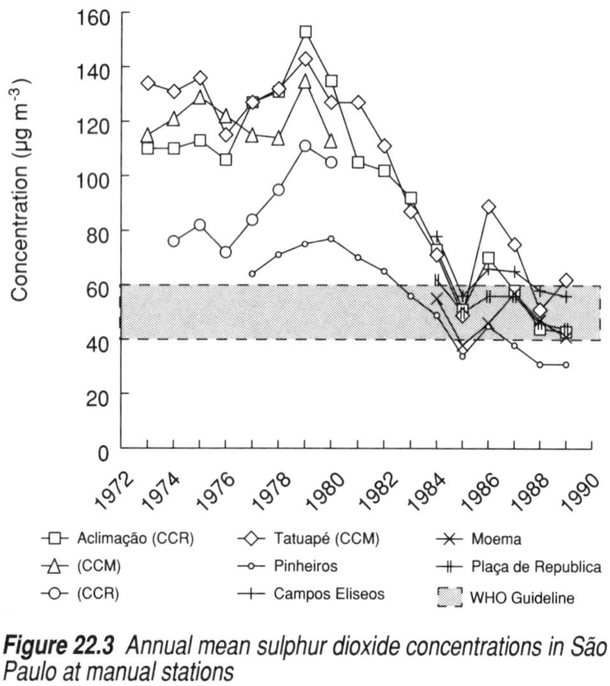

***Figure 22.3** Annual mean sulphur dioxide concentrations in São Paulo at manual stations*

Sources: CETESB, 1989; CETESB, 1990; GEMS data

met at all sites in 1989. The highest arithmetic mean of 62 μg m^{-3} is just above the WHO guideline value of 60 μg m^{-3}, while all 98 percentiles are lower than the corresponding 150 μg m^{-3} upper limit of the WHO guidelines. The GEMS sites appear to be representative of the manual network as a whole.

Results of the automatic network show a similar picture. A significant decrease in the levels from 1981 to 1984 was followed by a prolonged stable period at levels well below the national standards and under or within the WHO guideline. A more detailed analysis of the data shows that the automatic network results are comparable with, or lower than, the levels recorded in the manual network.

Suspended particulate matter

Emissions In 1988 estimated emissions of SPM totalled 68,000 tonnes per annum. In the early 1980s 162 stationary SPM sources were responsible for 96 per cent of all the known SPM emissions. These sources were given five years to comply with strict regulations aimed at large emission reductions. Although imposed reductions have now been met, SPM problems have not been solved (CETESB, 1990).

Emission abatement in the past was focused on stationary sources (using the Ringelmann scale as the control chart), but the enormous fleet of badly maintained and not properly controlled diesels probably cause the remaining smoke problems. CETESB has started a smoke control and abatement programme based on (CETESB, 1990):

- spot control of diesel vehicles using Ringelmann charts;
- certification of distributed diesel oils;
- a training and awareness campaign especially for truck, taxi and bus owners.

The local authorities hope that the problem will be solved by 1997 when the Programme for Controlling Air Pollution due to Motor Vehicles (PROCONVE) will have its full effect. A corner-stone of this programme are the emissions regulations for diesels (Table 22.1) (CETESB, 1990; Szwarc and Branco, 1987).

Ambient Concentrations Total suspended particulates are measured on a one-out-of-six-days basis at nine different sites (eight of the automatic network and one of the manual network, Figure 22.1) by high-volume samples. Figure 22.4 displays the annual mean results for the past six years. The main conclusions are:

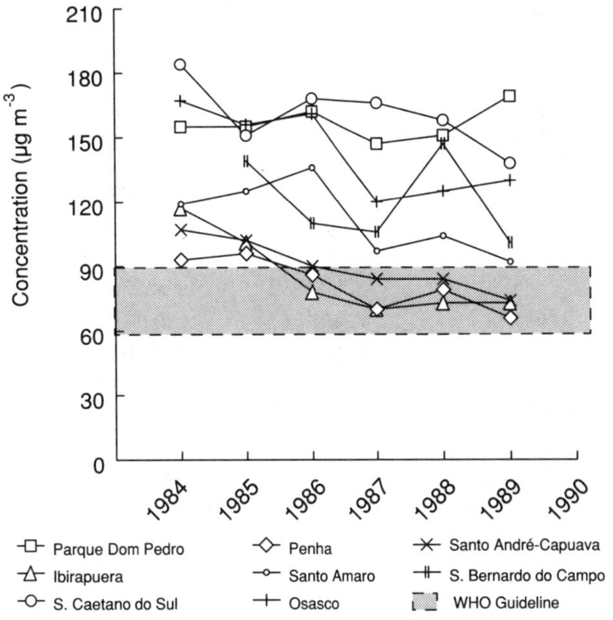

Figure 22.4 *Annual mean suspended particulate matter (TSP) concentrations at selected stations*

Sources: CETESB, 1989; CETESB, 1990

- no clear trend over the past six years;
- systematic exceedences of the national air quality standards at the majority of sites.

It is clear that under those conditions, WHO guidelines are also exceeded.

A second set of SPM measurements are the atomic-absorption spectrophotometric determinations in all the automatic monitoring network sites. A downward trend in those time series was notable and in 1989 the NAQS were met in all sites. The WHO guidelines for the arithmetic mean and the 98 percentile of the daily means were also met. A more detailed analysis of the TSP data obtained simultaneously at the same sites, by high-volume sampling and atomic absorption, shows that one is basically a measuring a different fraction of the SPM. Some statistics do agree while others differ by a factor of two or more, with atomic absorption values always being lower than those obtained by high-volume sampling. Agreement or disagreement also changes as a function of time and is different from one year to another.

With respect to the inhalable SPM, the 1987/1988 measuring campaign in four sites of the automatic monitoring network showed that:

Table 22.1 *Emission limits for heavy duty vehicles*

Type of Emission	Effective from	Vehicle type	Emission Limits			
			Smoke (K)*	CO (g km^{-1})	NO_x (g km^{-1})	HC (g km^{-1})
Exhaust	1 Oct 1987	Urban diesel bus	2.5			
	1 Jan 1989	All diesel vehicles	2.5			
	1 Jan 1993	All diesel vehicles	2.5	11.2	2.8	18
	1 Jan 1995	All diesel vehicles	2.5	11.2	2.8	14.4
Crankcase	1 Jan 1988	Urban diesel bus	Zero emissions under any operating conditions			
	1 Jan 1989	Otto cycle engines	Zero emissions under any operating conditions			
	1 July 1989	Naturally aspirated diesel engines	Zero emissions under any operating conditions			
	1 Jan 1993	Turbocharged diesel engines	Zero emissions or in HC emissons of exhaust			
Evaporative	To be proposed	Otto cycle engines	To be proposed			

* $K = \frac{c}{\sqrt{G}}$ where c = carbon concentration (g m^{-3}) and G= nominal air flow (l s^{-1})

After Murley, 1991

- at three of the four sites the mean levels of particulate matter < 10 μm (PM_{10}) exceeded the 50 μg m^{-3} limit of the NAQS with values of 85, 79 and 62 μg m^{-3};
- in all four sites the WHO daily guideline (150 μg m^{-3}) was exceeded more than once, with maxima of 285, 222 and 180 (twice) μg m^{-3}.

As with SO_2, major efforts have been made over the past years to lower the ambient smoke levels in São Paulo. Although there has been a net decrease in the yearly mean and in the extreme daily levels, it is clear that the net result is less successful than for SO_2 (Figure 22.5). Figure 22.5 shows that WHO guidelines and NAQS are still exceeded, with high smoke levels occurring especially in winter (May–August).

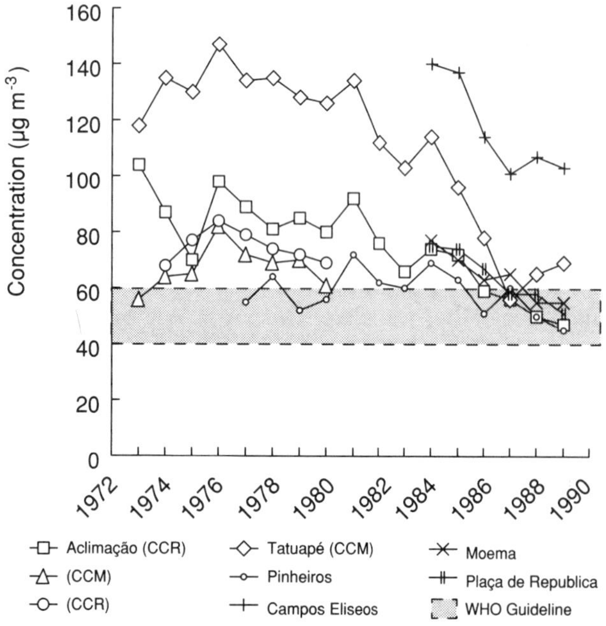

Figure 22.5 *Annual mean suspended particulate matter (smoke) concentrations*

Sources: CETESB, 1989; CETESB, 1990; GEMS data

Lead

Emissions The National Alcohol Programme (de Oliveira, 1991), which was launched in 1975 to reduce oil imports, has to a large extent been responsible for a large reduction in Pb emissions and ambient concentrations. Lead in petrol was first replaced by a blend of 13–22 per cent anhydrous ethanol and petrol, referred to as "gasohol". Lead is nevertheless still added to petrol in a number of places and in varying concentrations, although the Brazilian Standard NBR-8669 of December 1984 limits this to 0.4 ml tetraethyl lead per litre petrol.

The second effect of the National Alcohol Programme was the progressive introduction of (neat or hydrous) ethanol as fuel for redesigned alcohol-based Otto cycle engines. The first ethanol-fuelled cars appeared in 1980. By 1990 more than 90 per cent of the cars had alcohol-based engines. Sixty per cent of São Paulo's light-vehicle population - with life of at least 12 years - runs on petrol or "gasohol" and almost 40 per cent runs on ethanol.

Ambient Concentrations In 1976, Orsini and Boveres (1982) determined ambient Pb levels at two different locations; downtown São Paulo city and at a distance of approximately 7 km from the city. Based on a limited number of samples, mean Pb levels were shown to be in the vicinity of 1.0–1.1 μg m^{-3}. The

Table 22.2 *Three-monthly mean lead concentrations (µg m^{-3})*

Site	1978	1983	1987
Republica	0.8–1.4		
Pinheiros	0.9–1.2		
Parque Dom Pedro II		0.1–0.3	0.2–0.4
Ibirapuera		0.2–0.6	0.1–0.4
Osasco		0.1–0.2	0.1–0.2
São Caetano do Sul	0.8–1.6	0.2–0.4	0.4–0.5

After CETESB, 1989

results of more recent and more systematic monitoring of the daily Pb levels at different sites are summarized in Table 22.2. The observed levels are below WHO guideline values. Despite the continuous and rapid increase in the number of vehicles - from less than 1.5 million in 1976 to almost 4 million in 1990 - the expected Pb problem has been avoided.

Carbon monoxide

Emissions Figure 22.2 shows the contribution of motor vehicles to estimated total anthropogenic carbon monoxide (CO) emissions. In an effort to address the CO problem, and the other traffic-related emissions (e.g., oxides of nitrogen (NO_x), hydrocarbons (HC) and ozone (O_3) levels), the National Environmental Council (Consello Nacional do Meio Ambiente) enacted Resolution No. 18 on 6 May 1986, establishing the automotive emission control programme nation-wide under the name PROCONVE. Emission limits that become progressively more stringent are part of this programme and they are summarized in Table 22.1 for heavy duty vehicles and in Table 22.3 for light duty vehicles.

Ambient Concentrations The CO situation in São Paulo (city) gives great cause for concern and no improvement has been seen over the past years. The WHO guidelines and NAQS levels are exceeded in four of the five monitoring sites (Figure 22.6). The number of days with exceedences of the attention level of 15 ppm (eight-hour mean value) highlights the gravity of the situation (Table 22.4).

Oxides of nitrogen

Emissions Estimated anthropogenic emissions of NO_X totalled 245,000 tonnes in 1988. Of this total, 82 per cent was attributed to motor vehicles (Figure 22.2).

Ambient Concentrations The ambient nitrogen dioxide (NO_2) levels at four sites of the automatic monitoring network are summarized in Figure 22.7. The mean NO_2 levels in the Congonhas site fluctuate around the 100 µg m^{-3} level which is the NAQS. The Brazilian one-hour standard (320 µg m^{-3}) and the WHO guideline (400 µg m^{-3}) are exceeded at 3 stations with peak values ranging from 600 to 1500 µg m^{-3}.

Ambient nitric oxide (NO) levels in São Paulo are

Table 22.3 *Emission limits for Brazilian alcohol- and petrol-fuelled light duty vehicles*

			Emission Limits				
Type of Emission	Effective from	Remarks	CO (g km^{-1})	NO_X (g km^{-1})	HC (g km^{-1})	Aldehydes (g km^{-1})	Idle CO (%)
Exhaust	1 Jun 1988	Brand new vehicle configurations	24	2	2.1		3
	1 Jan 1989	Minimum of 50% of sales	24	2	2.1		3
	1 Jan 1990	100% of sales except light duty trucks	24	2	2.1		3
	1 Jan 1992	Only light duty trucks	24	2	2.1	0.15	3
	1 Jan 1992	100% of sales except light duty trucks	12	1.4	1.2	0.15	2.5
	1 Jan 1997	All light duty vehicles	2	0.6	0.3	0.15	0.5
Evaporative	1 Jan 1990	All light duty vehicles			6 g/test		
Crankcase	1 Jan 1988	All light duty vehicles	Emissions shall be nil under any engine operating conditions				

After Murley, 1991

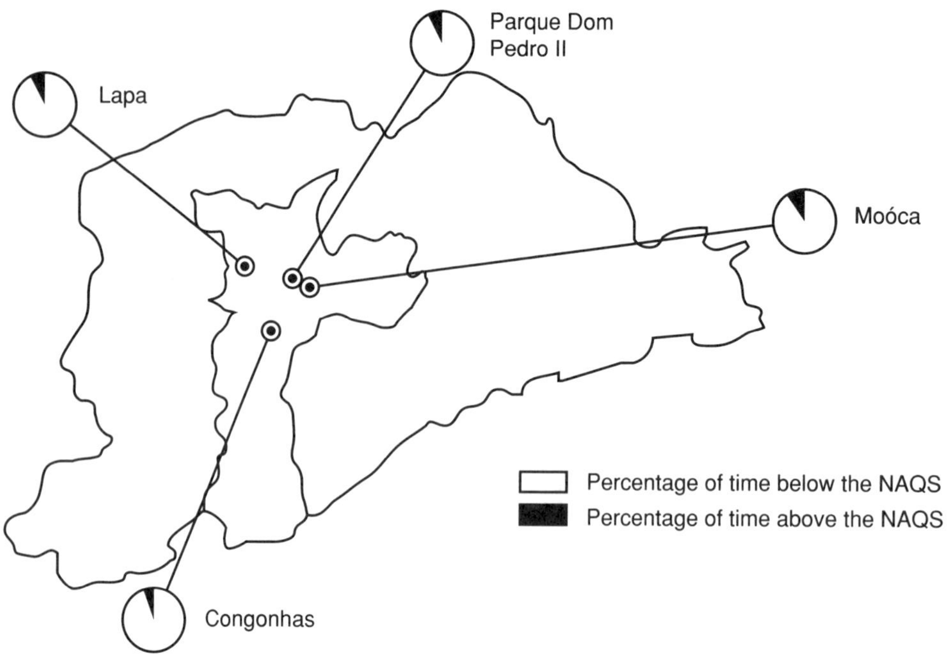

Figure 22.6 *Percentage of time that the NAQS for carbon monoxide was exceeded in 1989*

After CETESB, 1990

Table 22.4 *Number of days above the carbon monoxide attention level (15 ppm – 8 h)*

Site	1981	1982	1983	1984	1985	1986	1987	1988	1989
Correio	98	37	34	12	10	120	136	9	11
Parque Dom Pedro II	4	0	0	0	0	0	0	1	0
Moóca	0	0	0	0	0	0	0	0	0
Congonhas	9	8	0	0	2	0	14	4	3
Cerqueira César	7	2	5	4	1	3	8	5	2

After CETESB, 1990

relatively high which is not surprising considering the predominant role played by the motor vehicle fleet as an emissions source.

Ozone

Emissions High NO_x:HC ratios, high ambient temperatures and intense insolation form the ideal mixture to produce O_3 at ground level. All these conditions are frequently met in São Paulo.

Ambient Concentrations São Paulo's O_3 problem is summarized in Figures 22.8 and 22.9. The gravity of the problem is further illustrated by Table 22.5 which shows the number of days during which the attention level of 200 μg m^{-3} has been reached in past years.

22.4 Conclusions

Air Pollution Situation São Paulo has done much to improve its air quality in recent years, especially with respect to SO_2 and Pb. However, levels of SPM, CO and O_3 are still of great concern, as is NO_2 due to its effects upon health and the environment and its O_3-forming potential. Traffic is the greatest emissions source in the GSPA and although the National Alcohol Programme has helped to reduce Pb emissions,

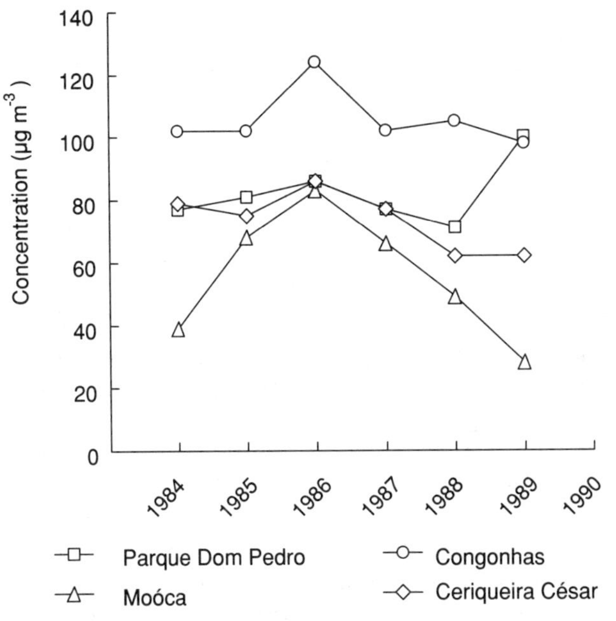

Figure 22.7 *Annual mean nitrogen dioxide concentrations at selected sites*

After CETESB, 1990

other traffic-related pollution problems remain. It is hoped that more stringent emission limits will help to stabilize emissions and ambient concentrations in the future. However, an overall decrease in motor vehicle emissions will be difficult to achieve because of increasing motor vehicle registrations and distances travelled.

Health Effects The first epidemiological study on short-term health effects of air pollution in the GSPA was carried out by Mendes and Wakamatsu in the mid-1970s (Mendes and Wakamatsu, 1976). A systematic analysis of the 1973 death records, and of local SO_2 and smoke measurements, suggested that during a two-week air pollution episode with average SO_2 levels above 200 µg m^{-3} and peak daily values of 450 µg m^{-3} (1 August 1973, Capuava site) an increase in the daily mortality rate could be related to concentrations. The increase in mortality, mainly through circulatory and respiratory diseases, occurred in the over 65 and the less than one-year age groups. No recent studies on the health effects of motor vehicle-related pollutants have been reported.

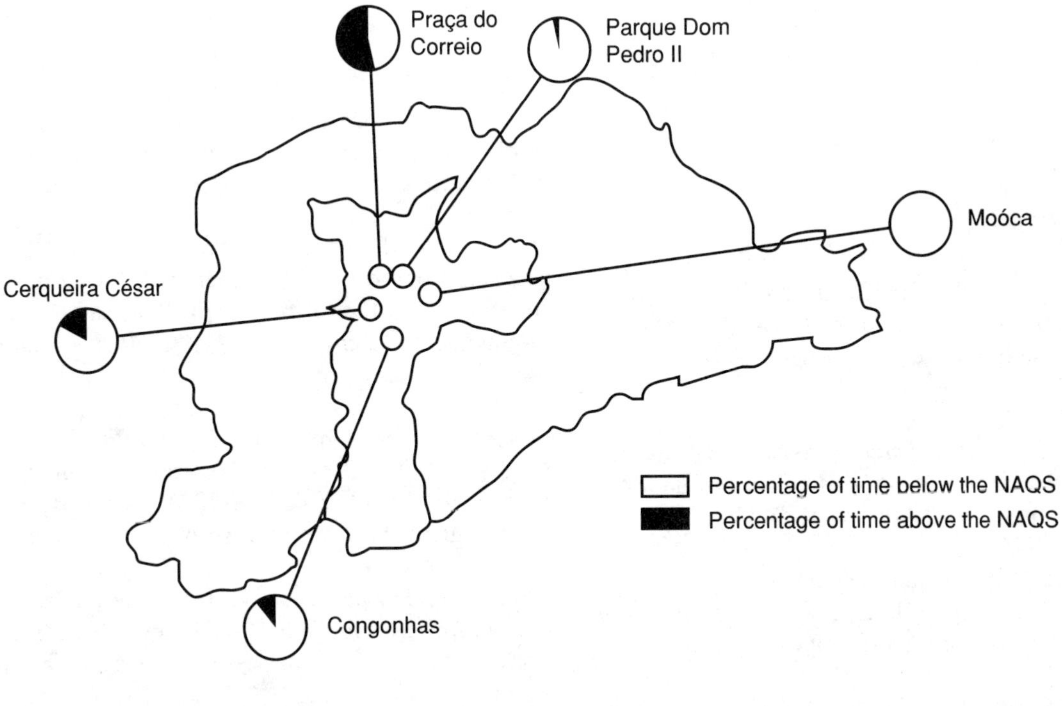

Figure 22.8 *Percentage of time the NAQS for O_3 was exceeded in 1989*

After CETESB, 1990

Table 22.5 *Number of days above the ozone attention level (400 μg m^{-3} – 1h)*

Site	1981	1982	1983	1984	1985	1986	1987	1988	1989
Parque Dom Pedro II	8	10	1	3	4	8	14	2	10
Moóca	19	4	9	5	4	24	23	14	15
Congonhas	3	2	7	4	2	14	3	0	7
Lapa	10	6	2	19	10	13	12	12	7

After CETESB, 1990

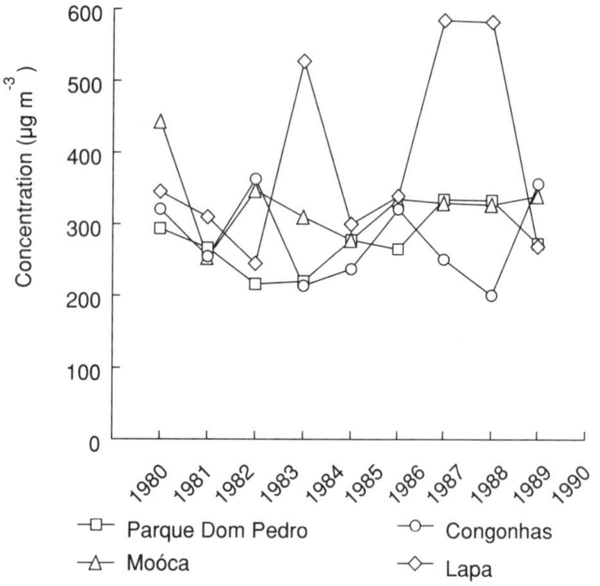

Figure 22.9 *Maximum hourly ozone concentrations at selected sites*

After CETESB, 1990

22.5 References

CETESB 1985 *Air pollution control programme and strategies in Brazil: São Paulo and Cubatao areas,* Companhia de Tecnologia de Saneamento Ambiental, São Paulo.

CETESB 1989 *Relatório de qualidade do ar no Estado de São Paulo - 1988,* Série Relatórios, Companhia de Tecnologia de Saneamento Ambiental, São Paulo.

CETESB 1990 *Relatório de qualidade do ar no Estado de São Paulo - 1989,* Série Relatórios, Companhia de Tecnologia de Saneamento Ambiental, São Paulo.

CESP 1989a *Estatisticas Energéticas Municipais e Regionais: Estado de São Paulo 1988,* Companhia Energética de São Paulo, São Paulo.

CESP 1989b *Balanço Energético do Estado de São Paulo 1986,* Companhia Energética de São Paulo, São Paulo.

de Oliveira, A. 1991 Reassessing the Brazilian alcohol programme, *Energy Policy,* **19**, 47–55.

Mendes, R. and Wakamatsu, C. 1976 A study of the relation of air pollution to daily mortality in São Paulo, Brazil, 1973. In: *Proceedings of the 4th International Clean Air Congress,* Tokyo, 76–80.

Murley, L. (Ed.) 1991 *Clean Air Around the World: National and International Approaches to Air Pollution Control,* International Union of Air Pollution Prevention Associations, Brighton.

Nefussi, N., Guimaraes, F. A. and Oliveira, G. 1977 Air pollution control programmes in the state of São Paulo, Brazil. In: *Proceedings of the 4th International Clean Air Congress,* Tokyo, 942–946.

OPS 1982 *Red Pan Americana de Muestro de la Contaminacion del Aire, Informe Final 1967–1980,* Publication CEPIS 23, Organizacion Pan Americana de la Salud, Lima.

Orsini, C. M. Q. and Boveres, L. C. S. 1982 Investigations on trace elements of the atmospheric aerosol of São Paulo, Brazil. In: *Proceedings of the 5th International Clean Air Congress 1980,* Argentina, 247–255.

Szwarc, A. and Branco, G. M. 1987 *Automotive Emissions: The Brazilian Control Programme,* SAE Technical Paper No. 871073, Companhia de Tecnologia de Saneamento Ambiental, São Paulo.

UN 1989 *Prospects of World Urbanization 1988,* Population Studies No. 112, United Nations, New York.

UNEP 1989 *United Nations Environment Programme Environmental Data Report,* 2nd edition, Blackwell, Oxford.

23

Seoul

23.1 General Information

Geography The capital of the Republic of Korea (Korea), Seoul is located on the western side of the Korean Peninsula (Latitude 37°33'N, Longitude 127°00'E). The centre of Seoul is 30 km inland from the Yellow Sea (Huang Hai) and is situated on the north side of the Han River (Han-gang), which originates in the hills south-east of the city (Figure 23.1). Korea is a relatively mountainous country. This feature is readily apparent from Seoul. While the centre of the city is quite flat and less than 100 m above sea level, elevations rise rapidly towards the north where the hills can exceed 800 m. In a southerly direction the elevations also rise significantly with 600 m hills within 20 km of the urban centre.

Demography Almost 25 per cent of the Korean people live in Seoul which had an estimated population of 11.33 million in 1990 (UN, 1989). The Seoul Area (600 km^2) is one of the most densely populated regions in the world (18,000 people per km^2). People have settled in the region due to its status as the centre of political and economic decision-making and education and the area has experienced rapid population growth. Between 1950 and 1990 Seoul's population increased tenfold. It is estimated that Seoul's population will be 12.97 million by the year 2000 (UN, 1989). The Greater Seoul Area (GSA) has an area of 1,650 km^2 and a population of 15 million people.

Climate Seoul has a continental warm summer climate. The annual mean temperature is 10°C. Winters are cold and dry with temperatures averaging approximately -5°C. Summers are hot and humid with a mean temperature of 25°C. The mean precipitation in a year is 1,200 mm.

Industry The largest proportion of the economically active population are employed in manufacturing. The main products are electrical machinery (electronics), textiles (clothing), chemicals (refining), motor vehicles and food products. Between 1980 and 1987 there was a mean annual increase of 10.8 per cent in industrial production in Korea. The trade and service industries and construction generate a significant amount of economic activity. Agriculture, forestry and fishing are also important components, but are more prevalent outside Seoul.

In conjunction with the increase in population and economic activity, Korean energy consumption has risen sharply. Between 1973–1987, consumption almost tripled and industrial and domestic/commercial activities accounted for a majority of the energy used. Economic growth and energy use began to increase rapidly in the early 1960s, after implementation of Korea's development plans. Since then, power generation capacity has expanded by about 42 times, from 434 MW in 1962 to 18,060 MW in 1986. By 1987, 53, 23, 13 and 4 per cent of the total electricity generated was derived from nuclear power, coal, natural gas and petroleum respectively. While nuclear power dominated in the production of electrical energy, coal and oil accounted for roughly 82 per cent of total energy use in 1987 (ADB, 1989). Bunker-C oil and diesel fuel were, and are, common energy sources for industrial facilities, but for heating purposes anthracite (coal) is the dominant fuel. The national pattern of fuel use is also typical of Seoul. For heating, anthracite and bunker-C oil predominated in 1988, while industry and power generation in Seoul depended mainly on diesel fuel, bunker-C oil and anthracite.

Transport Seoul is connected to other principal cities by a relatively modern network of motorways. The most important of these are the Seoul-Pusan motorway, the Yeongdong motorway and the Chungpu motorway. The number of motor vehicles in South Korea has increased very rapidly over the past 30 years. There was nearly a 100-fold increase from 1960 (29,000) to 1989 (2,660,000) with the number of passenger cars rising from approximately 13,000 in 1965 to 1,118,000 at the end of 1988. The rate of increase has remained rapid during the latter half of the 1980s. From 1987 to 1989 there was a 40 per cent increase in the total number of motor vehicles in use and nearly a 50 per cent rise in the number of cars. The number of vehicles in use for commercial transit of goods and

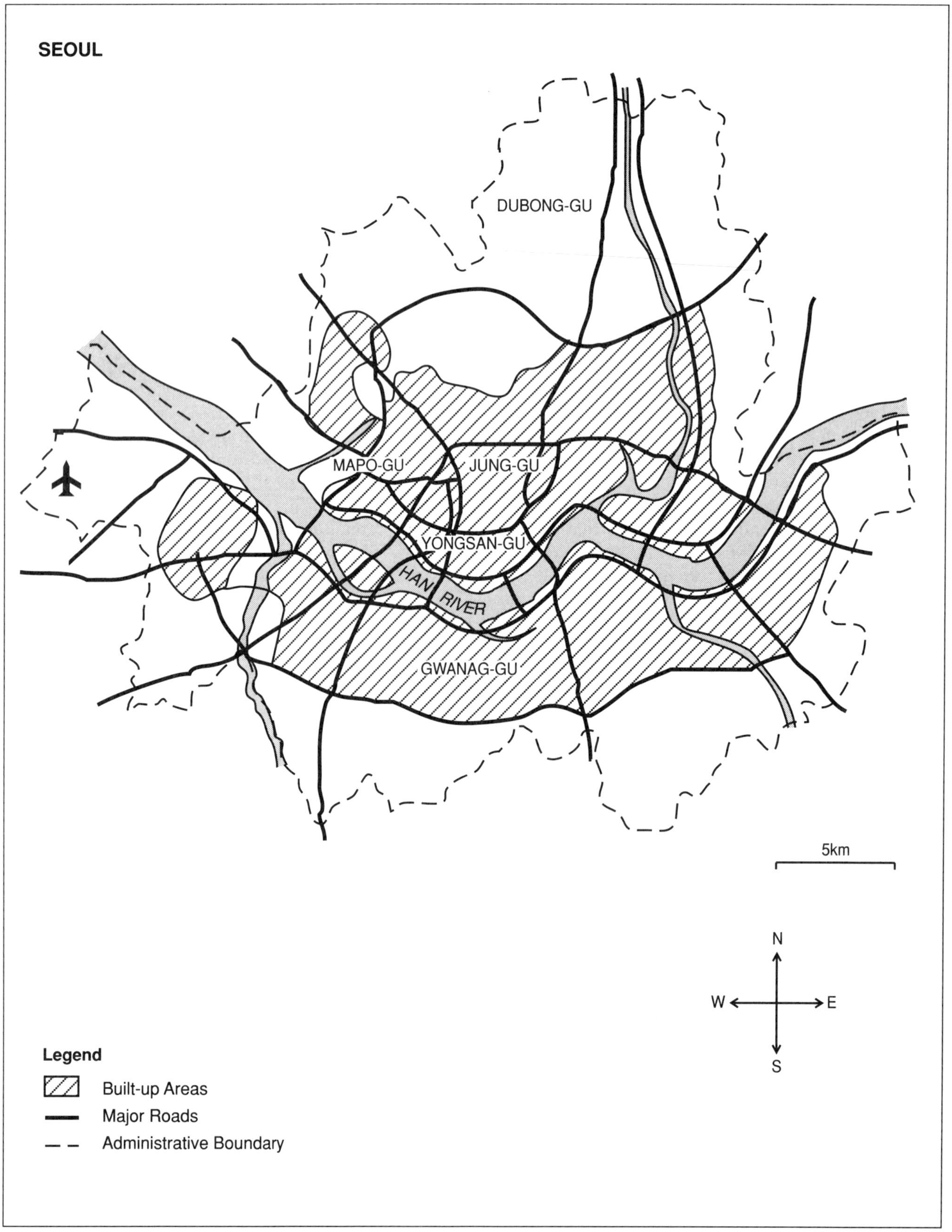

Figure 23.1 *Sketch map of the Greater Seoul Area*

people has also increased over this time (35 per cent) as has the number of motorcycles and mopeds. Seoul, which has about 40 per cent of the registered vehicles in Korea, has problems of traffic congestion and air pollution related to motor vehicles. Rail has been a popular form of transport but its use remained steady from 1986 to 1988. Within the Seoul area the metropolitan subway, which has 120 km of track, accounted for 17 per cent of all passenger trips (Kim and Cho, 1986).

23.2 Monitoring

In the past, environmental policy was the responsibility of the Environment Administration, which came under the Ministry of Health and Social Affairs.

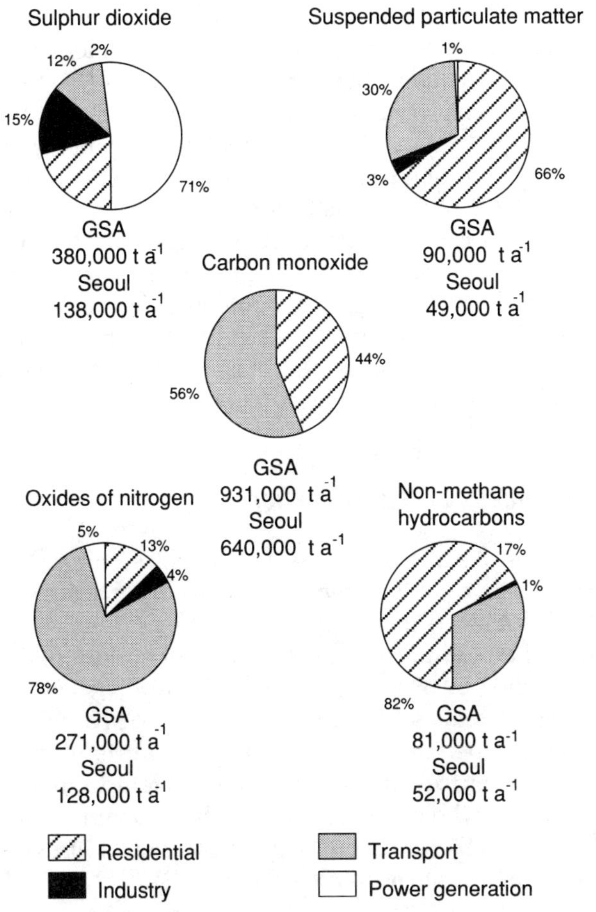

***Figure 23.2** Pollutant emissions in the Greater Seoul Area in 1990*

Source: Rhee, 1991

However, this structure was not well suited to the development and implementation of national environmental policies. Thus in 1990, the Administration was upgraded to the Ministry of Environment. It is responsible for policy formulation, programme development and implementation of environmental protection activities. The National Institute of Environmental Research which provides technical guidance, is also part of the Ministry of Environment.

The past and present environmental agencies have built up a national air monitoring network. As of 1990 it consisted of 62 automated stations where sulphur dioxide (SO_2), oxides of nitrogen (NO_x), ozone (O_3) and hydrocarbons (HC) are measured continuously and also had 29 high-volume air samplers for the collection of total suspended particulates (TSP). In addition, there are 28 acidic deposition monitoring stations spread across the country. Within Seoul there were 29 monitoring sites in 1990. These included: 20 sites for the measurement of SO_2, NO_x, carbon monoxide (CO), O_3, TSP, lead (Pb) and HC; and 9 precipitation chemistry stations. The Global Environment Monitoring System (GEMS) includes 6 sites in Seoul for monitoring SO_2. A gas bubbler (hydrogen peroxide) has been used at each site. The GEMS network has not been collecting information on TSP in Seoul.

23.3 Air Quality Situation

Sulphur dioxide

Emissions In 1990 anthropogenic emissions of SO_2 were estimated to total 380,000 tonnes per annum in the GSA compared with 138,000 tonnes per annum in the city of Seoul. The residential sector accounted for 71 per cent of total emissions due to the burning of anthracite briquettes as a domestic fuel (Figure 23.2). Industry and transport are responsible for 15 and 12 per cent respectively and power generation within the city produced only 2 per cent. In 1990 the municipalities of Inchon and Kyung-Gi contributed 86,000 tonnes and 156,000 tonnes respectively to the GSA burden (Rhee, 1991).

Current projections indicate that by 1993 desulphurization facilities will be complete (Rhee, 1990) and a commensurate reduction in SO_2 emissions is therefore expected. However, coal is a major source of SO_2 and its consumption is projected to increase by 20 per cent during the 1990s.

Ambient Concentrations Thirteen years of SO_2 data collected by various domestic agencies are currently

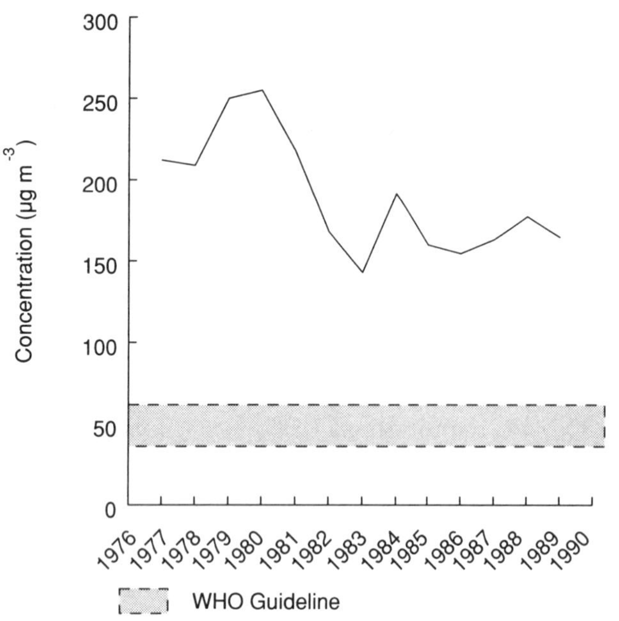

***Figure 23.3** Annual mean sulphur dioxide concentration*

After Ministry of the Environment, 1991

available. This record extends from 1977 to 1989. In contrast, only a very limited amount of the GEMS data have been reported. These measurements are particularly interesting because they represent SO_2 levels in industrial, commercial and residential areas. Even though the GEMS information is not recent, it provides alternate measurements for comparison with the Korean network results.

All the reported GEMS measurements exceeded the WHO annual guideline of 40-60 μg m^{-3}. The higher values correspond with the industrial site. The 1983 values were lower than the levels reported in the other two years. The 98 percentiles, which ranged from 456–713 μg m^{-3}, also exceeded the WHO daily guideline (100–150 μg m^{-3}). The GEMS annual mean for 1983 is in agreement with the mean value of 131 μg $m^{-3}$from the national network in Seoul.

The annual mean SO_2 concentrations in Seoul from 1977 to 1989 are shown in Figure 23.3. The 1983 value, discussed above, is shown on this plot. The main feature is the decline in the levels between 1981 and 1983. From 1977 to 1981 the annual mean concentrations were relatively steady at around 236 μg m^{-3}. From 1983 to 1989 they were closer to 157 μg m^{-3}. The significant decline in concentrations between 1981 and 1983 is coincident with the increased use of low-sulphur oil by industry reported by Rhee (1990). However, levels still exceed both the WHO guidelines and the Korean air quality standard. The long-term exposures to elevated SO_2 concentrations suggest that there was a high potential for health-related problems including an increased risk of respiratory illness in children and respiratory symptoms throughout the population.

Owing to the distribution of industry, population and topography, SO_2 levels vary across the region. In 1988 and 1989, the mean annual concentrations were highest in the eastern and south-western sections of the Seoul area. They tended to be lower towards the north, north-west and south to south-east. This pattern is consistent with the orientation of the river valley and the corresponding wind flow. Relatively clean air, originating from rural areas and the sea, would tend to move in from the north-west, collect pollutants from the city and concentrate them over eastern sections of Seoul. For example, the upwind (NW) mean annual SO_2 concentration (1988 and 1989 combined) was 134 μg m^{-3} and the downwind (E) concentration was 223 μg m^{-3}. From the most polluted to the cleanest site the annual average concentration ranged from 225 μg m^{-3} (SW) to 123 μg m^{-3} (S). This spatial pattern was also apparent on a monthly basis. The highest concentrations in 1988 and 1989 were reported in winter months to the south-west of Seoul. In the winter, concentrations were greater than 262 μg m^{-3} while in the summer they were less than 53 μg m^{-3}. This variation was largely a function of the fluctuation in the demand for heating. Coal use, particularly briquettes, increases as temperature declines. Combining this factor with a decrease in the height of emissions (domestic sources), and with a decrease in vertical mixing due to lowering inversion heights, leads to a greater potential for high SO_2 levels. The concentrations were likely to have been much higher than measured in some locations, given the local terrain and the population density.

Suspended particulate matter

Emissions As with SO_2 the residential sector is the major source (66 per cent) of suspended particulate matter (SPM) in Seoul (Figure 23.2). In 1990 emissions totalled an estimated 90,000 tonnes per annum in the GSA and 50,000 tonnes per annum in Seoul. Transport is the second major source of SPM accounting for 30 per cent of the total. The contribution of emission sources in Seoul is very different to that in Korea as a whole, where industry is the main source of SPM and SO_2. In Seoul only 3 per cent of emissions are attributable to industry (Rhee, 1991).

Ambient Concentrations The SPM (measured as TSP) concentration in the Seoul atmosphere is very high. The annual mean concentrations have consistently exceeded the national standard of 150 μg m^{-3}.

Figure 23.4 shows annual mean TSP concentrations from 1984 to 1989, as reported by the Ministry of the Environment (1991) and shows an apparent decrease in concentrations in that time, consistent with the reduction in emissions. However, the concentration decrease is much less than the reported decrease in TSP emissions. The problem of high TSP concentrations is recognized by national officials and there have been attempts to reduce these levels. However, it is not known how much of the decline is a result of control or of natural variability.

The seasonal pattern of variation for TSP is similar to that for SO_2, indicating that similar sources are responsible. Spatially, there is a slight difference between the locations of maximum TSP and SO_2. Total suspended particulate concentrations are highest over the east of the region with a secondary maximum in central areas and to the south-west. As with SO_2, the cleanest areas are in the north to north-west and to the south. This provides a clear indication of a common origin of TSP and SO_2.

The results of two special studies provide some additional information on the characteristics of TSP in Seoul. Based upon six months of data, from January to June 1985, Chung et al. (1987a) found that on average, 73.5 per cent of the particle mass was in the fine (< 2.5 μm) fraction. The fine to coarse ratio varied slightly from 70 per cent in the winter to 75–79 per cent in the spring. These results suggest that the TSP problems in Seoul are primarily a result of anthropogenic activities which are the chief source of fine aerosols. In addition, a large portion of the particles in this size range tend to be secondary pollutants, such as sulphate.

Hashimoto et al. (1987) examined the composition of TSP in Seoul from May 1986 to March 1987. During this time the average concentration was 142 μg m^{-3}. Their analyses indicated that carbon was the major component, comprising 20–30 per cent by weight of the Seoul dust. Sulphate was found to be the second most important component, accounting for approximately 10 per cent. Nitrate and chlorine ions were next, followed by iron, lead and a variety of other metals. About 55 per cent of the total mass was not accounted for by their analyses. Results from Hashimoto et al. (1987) and Chung et al. (1987a) both suggest that anthropogenic activities are an important source of aerosols. Hashimoto et al. (1987) showed that the fraction due to carbon, both elemental (20 per cent) and organic (6 per cent), was relatively high. Their findings suggest that diesel fuel had a significant impact on the TSP burden in Seoul.

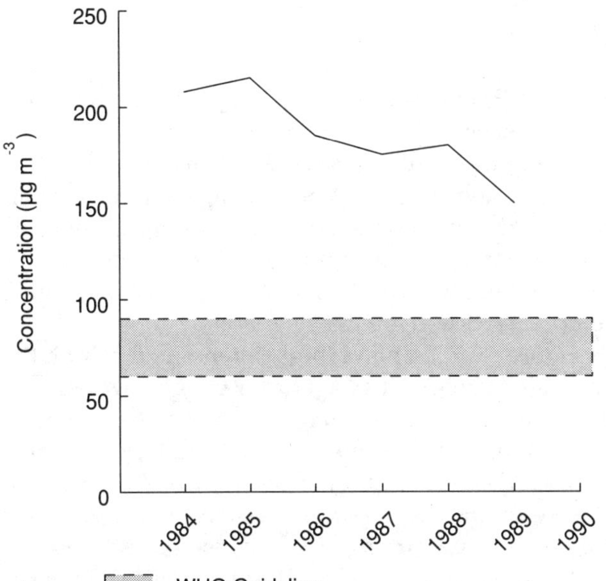

Figure 23.4 *Annual mean suspended particulate matter (TSP)*

After Ministry of the Environment, 1991

Lead

Ambient Concentrations No ambient Pb concentrations are reported from the co-ordinated monitoring network. However, Pb levels in TSP were determined by Hashimoto et al. (1987). In 1986 there were no limits on the Pb content of petrol in Korea. This explains the high ambient Pb levels observed. Current Pb concentrations are expected to be significantly lower than those in 1986 because of the introduction of unleaded fuels in July 1987.

Carbon monoxide

Emissions Over 930,000 tonnes of CO were emitted in the GSA in 1990 and over 68 per cent of this total was emitted in Seoul. Emission sources are split between residential (mainly incomplete combustion in domestic use of coal briquettes) which accounts for 44 per cent, and transport (56 per cent).

Ambient Concentrations Carbon monoxide levels in Seoul present a significant air pollution problem. The monthly mean concentrations based on the 1984–1989 time period are shown in Figure 23.5. Annual mean concentrations varied by less than 1 ppm (1.25 mg m^{-3}) with an overall mean of about 3 ppm (3.75 mg m^{-3}). There was no trend during the six-year period shown. Monthly concentrations display the same seasonal cycle as SO_2 and SPM but there is more disparity between the winter and summer levels of CO. They range from 5.0 ppm

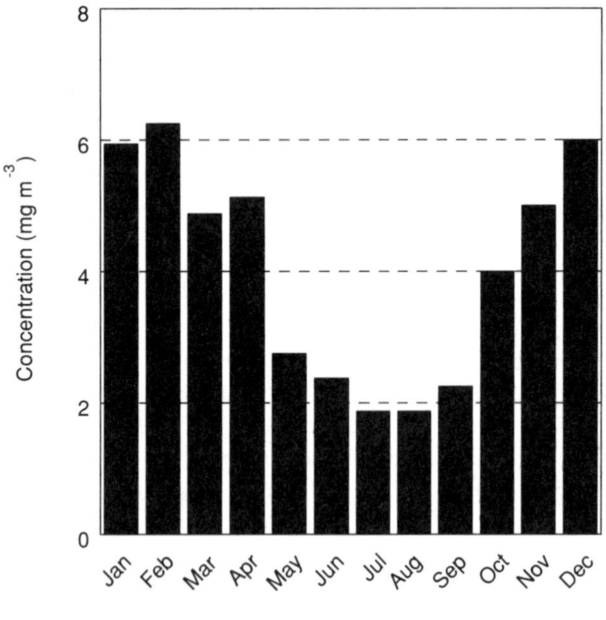

Figure 23.5 *Monthly mean CO concentrations (based on 1984-1989 data)*

After Ministry of the Environment, 1991.

(6.25 mg m^{-3}) to 1.5 ppm (1.875 mg m^{-3}). The enhanced seasonal cycle is consistent with the increased need for home heating in the cold season.

No short-term concentration data are available for comparison with the episodic standards of South Korea or WHO guidelines. It appears that there is no serious problem in Seoul. However, at individual stations, the standard was frequently exceeded during December and January 1984–89. The highest levels were reported in central and south-western districts. Given the magnitude of the monthly values, the prevalence of domestic coal combustion and the meteorological conditions in the winter and at night, it is probable that the short-term concentrations were close to, or in excess of, the national standards. Chung et al. (1987b) reported a maximum CO concentration of 13.8 ppm (17.25 mg m^{-3}) during January-February 1986 (the measurement is assumed to have been made over a 24-hour period).

Oxides of nitrogen

Emissions The emissions of NO_X increased markedly from 1984 to 1989. An increase in motor vehicle traffic was probably the main contributor to this change. The transport sector is dominant as a source of NO_X (78 per cent) (Figure 23.2). Total man-made NO_X emissions in 1990 were estimated to be 270,000 tonnes per annum in the GSA and 130,000 tonnes per annum in Seoul (Rhee, 1991).

Ambient Concentrations The mean annual nitrogen dioxide (NO_2) concentrations in Seoul have not risen in line with emissions. The data for ambient NO_2 are incomplete for some monitoring stations and the actual levels may be different to those reported.

The mean concentration increased slightly from 1984 to 1986–88, but dropped noticeably in 1989 (Figure 23.6). Throughout this time the levels were generally around 61 μg m^{-3}. This is well below the national air quality standard (NAQS). Some seasonal structure to the concentrations was apparent but it is difficult to explain the winter maximum based upon the temporal variation in emissions. This is because a dramatic shift in transportation behaviour is not expected from season to season. This suggests that meteorological factors had a significant influence. Lower mixing heights may have served to concentrate NO_2 during the winter, as well as a reduction in photochemical activity. Unlike SO_2, TSP and CO, there is no strong spatial pattern in NO_2 levels across Seoul because their sources are more diffuse and mobile.

While NO_2 did not appear to be a problem in 1989, continued increase in the motor vehicle population could lead to future problems. These could surface directly as high NO_2 levels or as secondary pollutants such as O_3 and nitric acid (HNO_3). Fortunately, South Korea has relatively strict emission standards for motor vehicles. These standards, and the use of catalytic converters, have been in effect since 1988 (Rhee, 1990) and should prove beneficial as older vehicles are replaced by newer, "cleaner" ones.

Ozone

Ambient Concentrations The six-year trend in annual mean concentrations is shown in Figure 23.7. As with NO_2, the measurements were not made consistently at all the monitoring stations during the period. Thus, the concentration data provided may not be entirely representative. Concentrations are higher in the warmer months due to increased insolation, higher temperatures and greater stagnation.

Figure 23.7 displays high interannual variability with 1985 being a "bad" year for O_3. One would conclude that this was because of a higher frequency of meteorological conditions favourable to O_3 production which implies that meteorology had a strong influence on the annual mean O_3 concentrations. This is true with respect to high O_3 episodes and is borne out by the seasonal variability. The site-to-site differences apparent in the data are in contrast to the variations for most of the other pollutants, particularly SO_2 and TSP. Annual mean and monthly O_3 concentrations were consistently lower in the east and south-west

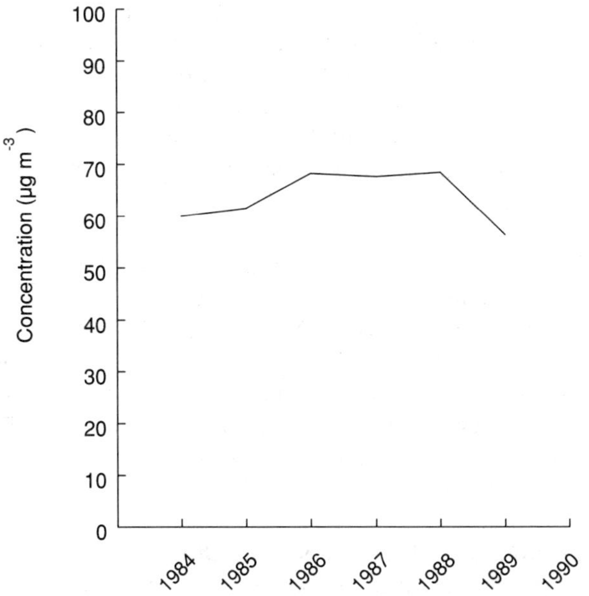

Figure 23.6 *Annual mean nitrogen dioxide concentrations*

After Ministry of the Environment, 1991

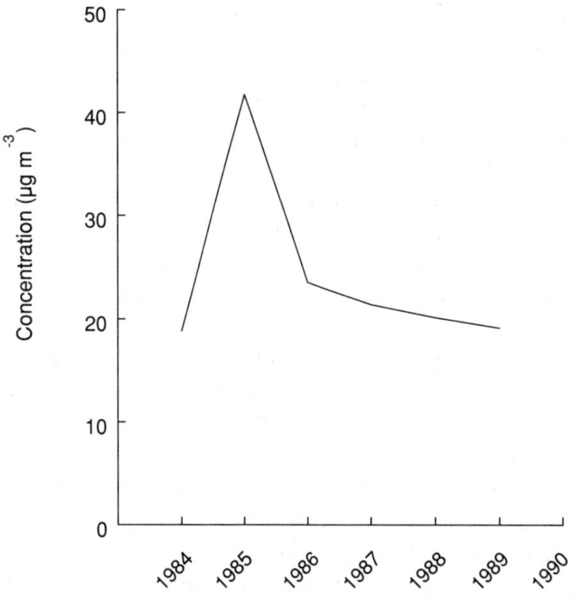

Figure 23.7 *Annual mean ozone concentrations*

After Ministry of the Environment, 1991

and higher in the south, north-east and, to a lesser degree, over central Seoul.

It can be seen from Figure 23.7 that concentrations since 1985 were within the national standard of 0.2 ppm (40 µg). However, data on the frequency of episodes are generally more appropriate for analyses of O_3 behaviour and in the assessment of the potential for health effects but this information was not reported by the Korean Environmental Administration. However, Chung et al. (1987b) examined the factors influencing O_3 in Seoul in September–October 1983 and 1984. They suggested that O_3 levels could rise in the future due to increasing motor vehicle traffic. Acting on this hypothesis, Chung et al. analysed the data collected at five monitoring stations in order to obtain a baseline for future studies. In their report they presented some values for the maximum hourly O_3 concentrations which ranged from 0.028 ppm (60 µg m^{-3}) to 0.087 ppm (186 µg m^{-3}) which were below the 0.10 ppm one-hour standard set by the national government. However, 0.087 ppm is a relatively high value, which could cause health problems for certain members of the population. Increased NO_x emissions and favourable meteorology could easily result in more O_3 episodes in the future.

23.4 Conclusions

Air Pollution Situation Industrialization and urbanization, in combination with a rise in living standards which has resulted in a rapid increase in energy consumption, have led to air pollution problems in Seoul. High SO_2 and TSP levels are the most serious problems. The national ambient air quality standards and WHO guidelines are frequently exceeded on an annual and episodic basis. In the winter, the levels tend to be at their highest due to increased heating demand and decreased vertical mixing. A major portion of the air pollution comes from the combustion of oil and coal. More specifically, the use of anthracite briquettes for home heating and cooking, bunker-C oil for industry and power plants, and motor vehicles are the predominant causes of the air quality problems. There is generally no record of chronic violation of the air quality standards for NO_2, CO and O_3. However, given the source characteristics and rate of emissions of some of these species, there is a potential for local episodic problems. In addition to the spatial distribution and rate of emissions, the local terrain around Seoul can have an effect on the air pollution levels. The mountain-valley combination has a strong influence on the prevailing winds through a direct funnelling effect and through more complex thermal circulations. The mountains also serve as a barrier which can contain the pollutants and allow their levels to build up over time.

Control Measures The actions of the Korean government (Rhee, 1990), are progressive and have potential to improve the air quality in Seoul. However, the projected increases in development and fuel consumption will significantly hinder their programmes.

The effects of the current policies and of new growth are not well understood.

The greatest need and potential for improvement in air quality are for SO_2 and TSP (Cho et al., 1986). Past reductions in the sulphur content of the fuel oils resulted in a decrease in SO_2 levels and plans for reduction should be adhered to in order to secure future gains. Industries are the main user of the high-sulphur oils and thus a major source of SO_2. Consequently, increased industrial activity could negate any air quality improvements due to decreases in fuel sulphur content. Stricter controls on industrial emission rates are likely to be needed if significant reductions in SO_2 levels in the industrial regions are to be made.

Increased emissions from coal combustion have the potential to offset any benefits as a result of changes in oil sulphur content so strategies to control SO_2 emissions due to coal are crucial. Since heating is the dominant use for coal in Seoul, new policies aimed at heating practices are necessary. While many of the larger facilities have been converting to liquified natural gas (LNG), it will be important for similar measures to be adopted for domestic heating. It is probably not feasible to control the emissions from the use of anthracite briquettes because of the large number of sources involved, and cleaner fuels are the most realistic option. These measures are doubly important because they will also serve to reduce CO emissions.

Motor vehicles are increasingly important as a source of air pollution, for example of TSP. Almost 50 per cent of the vehicles use diesel fuel. The current policy nationally is to reduce the number of diesel vehicles and to continue examining the potential for "cleaner" engines. The control of emissions from all motor vehicles is important. However, the current standards for cars are already relatively strict (equivalent to the US requirements for new cars). In addition to standards, routine inspection and maintenance have been incorporated into the programme. The final outcome of control measures is not known because the total vehicle population is increasing.

23.5 References

ADB 1989 *Energy indicators of developing member countries of Asian Development Bank,* Asian Development Bank, Manila.

Cho, B. H., Choo, S. Y. and Tamplin, S. A. 1986 *Air pollution control program development and policymaking in the Republic of Korea,* 79th Annual Meeting of the Air Pollution Control Association, Minneapolis.

Chung, Y., Jang, J. Y., Kwon, S. P. and Shin, D. C. 1987a Suspended heavy metals in the ambient air in Seoul. In: *Proceedings of the Third Joint Conference of Air Pollution Studies in Asian Areas,* Japan Society of Air Pollution, Environment Agency of Japan and United Nations Environment Programme Regional Office for Asia and the Pacific, Tokyo, 273–289.

Chung, Y., Jang, J. Y., Kwon, S. P. and Shin, D. C. 1987b An analysis of influencing factors on ozone concentration in the ambient air in Seoul. In: *Proceedings of the Third Joint Conference of Air Pollution Studies in Asian Areas,* Japan Society of Air Pollution, Environment Agency of Japan and United Nations Environment Programme Regional Office for Asia and the Pacific, Tokyo, 124–132.

Hashimoto, Y., Kim, H. K. and Otoshi, T. 1987 Air quality monitoring at Seoul as a part of JACK (Japan-China-Korea) network plan. In: *Proceedings of the Third Joint Conference of Air Pollution Studies in Asian Areas,* Japan Society of Air Pollution, Environment Agency of Japan and United Nations Environment Programme Regional Office for Asia and the Pacific, Tokyo, 87–100.

Kim, Y.-K. and Cho, K.-R. 1986 The motor vehicle emissions control programme in Korea, *The Second Joint Conference of Air Pollution Studies in Asian Areas,* Japan Society of Air Pollution, Tokyo, 100–111.

Ministry of the Environment 1991 *Korean Environmental Yearbook 1990, Part I: Air Quality, and Part II: Air Quality Management,* Ministry of the Environment, Seoul.

Rhee, D-G. 1990 Environmental Protection in Korea In: *Workshop on Energy Efficiency and Environmental Institutions in East and Southeast Asia,* 29 November 1990, Seoul, Ministry of Environment, Seoul.

Rhee, D-G. 1991 Air Pollution in the Megacity of Seoul, Ministry of Environment, Seoul. (Unpublished.)

UN 1989 *Prospects of World Urbanization 1988,* Population Studies No. 112, United Nations, New York.

24

Shanghai

24.1 General Information

Geography Shanghai is on the east coast of China at Latitude 31°10′N and Longitude 122°26′E. The Yangtze River (Changjiang) is to the north and Hangzhou Bay is to the south. Two main rivers, the Huangpu and the Suzhou, flow through the city, which is at an elevation of 4.5 metres above mean sea level (Figure 24.1). Shanghai has an intricate network of canals and waterways providing westward connections to the Taihu region.

Metropolitan Shanghai covers a total area of approximately 6,300 km^2, and consists of a main city area and nine surrounding counties including an agricultural hinterland. A number of nearby islands in the mouth of the Yangtze River and in the East China Sea are also considered part of the municipality on the province level.

Demography Shanghai is the most populated urban area in China. The 1982 census counted 6.3 million people in the city proper and 11.2 in the entire urban area. The official 31 December 1988 population estimate was 12.6 million for the municipality, and the July 1990 census count was 13.3 million. The UN estimates project the population of Shanghai to grow to 14.69 million by the year 2000 (UN, 1989).

Shanghai is one of the world's most densely populated cities. The average population density is about 2,100 persons per km^2, but in some areas population density exceeds 48,000 persons per km^2. The densely populated central city is surrounded by transitional zones and a rural hinterland which is the most densely populated rural area in the world.

Climate Shanghai has a humid sub-tropical climate, with an annual mean temperature of 15.7°C. The monthly minimum is approximately 3°C in January and the monthly maximum is about 28°C in August.

The city lies on a flat deltaic plain over which wind speeds average 3.7 m s^{-1}. The annual mean precipitation is 1,040 mm, with rainfall heaviest in June and lightest in December. The mean atmospheric pressure at Shanghai is 102 kPa. Acid precipitation problems have been noted, with an annual mean pH value of 4.7 (Ning et al., 1987).

Industry Shanghai is a major industrial and commercial centre with one of the largest ports in the world. The principal industries in the city are machine tools, iron and steel, and chemical plants. The chemical industries produce plastics, fibres, paint, drugs, pesticides, fertilizers, detergents and petroleum products. There are over 160 textile mills in the municipality in addition to many plants producing consumer goods. The city has 10,565 "factories and enterprises", with 8,000 steam boilers (Rukang, 1989). Owing to an absence of industrial zoning, many small industrial facilities are located in residential areas, with the accompanying direct personal exposure to industrial air pollutants.

Shanghai is a net exporter of electrical energy. Coal-fired power plants generate 500,000 kilowatts per year. Total coal consumption is estimated at 25 million tonnes per annum which leads to a waste gas volume of about 800 million m^3 per day.

Over 70 per cent of the coal burned in Shanghai is for industrial purposes, while the remaining 30 per cent is for domestic purposes. Shanghai has been given permission to purchase coal with a sulphur content of 1.2 per cent, which is lower than that of the average coal in China. Coking gas production is approximately one million m^3 per day. In 1990 the Shanghai portion of China's GNP was Renminbi (RMB) 50 billion per year (1 $US = 3.72 RMB).

Transport In 1990 there were over 147,700 civilian motor vehicles, including 54,586 passenger vehicles and 79,192 trucks. Buses and trolley-buses are the main means of urban passenger transport.

24.2 Monitoring

In the framework of the GEMS/Air programme, sulphur dioxide (SO_2) and suspended particulate matter (SPM) measurements have been made since 1981. Daily data for the period 1981-1989 are available. The

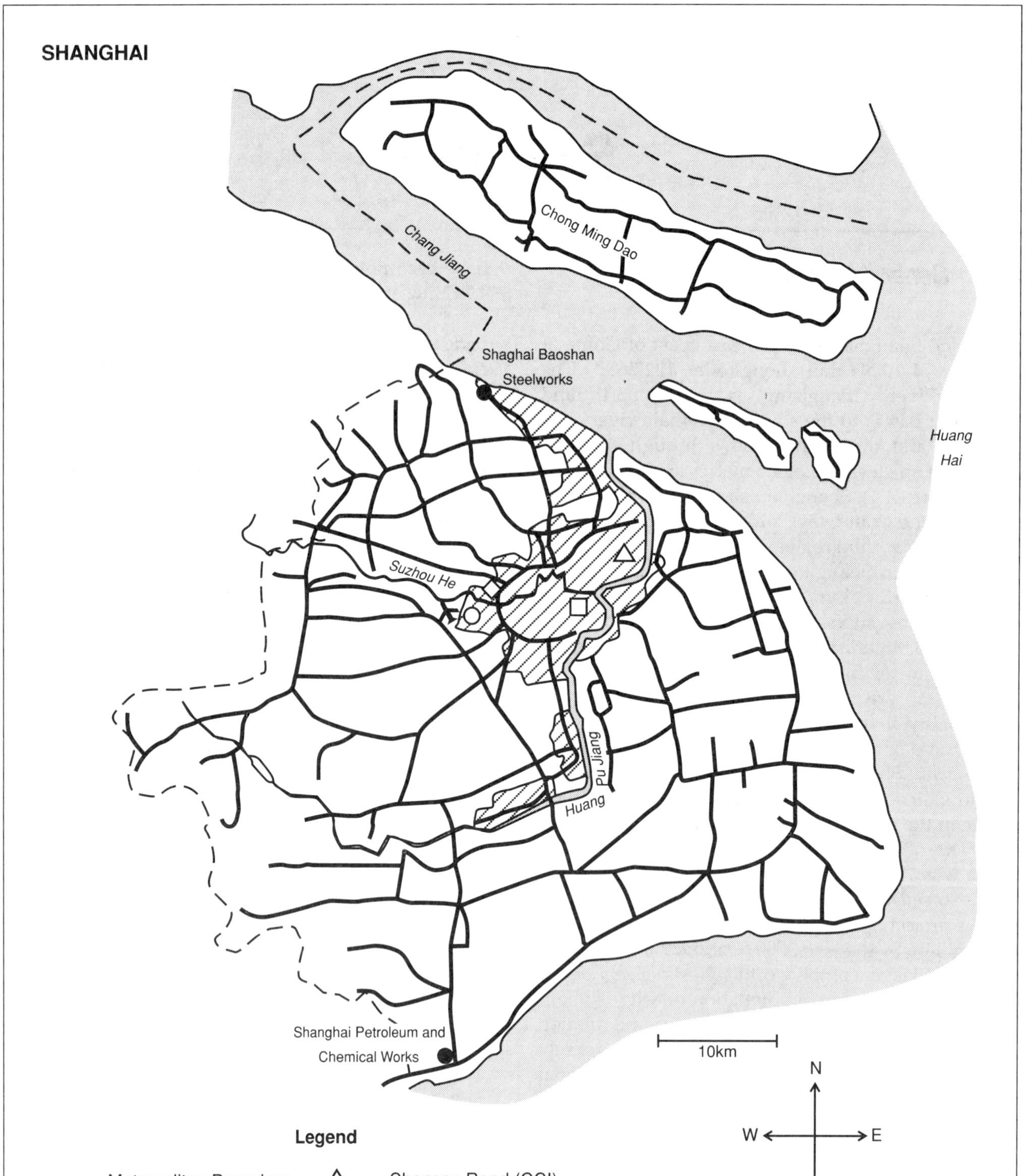

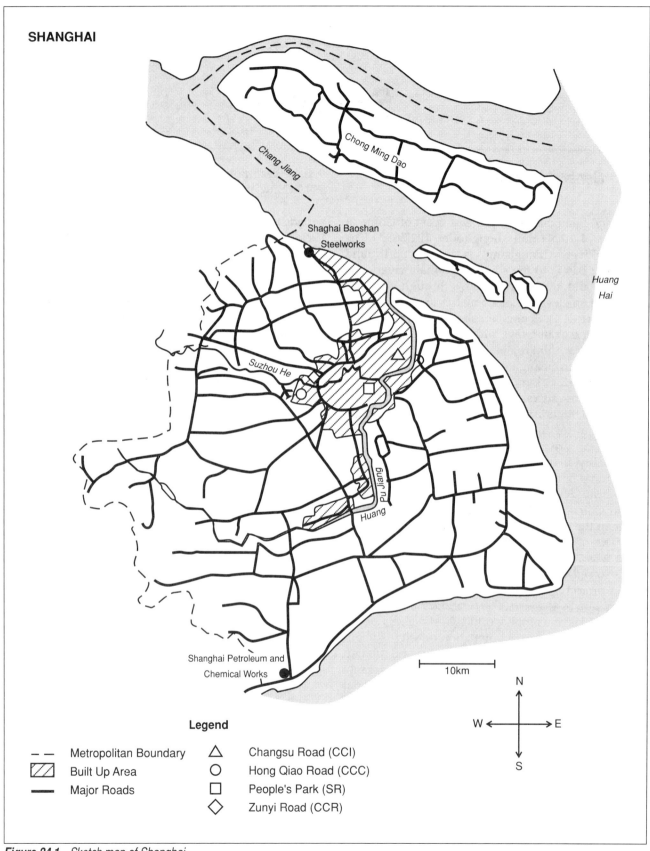

Figure 24.1 *Sketch map of Shanghai*

Table 24.1 *Comparisons of EPB and GEMS sulphur dioxide data ($\mu g\ m^{-3}$), 1985*

Area	Network	Median (50%)	Arithmetic Average	98 Percentile	Maximum
Industrial	GEMS	39	60	193	380
	EPB	116	162	436	568
Commercial	GEMS	35	55	207	346
	EPB	92	104	268	572
Residential	GEMS	97	101	208	264
	EPB	91	93	220	400
Suburban	GEMS	6	13	50	63
	EPB	16	26	129	233

Source: Mage, 1988

four Shanghai GEMS/Air monitoring stations are classified as suburban residential (SR), centre city commercial (CCC), centre city residential (CCR), and centre city industrial (CCI).

In addition to the GEMS/Air network, the Shanghai Environmental Protection Bureau (EPB) runs an environmental monitoring centre for air pollution (Ning et al., 1987; Rukang, 1989). Data on all the major pollutants are monitored continuously at six locations and are telemetered by radio to the central stations where they are recorded. Manual samples for SPM, SO_2 and oxides of nitrogen (NO_x) are collected by EPB at an additional 32 sites in Shanghai city and the surrounding metropolitan area.

During a WHO site visit (Mage, 1988), SPM and SO_2 monitoring data from GEMS stations were compared with data from the Shanghai EPB stations. Whereas SPM data from both monitoring networks were comparable, the 1985 SO_2 data from EPB appeared to be 100 per cent higher than those from the GEMS stations at similar locations (Table 24.1). The cause of these differences is unknown as intercomparison studies have not been conducted. The EPB data for other years have not been made available for this report so the data for ambient air pollutant concentrations given in the subsequent chapter refer to the GEMS data only.

Table 24.2 *Emissions from industrial and domestic fuel combustion for base year 1983*

Pollutant	Emissions ($t\ a^{-1}$)
SO_2	267,000
SPM	324,000
CO	229,000
NO_x	127,000

Note: There are additional emissions resulting from motor vehicle exhaust (CO, NO_x, SPM) and from industrial processes (SO_2, SPM)
After Zhao and Zhao, 1985 as tabulated by Ning et al., 1987

The Shanghai EPB has developed a low-cost temperature controlled bubbler system for SO_2 monitoring using the modified Schiff reaction (formalin-pararosanaline) as produced by several Chinese companies.

24.3 Air Quality Situation

Sulphur dioxide

Emissions There is no recent SO_2 emissions inventory available for Shanghai. Taking 1983 as a base year, SO_2 emissions from industrial and domestic fuel combustion have been estimated to be 267,000 tonnes per annum (Zhao and Zhao, 1985 as tabulated by Ning et al., 1987, Table 24.2). When assessing these figures, it must be considered that there are additional SO_2 emissions resulting from industrial processes. Furthermore, emissions will certainly have risen considerably since 1983. However, in the absence of relevant data no 1991 estimate can be produced.

It was reported that inefficient household stoves burning high-sulphur coal (with an average sulphur content of 2-3 per cent) contribute approximately 35-40 per cent of the SO_2 released into the atmosphere in China (Ning et al., 1987).

Ambient Concentrations Sulphur dioxide has been monitored at the Shanghai GEMS/Air sites since

1981. As stated above, the Shanghai EPB runs an additional monitoring network for six locations.

As shown in Figure 24.2, annual mean SO_2 concentrations have remained relatively constant during the past four years. Only one of the four monitoring stations, the SR site, was below the WHO annual mean guideline value of 40–60 μg m^{-3} for SO_2 during the whole monitoring period. Annual means at the CCI site were usually within the WHO guideline range, although there was an increasing trend in recent years; in 1989, the annual mean (63 μg m^{-3}) slightly exceeded the WHO annual guideline. Sulphur dioxide concentrations at the CCC site exceeded the WHO guidelines concentration by 17–25 per cent from 1986 to 1988, but concentrations were decreasing steadily and the 1989 annual mean value was 59 μg m^{-3} which is just below the WHO guideline value. Annual mean SO_2 concentrations at the CCR site increased steadily from 1981 to 1987 up to about 100 μg m^{-3} which is about 70 per cent above the upper value of the WHO guideline. Since 1987 annual mean SO_2 levels have remained high at the CCR station.

As documented in Table 24.1, GEMS data are considerably lower than the EPB data. Whatever the cause of these discrepancies, the air pollution situation as described by the GEMS network is probably optimistic.

Daily SO_2 mean concentrations were monitored on 86–180 days per year. As the short-term WHO 98 percentile guideline for SO_2 is set at 150 μg m^{-3}, not to be exceeded on more than 2 per cent of the monitored days, the available exceedences were standardized for the full 365 days per year. Figure 24.3 shows the calculated number of days per year on which the WHO 98 percentile guideline was expected to be exceeded. Only the SR monitoring site met the WHO short-term guideline in all the years. At the CCC site, the guideline was met in 1984, 1987 and 1989. At two (CCR, CCI) of the four monitoring sites, the short-term guideline was not met in the last five years with about 50–60 days per year exceeding 150 μg m^{-3} as a daily average. Especially at the CCR site, and the number of days above 150 μg m^{-3} has been generally increasing since 1981.

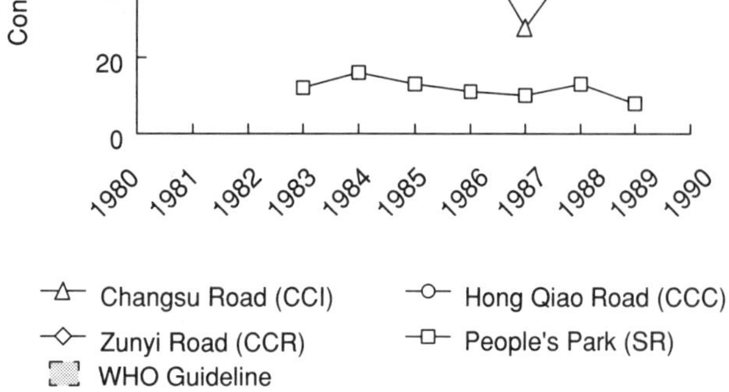

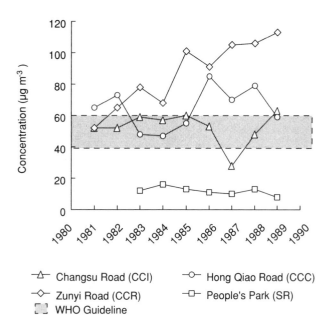

Figure 24.2 *Annual mean sulphur dioxide concentrations in Shanghai*

Source: GEMS data

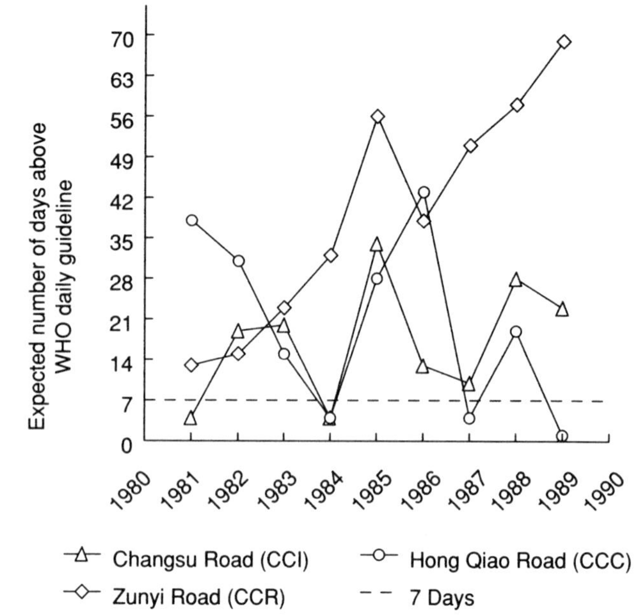

Figure 24.3 *Number of days on which 98 percentile sulphur dioxide guideline concentration was expected to be exceeded*

Note: SR station is not included because the guideline was always met
Source: GEMS data

Maximum daily SO_2 concentrations were around 150-300 μg m^{-3} at the city centre sites and around 50 μg m^{-3} at the suburban site. The highest 24-hour mean SO_2 concentration was 479 μg m^{-3}, recorded at the CCR site.

However, indoor SO_2 air pollution can be much more serious than ambient levels. For instance, Rukang (1989) reported that SO_2 levels in Shanghai kitchens have been as high as 2,440 μg m^{-3}, which is 16 times the WHO daily guideline value (the WHO 10-minute guideline is 500 μg m^{-3}). This indicates that indoor SO_2 pollution in Shanghai is cause for significant concern.

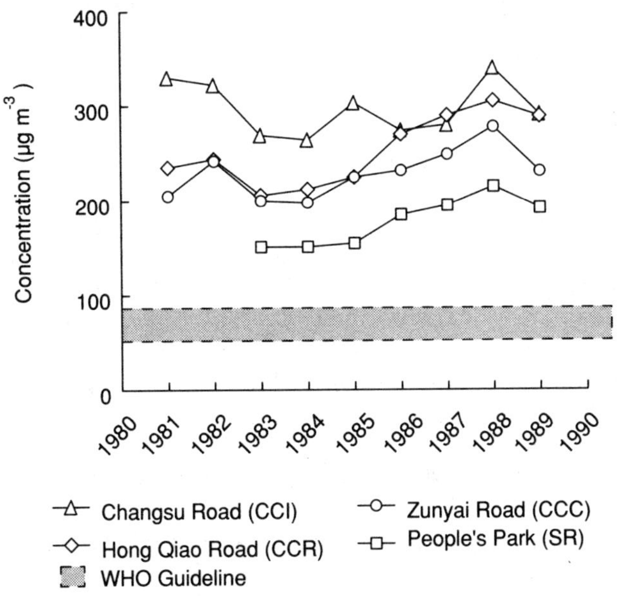

Figure 24.4 *Annual mean suspended particulate (TSP) concentrations*

Source: GEMS data

Suspended particulate matter

Emissions As with SO_2, there is no recent SPM emission inventory available for Shanghai. It has been estimated that in 1983 SPM emissions from industrial and domestic fuel combustion sources were 324,000 tonnes per annum (Zhao and Zhao, 1985, as tabulated by Ning et al., 1987, Table 24.2). When assessing these figures, it must be considered that there are additional SPM emissions resulting from industrial processes and motor vehicles, as well as from natural sources. As most natural sources of SPM are not quantifiable, SPM emission inventories are difficult to establish. Since 1983, SPM emissions from fuel combustion have probably risen in line with the increase in population. However, as with SO_2, no estimate of probable 1991 emissions can be produced.

Ambient Concentrations Suspended particulate matter has been monitored in Shanghai by the GEMS/Air network since 1981. The high-volume samplers capture total suspended particlulates (TSP) including respirable particles (less than 10 µm in aerodynamic diameter), and non-respirable particles (greater than 10 µm). A comparison of SPM data from the GEMS and the EPB networks showed that the two were comparable.

Annual mean SPM (TSP) levels at the four GEMS/Air monitoring sites are shown in Figure 24.4. The WHO annual mean guidelines of 60–90 µg m^{-3} were exceeded at each site during the whole monitoring period by a factor of two to three. After a decrease in SPM levels from 1981 to 1983, annual mean SPM concentrations increased again after 1984. Suspended particulate matter values were lowest at the SR site, increasing from about 150 µg m^{-3} in 1985 to 200 µg m^{-3} in 1988. The highest annual SPM values were measured at the CCI site, where the minimum annual value was 264 µg m^{-3} in 1984 and the maximum was 340 µg m^{-3} in 1988 (EPB, 1991). The CCR site shows similar SPM pollution to the CCI site. Suspended particulate matter levels at the CCC site are lower than at other city centre sites: annual mean SPM concentrations increased from about 200 µg m^{-3} in 1984 to about 280 µg m^{-3} in 1988. From 1988 to 1989, annual SPM concentration values fell at all monitoring sites by 10–17 per cent. These data are consistent with a report (Ning et al., 1987) which noted an annual mean SPM (TSP) concentration of 200 µg m^{-3} in Shanghai for 1983. These annual SPM values even exceed the Shanghai interim air quality goal of 250 µg m^{-3}.

With respect to daily mean SPM levels, the WHO 98 percentile guideline of 230 µg m^{-3} (which should not be exceeded on more than 2 per cent of the days monitored, in this case on 7 out of 365 days) was exceeded in all stations during the whole monitoring period. Figure 24.5 shows the calculated number of days per year on which this WHO daily guideline was exceeded. This was the case on approximately

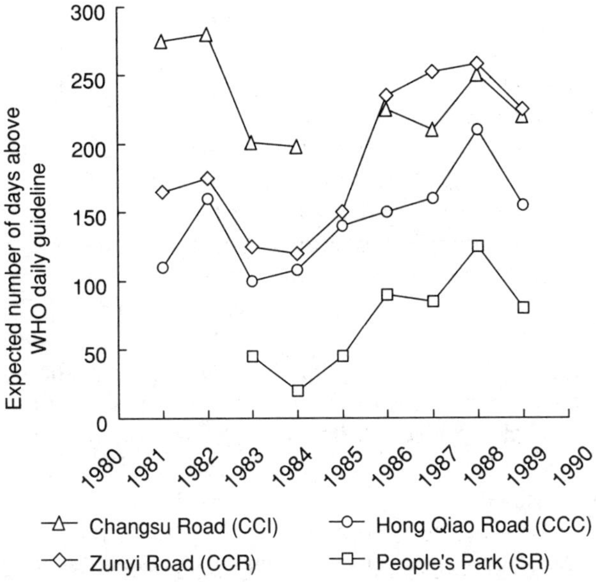

Figure 24.5 *Number of days on which the 98 percentile suspended particulate matter guideline concentration was expected to be exceeded*

Source: GEMS data

200 days per year at the CCR and CCI sites, on approximately 150 days at the CCC site, and on approximately 75 days at the SR site.

According to the EPB, the mean concentration of inhalable particulate matter (IP) has fallen from 280 $\mu g\ m^{-3}$ in 1978 to 160 $\mu g\ m^{-3}$ in 1984 due to the implementation of effective control actions. In 1987 IP levels had risen again to 190 $\mu g\ m^{-3}$ (which is above the Shanghai interim goal of 175 $\mu g\ m^{-3}$) because of increased urban construction and expansion of industrial production.

Lead

Emissions There are no estimates of lead (Pb) emissions in Shanghai. It is not even possible to estimate the order of magnitude because there are also no traffic statistics available. However, it can be assumed that Pb emissions from motor vehicle traffic are probably relatively small due to the still low density of motor vehicles.

Ambient Concentrations No data for Pb concentrations in ambient air are collected by the GEMS/Air network. Shanghai EPB data on Pb have not been made available for this report.

Carbon monoxide

Emissions As with SO_2 and SPM, there is no recent carbon monoxide (CO) emission inventory available for Shanghai. Taking 1983 as a base year, CO emissions from industrial and domestic fuel combustion have been estimated to be 229,000 tonnes per annum (Zhao and Zhao, 1985 as tabulated by Ning et al., 1987, Table 24.2). This figure would seem to be rather low, especially if it is taken into account that most combustion occurs in low-efficiency small or medium boilers which emit high amounts of CO. Furthermore, there are additional CO emissions from motor vehicle exhausts. Since 1983, population and fuel use have increased which should have led to increased CO emissions. At the same time, improved combustion efficiency probably offset part of the increase in emissions. Thus no estimate of 1991 emissions can be produced.

Ambient Concentrations Ambient CO concentrations are not measured by the GEMS/Air network. The Shanghai EPB measures ambient CO routinely, but these data are not available.

Indoor CO levels as high as 21.4 mg m^{-3} have been reported by Rukang (1989) which are well above the WHO recommended maximum eight-hour mean value of 10 mg m^{-3}. This indicates high personal exposure for the most seriously exposed persons.

Oxides of nitrogen

Emissions As with the other major air pollutants, there is no recent emissions inventory for NO_x available for Shanghai. For the base year 1983, NO_x emissions from industrial and domestic fuel combustion have been estimated to be 127,000 tonnes per annum (Zhao and Zhao, 1985 as tabulated by Ning et al., 1987, Table 24.2). Additional NO_x emissions come from motor vehicle exhausts.

Since 1983 NO_x emissions will have certainly increased considerably, as more efficient high-temperature combustion processes normally lead to increased NO_x emission rates. Thus it may be assumed that 1991 NO_x emissions are probably more than 50 per cent higher than the 1983 figure.

Ambient Concentrations Ambient NO_x concentrations are not measured by the GEMS/Air network. The only available data are from Ning et al. (1987) who noted that annual mean concentrations of nitrogen dioxide (NO_2) in Shanghai in 1983 were around 50 $\mu g\ m^{-3}$. This would be considerably below the WHO one-hour guideline (400 $\mu g\ m^{-3}$) and the WHO 24-hour guideline (150 $\mu g\ m^{-3}$).

Indoor NO_2 levels seem to be higher than those in outdoor air. Oxides of nitrogen levels as high as 90 $\mu g\ m^{-3}$ have been reported in Shanghai kitchens using coal (Rukang, 1989). The sampling time for these measurements is not clear, but the concentrations are below the WHO 24-hour guideline. These low levels of NO_x are due to low-temperature combustion in inefficient stoves.

Ozone

Ambient Concentrations Ambient ozone (O_3) concentrations are not measured by the GEMS/Air network. The Shanghai EPB measures ambient O_3 routinely, but these data were not available for this report.

24.4 Conclusions

Air Pollution Situation The main ambient air pollution problem in Shanghai results from the very high

SPM and SO_2 levels. Short-term as well as long-term averages regularly exceed the WHO guidelines in large parts of the urban area. Although Shanghai has low SPM values compared with some other major Chinese cities, it is possible that the chemical composition may be more toxic owing to the higher proportion of industrial SPM emissions.

For the other pollutants, a lack of available routine monitoring data makes an assessment difficult. There are no data at all on Pb, CO, and O_3 concentrations. A report from 1983 suggests that NO_2 concentrations may be low in most urban areas.

Indoor air pollution in Shanghai can be very high because of inefficient household stoves burning high-sulphur coal. Sulphur dioxide levels in coal-fired kitchens were found to be as high as 16 times the WHO short-term guideline. CO levels in kitchens also seem to be above WHO short-term guidelines.

Main Problems From the monitoring data it must be concluded that the main pollution problem in Shanghai stems from combustion of high-sulphur coal in inefficient boilers. Although there are no data available, urban transport probably contributes to a lesser extent to overall air pollution in Shanghai. However, motor vehicle exhausts certainly have considerable effects on roadside air pollution, and motor vehicle traffic is likely to increase greatly in the next decade. Shanghai has significant problems with control of SO_2 and SPM because the industrial and residential zones are not separated. Industry is dispersed throughout the city and residences surround the factories. The close proximity of residences and industry means that the exposure of the population to air pollution may be very high.

Control Measures Air pollution controls have been implemented in Shanghai over the past decade by the EPB. As part of an effort to ameliorate the SO_2 pollution situation, Shanghai has been granted permission to purchase low-sulphur (1.2 per cent) coal. In 1982–1984, black smoke emissions were reduced significantly through a campaign of furnace and boiler renovation in which 95 per cent of these units were modified. Improvements in SPM and dust controls were instituted by the introduction of a new type of cyclone separator that increased capture efficiency from less than 70 per cent to 85 per cent.

According to the EPB, SPM control actions were effective as the average concentration of inhalable particulate matter (IP) had fallen by about 40 per cent from 1978 to 1984. However, until 1987, levels of IP had risen again by 20 per cent. The reasons for this recent increase in IP levels were:

- Increased construction in Shanghai: the population boom in Shanghai has led to growth in domestic housing and new commercial activities;
- Expansion of industrial production in Shanghai: recent economic developments led to demand for expansion of local industry; and
- Ageing of existing facilities causing operation efficiencies to decrease: some industry has emission controls, but older equipment does not operate at design efficiency.

Shanghai has instituted SPM (TSP) emission regulations of 200–600 mg m^{-3}, depending on the size of the facility. There is concern as to whether the Shanghai interim air quality goals of 250 $\mu g\ m^{-3}$ SPM (TSP) or 175 $\mu g\ m^{-3}$ can be met because coal use will increase from the present 25 million tonnes per annum to 50 million tonnes per annum by the year 2000. A portion of this additional demand will be supplied through high-sulphur coal, which again would greatly increase the SO_2 emissions. At the present time the high cost of flue gas desulphurization (FGD) would cause a great economic burden for Shanghai if SO_2 levels were to be reduced to meet the applicable standards. The recommendations of a Japan International Co-operation Agency (JICA) study on Shanghai SO_2 pollution in 1986 and 1987 (Mage, 1988) were:

- Change over to coal gas from coal briquette burning in homes. At present 55 per cent of the people use coal gas and the goal should be 80 per cent;
- Institute central heating and co-generation of electricity for groups of industries;
- Improve the coal briquetting process by adding lime (CaO) in small-scale plants;
- Develop more economical FGD processes and build a pilot plant for demonstration;
- Relocate major industries from residential areas in Shanghai to industrial areas outside the town.

The Shanghai EPB is currently working on a master plan that parallels in many instances the JICA proposal. The planning model is based on the Gaussian plumes computed from a detailed emission inventory using individually determined emission factors. This model, which is used to predict concentrations up to the year 2000, has been calibrated with recent data and it is being used to predict the effects of different control strategies on the various sources.

Table 24.3 *Comparison of dustfall and benzo-[a]-pyrene (BaP) in various urban functional zones in Shanghai*

Urban Areas	Dustfall ($g\ m^{-2}\ d^{-1}$)	BaP in dust ($\mu g\ g^{-1}$)	BaP deposition ($\mu g\ m^{-3}\ d^{-1}$)
Commercial district	1.02	7.35	7.78
Industrial district	1.15	5.46	6.60
Residential district	0.70	5.61	3.92
Suburban district	0.22	1.31	0.34

After Rukang, 1989

In summary, Shanghai has developed control strategies for meeting environmental standards. The costs of the control measures are the main constraints since most of the technology is already available. This cost problem will require adaptation of low-cost pollution control systems to the prevailing conditions for local industry.

Health Effects The current SO_2 and SPM levels in Shanghai are of significant concern from both acute and chronic perspectives. Observed levels are associated with chronic pulmonary insufficiency in ageing populations and in susceptible individuals. The monitored extreme maximal SPM values are associated with acute morbidity in older age groups and premature mortality for persons most at risk.

Rukang (1989) noted that "male cancer mortality in Shanghai is the highest in China" and that from 1963 to 1985 lung cancer mortality increased from 20.9 per 100,000 to 44.3 per 100,000. It has been observed that lung cancer incidence is highest within urban districts and lowest in the outermost regions, and higher in industrial and commercial regions than in residential areas of Shanghai (Rukang, 1989). In relating air pollution to lung cancer mortality, it was shown that concentrations of the highly carcinogenic compound benzo-[a]-pyrene (BaP) were highest in commercial and industrial districts where particulate matter (expressed as dustfall in g m^{-2} per day) was also highest (Table 24.3). The high incidence of lung cancer in Shanghai has been mapped in detail (Editorial Committee for the Atlas of Cancer mortality in the People's Republic of China, 1979).

In one area of Shanghai, where extremely high levels of hexavalent chromium have been measured, it was reported that all children who attended a particular elementary school have failed the military entrance physical examination and the people have a very high lung cancer rate compared with other communities.

24.5 References

EPB 1991 Personal communication, Shanghai Environmental Protection Bureau.

Editorial Committee for the Atlas of Cancer Mortality in the People's Republic of China 1979 *Atlas of Cancer Mortality in the People's Republic of China,* China Map Press, Shanghai.

Mage, D. T., 1988 *WHO Assignment Report,* Centre for Promotion of Environmental Planning and Applied Studies (PEPAS), Kuala Lumpur.

Ning, D., Whitney, J. B. and Yap, D. 1987 Urban air pollution and atmospheric diffusion research in China, *Environmental Management* **11**, 721–728.

Rukang, F. 1989 Environment and cancer in Shanghai, *Journal of Environmental Science (China)* **1**, 1–9.

UN 1989 *Prospects of World Urbanization 1988,* Population Studies No. 112, United Nations, New York.

Zhao, Z. X. and Zhao, Z. P. 1985 *Air pollution in Shanghai and approaches to its control,* Preprint No. 85-59-A7 of the 78th Annual Meeting of the Air Pollution Control Association, Detroit.

25

Tokyo

25.1 General Information

Geography Tokyo (Latitude 35°40′N, Longitude 139°45′E), the capital of Japan, is located roughly in the middle of the Japanese archipelago, in the south of the Kanto plain. The city of Tokyo lies at the mouth of the Sumida River (Figure 25.1). To its east are suburbs comprising the prefecture of Chiba which is separated from Tokyo by the Edo River. To the north of Tokyo City are the suburbs of Saitama prefecture. South of Tokyo, separated by the Tama River, is the large industrial centre of Kawasaki. The city of Yokohama (part of the urban agglomeration) is situated to the south-west.

In addition, bordering the city of Tokyo is Tokyo Bay which receives its water from the Pacific Ocean. The original terrain consisted mainly of river deltas, hills, and sand dunes. The present coastline is lined with warehouses, docks and industrial complexes.

Demography Tokyo covers an area of approximately 2,200 km^2, which is 0.6 per cent of Japan's total land area. It is one of the most populous cities in the world with about 12 million people, which accounts for about 10 per cent of the national total. The Tokyo/Yokohama agglomeration had an estimated population of 20.52 million in 1990 and at that time was the largest urban agglomeration in the world (UN, 1989). Owing to the importance of Tokyo as an economic and cultural centre, people have been settling in this area, resulting in a 50 per cent increase in population since 1980. The population density is 5,417 people per km^2. The projected population in 2000 is 21.32 million (UN, 1989).

Climate Tokyo is located in a humid sub-tropical climatic zone and has four seasons. The conditions in this region are generally mild, and the mean annual temperature is about 15°C. The winters have low temperatures averaging 5°C and summer temperatures average 26°C. Tokyo's climate is controlled by the summer and winter monsoons: in summer, masses of warm, humid air from the Pacific; in winter, a flow of cold, dry air from Siberia. There are two rainy seasons: June–July and September–October. Tokyo usually has two or three typhoons a year during the autumn rainy season. The annual mean rainfall is about 1,460 mm.

Industry Tokyo has a large number of manufacturing industries including textiles, toiletries, printing and publishing. It also produces goods requiring a large labour force, such as electrical products, cameras and automobiles. The neighbouring Yokohama-Kawasaki district is an area of heavy industry specializing in chemicals, machinery, metallurgy, petroleum refining, ships, motor cars and fabricated metal products. A centre for iron and steel, petroleum refining, petrochemicals, electric power, and other heavy industries is also located on the Chiba-Ichihara coast in the north-east of Tokyo.

Most of the metropolitan area is supplied with gas by the Tokyo Gas Company. Power plants are located mainly on the coast where domestic and imported coal, petroleum, liquid gas, and natural gas are brought in by ship. Petroleum and bottled propane gas are widely used for heating and cooking. The bulk of Tokyo's power comes from thermal stations but approximately 15 per cent comes from hydroelectric stations. Electricity is also generated by nuclear power stations. It is interesting to note that fossil fuel consumption in Japan between 1975 and 1987 has remained relatively stable, possibly with marginal increases.

Transport Tokyo is a national transport hub. The metropolitan area is spanned by a dense network of electric railways and subways, bus routes, and motor highways. Tokyo's system of motor vehicle expressways consists of a loop around the central business district and several radial lines connecting with the national expressways.

Motor vehicle traffic has increased rapidly over the last 30 years. While in 1960 the number of registered motor vehicles was about 500,000, it was more than 2 million by 1970 and reached 4.4 million in 1990, which is almost a ten-fold increase in 30 years. The number of motor vehicle registrations for 1975–1990 is summarized in Table 25.1. Nearly 80 per cent of all vehicles are regular and "very small" passenger cars (TMG, 1989a).

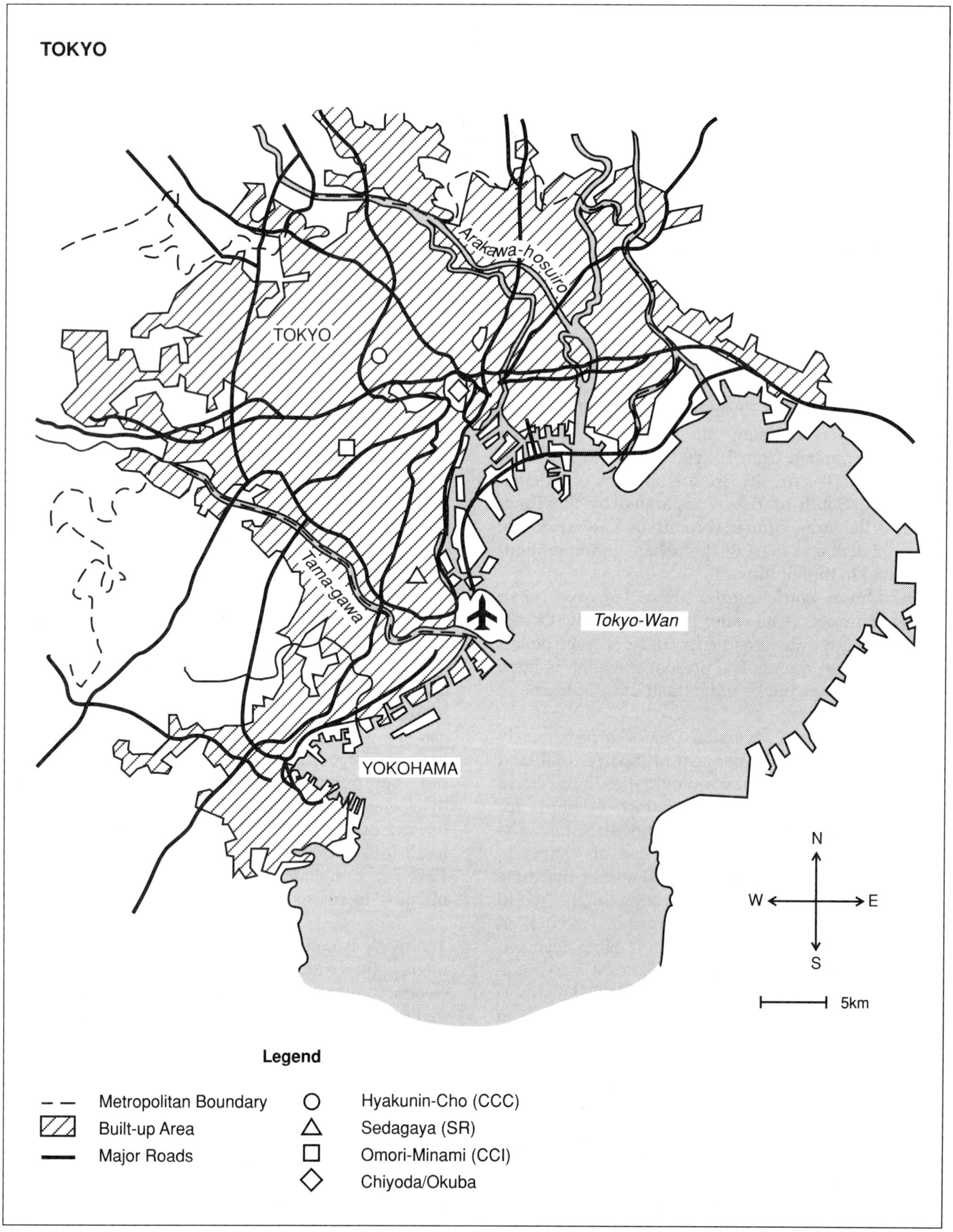

Figure 25.1 *Sketch map of Tokyo/Yokohama agglomeration with monitoring sites*

Table 25.1 *Motor vehicle registrations (thousands) in Tokyo Metropolitan Government area*

	1975	1980	1985	1990
Passenger vehicles	1,439	1,731	2,120	2,701
Very small vehicles	329	355	571	766
Buses	18	16	15	16
Small cargo vehicles	664	725	678	650
Heavy cargo vehicles	111	128	144	182
Other	49	55	65	80
Total	2,610	3,010	3,593	4,395

After TMG, 1989a

25.2 Monitoring

The Environment Agency of Japan has set up a national air monitoring network. The network is spread over 15 regions and contains over 1,000 stations (Government of Japan, 1988). Each station is equipped with several instruments for monitoring sulphur dioxide (SO_2), oxides of nitrogen (NO_x) and other air pollutants. National environmental background air monitoring stations are located at eight sites with the purpose of analysing atmospheric composition in unpolluted regions.

Within the Tokyo metropolitan area, there are 35 automated ambient air monitoring stations, 20 of which are in the city centre and 15 in the suburbs. In those stations SO_2, carbon monoxide (CO), suspended particulate matter (SPM), nitrogen dioxide (NO_2), nitric oxide (NO) and photo-oxidants are monitored continuously. In addition, there are 32 car exhaust monitoring stations measuring CO, NO_2 and NO. Some of the car exhaust monitoring stations also measure SO_2, total suspended particulates (TSP) and photo-oxidants. Besides the routine monitoring stations, there are two special monitoring sites: one measuring pollutant concentrations on a vertical tower, and the other in a countryside area.

There are three GEMS/Air monitoring sites in Tokyo measuring SO_2 and SPM. One site is located in the commercial city centre (Hyakunin-Cho, Shinjuku-Ku), one in an industrial city centre area (Omori-Minami, Ota-Ku), and one in a suburban residential area (Sedagaya, Sedagaya-Ku). Sulphur dioxide is measured with an instrumental conductimetric method, and SPM is measured using filter beta-ray absorption, nephelometer light scatter and high-volume gravimetric methods.

25.3 Air Quality Situation

Sulphur dioxide

Emissions Data on emissions of SO_2 in the Tokyo metropolitan area were not available for this report. National SO_2 emissions in Japan were 835,000 tonnes per annum in 1986. This equals a per capita emission of 6.9 kg SO_2 per person per annum which is one of the lowest per capita rates in industrialized countries (OECD, 1991).

Sulphur dioxide emissions in Tokyo and Japan have reduced dramatically since 1970. This was achieved through fuel desulphurization, and through the development and wide use of emission gas treatment technologies for mobile and stationary sources.

Ambient Concentrations Sulphur dioxide has been monitored in Tokyo since 1965. Data from the GEMS/Air sites are available from 1972 to 1988. Annual mean SO_2 concentrations were about 160 $\mu g\ m^{-3}$ in 1965 and 1966. With a strong initial effort to curb air pollution, those high levels were cut drastically from 1966 to 1973. From 1974 to 1979 SO_2 annual mean levels were about 50–80 $\mu g\ m^{-3}$ in the city centre and 50 $\mu g\ m^{-3}$ in the suburban areas. Sulphur dioxide levels continued to decrease and, in 1988, the annual SO_2 concentrations at the GEMS/Air sites ranged from 18–26 $\mu g\ m^{-3}$, which was well below the WHO annual mean guideline of 40–60 $\mu g\ m^{-3}$. The mean value from the Tokyo municipal monitoring stations (13 $\mu g\ m^{-3}$) was lower than the data from the GEMS/Air sites. While there were big differences between the SO_2 levels in the city centre and in the suburbs during the 1970s, these are now very low.

In Figure 25.2 annual mean data from the three GEMS/Air monitoring sites are presented and, for comparison, two data sets from other sources are included. These data sets agree with the GEMS data.

The 98 percentile values, which characterize the short-term high concentrations, ranged from 47-68 $\mu g\ m^{-3}$, which is also much below the WHO daily guideline (98 percentile) of 150 $\mu g\ m^{-3}$.

National environmental standards for SO_2 have been met at all the stations since 1985 and the levels have remained relatively low. Only in 1987 did one station measure a concentration in excess of the national annual mean standard.

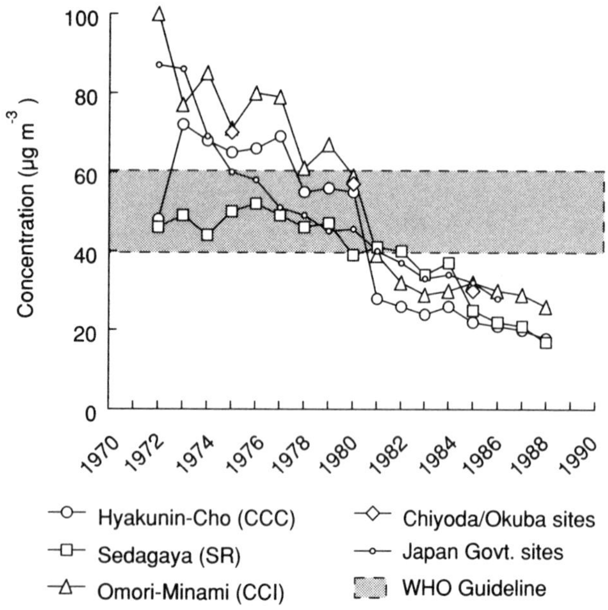

***Figure 25.2** Annual mean sulphur dioxide concentrations*

Sources: GEMS data; Government of Japan, 1988; Komeiji et al., 1990

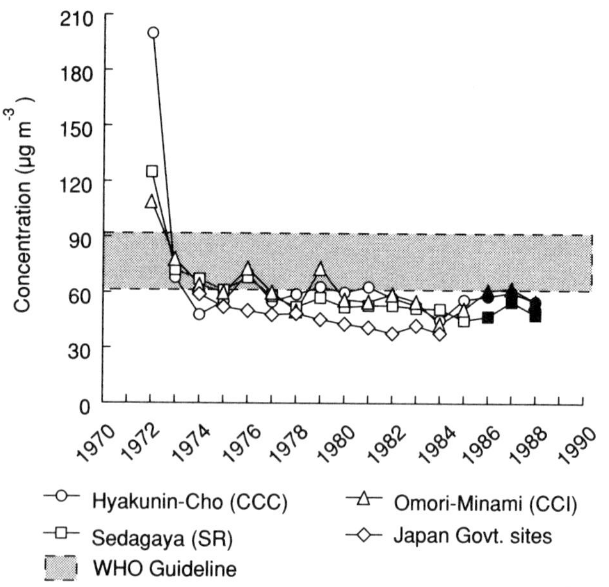

***Figure 25.3** Annual mean suspended particulate matter concentrations*

Note: Solid symbols indicate change in monitoring technique. Nephelometer light scatter 1972–1985. Filter β-ray absorption 1986 onwards

Sources: GEMS data; Government of Japan, 1988

Suspended particulate matter

Emissions Data on emissions of SPM in the Tokyo metropolitan area were not available for this report.

Ambient Concentrations There are various methods of measuring particulates in ambient air. The three GEMS sites use filter beta-ray absorption and nephelometer light scatter. One of the sites also uses a high volume sampler, which can collect particles of varying sizes.

Results from the GEMS monitoring sites are summarized in Figure 25.3. It has to be noted that because of changes in the monitoring methods, the time series from 1975–1984 and 1984–1989 are not directly comparable. Nevertheless, it is clear from the data that SPM decreased steadily from 1975 to 1989. While concentrations during 1975–1977 were about 55–75 $\mu g\ m^{-3}$, SPM levels were below 60 $\mu g\ m^{-3}$ at all monitoring sites in 1989. There were no significant differences between SPM pollution levels in the city centre and in the suburbs. The data confirm that SPM levels are below the WHO annual mean guidelines of 60–90 $\mu g\ m^{-3}$.

Data from GEMS and the Tokyo municipal sites correspond generally well (TMG, 1989b). However, the municipal stations indicate that after the lowest recorded SPM values in 1980 there was a slightly increasing trend in SPM which is not reflected in the data from GEMS/Air sites.

The 98 percentile, which is an indicator for high short-term pollution levels, was 131–148 $\mu g\ m^{-3}$ in the various stations. This is well below the WHO daily guideline (98 percentile) of 230 $\mu g\ m^{-3}$ (not be exceeded on more than 2 per cent of days).

Lead

Emissions In urban areas, the major source of lead (Pb) emissions is motor vehicle traffic. Lead compounds are added to petrol to improve octane rating and, after combustion, lead-containing particles are emitted with the exhaust gases into the air. In Japan, the maximum level of Pb additive in petrol is 0.15 $g\ l^{-1}$. Lead emissions were reduced dramatically when unleaded petrol was introduced. Most of the petrol used in Tokyo is unleaded (see Table 4.2, Chapter 4). Thus, at present, Pb emissions in Tokyo are very small.

Ambient Concentrations No regular reports on ambient levels of Pb were available for this report. Lead measurement is not part of the routine national air monitoring programme, or of GEMS/Air.

Usually, ambient Pb concentrations are closely related to SPM levels since Pb resides on fine particles. As Pb emissions are small in Tokyo, and SPM concentrations are low, ambient air Pb levels are thought to be low, too. This is supported by analyses of particulate samples at the Okuba station which have been carried out since 1966. These data indicate a downward trend for reported metals including Pb (Komeiji et al., 1990).

Carbon monoxide

Emissions Carbon monoxide emissions generally result from incomplete combustion processes. Thus in urban areas most CO emissions normally come from motor vehicles exhausts. A CO emission inventory for Tokyo was not available for this report. However, catalytic converters for the control of automobile exhaust pollutants were introduced in 1976, and thus CO emissions have been steadily reduced.

Ambient Concentrations Carbon monoxide levels have been monitored in Tokyo since 1965. There are 35 stations for monitoring ambient air and 32 other stations specifically designed to monitor automobile exhaust pollution at roadside locations. All stations use infrared analysers. Figure 25.4 summarizes the annual averages for CO concentrations at the ambient air monitoring sites operated by the Tokyo Metropolitan Government (TMG) (TMG, 1989b). The time series shows that annual average CO levels were high (6–8 mg m^{-3}) during 1965–1969. Since then, average CO concentrations have declined because of strict control legislation. While CO levels were still 2–3 ppm (2.3–3.5 mg m^{-3}) during 1971–1975, the average concentration in 1988-1989 was 0.9 ppm (1 mg m^{-3}). There were no significant differences of CO pollution between the city centre and the suburbs. The national standard was met at all monitoring stations.

No information on short-term CO concentrations was available for this report. However, from the very low long-term average data, it can be assumed that short-term CO values in Tokyo were below the WHO eight-hour guideline of 10 mg m^{-3}.

Oxides of nitrogen

Emissions Oxides of nitrogen include both NO and NO_2. Although NO is an order of magnitude less toxic than NO_2, both substances are normally included in emissions inventories because NO is ultimately converted to NO_2. The main sources of NO_x are high-temperature combustion processes and the chemical industry (nitric acid and fertilizer production). Oxides of nitrogen emissions in Tokyo were estimated to be 52,700 tonnes per annum in 1985. The

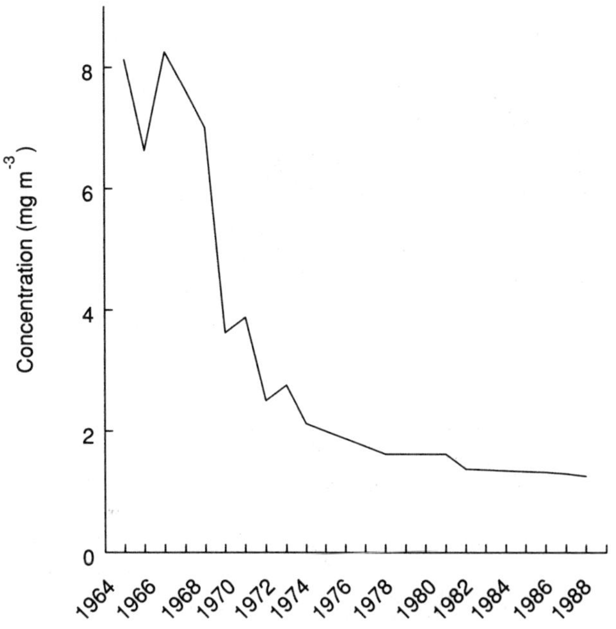

Figure 25.4 *Annual mean carbon monoxide concentrations*

After TMG, 1989b

majority of those emissions (35,400 tonnes per annum or 67 per cent) come from motor vehicle automobile exhausts. Eighteen per cent of the total NO_x emissions come from industrial sources, 9.5 per cent from domestic sources and 5 per cent from ships and aircraft (TMG, 1989a).

With respect to the NO_x emissions from motor vehicle sources, diesel-powered vehicles (especially cargo vehicles) are of particular importance: although they contribute only about 20 per cent of the total mileage travelled, they contribute more than 50 per cent of all NO_x emissions. Passenger vehicles with petrol engines are equipped with exhaust pipe emission control devices. They contribute more than 50 per cent of the total mileage travelled, but emit about only 21 per cent of NO_x (TMG, 1989a).

Ambient Concentrations In Tokyo, NO_x levels are monitored at the 67 stations operated by the TMG. The method of measurement is the Salzman automatic colorimeter.

Figure 25.5 shows the annual mean NO_2 concentrations at the ambient air monitoring sites operated by the TMG (TMG, 1989b). From the monitoring data it is clear that while control of other air pollutants was successful in Tokyo, NO_2 concentrations could not be reduced during the last 20 years. Since monitoring began, annual mean concentrations have been around 0.03 ppm (60 $\mu g\ m^{-3}$) as a city mean. Annual means were around 0.033 ppm (65 $\mu g\ m^{-3}$) in the city centre and 0.025 ppm (50 $\mu\ m^{-3}$)

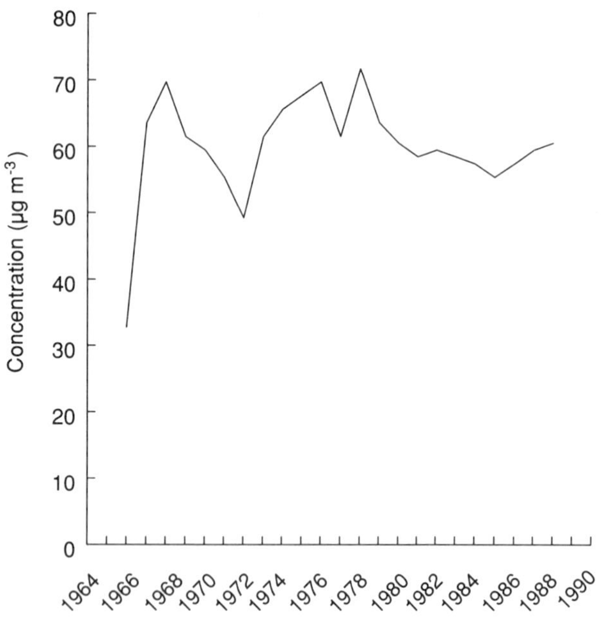

Figure 25.5 *Annual average nitrogen dioxide concentrations*

After TMG, 1989b

in the suburbs (TMG, 1989b).

The national environmental standard for NO_2 (as of July 1987) is 0.04 ppm (76 µg m^{-3}) as a daily mean, and a level of 0.06 ppm (115 µg m^{-3}) is considered as tolerable. These NO_2 guidelines are considerably lower than the WHO 24-hour guideline of 150 µg m^{-3}.

For the whole of Japan, the 98 percentile data indicate that the 0.06 ppm standard is exceeded at about 25 per cent of the automobile exhaust monitoring stations, which are located primarily in the large metropolitan areas such as Tokyo, Osaka and Kanagawa. In Tokyo during 1988 the 0.04–0.06 ppm daily average national standard was exceeded on several occasions. However, it seems unlikely that the present WHO 24-hour guideline of 150 µg m^{-3} is exceeded in Tokyo.

Ozone

Emissions Photochemical oxidants are a secondary pollutant and are a product of complex atmospheric reactions of NO_2 and reactive volatile organic compounds (VOC) under the influence of sunlight. As the formation of photochemical oxidants takes some time, the highest concentrations are normally found downwind of the major emission areas. Ozone (O_3) is one of the most important photochemical oxidants and is regarded as a marker substance for photochemical pollution.

Emissions of NO_x in Tokyo have been estimated to be about 52,700 tonnes per annum. For this report, no VOC emission inventory was available. Sources of VOC emissions are incomplete combustions, industrial processes (e.g., refineries), petrol and solvent evaporation, and natural sources (e.g., vegetation).

Ambient Concentrations In Tokyo, photo-oxidants are measured by all 35 ambient air monitoring stations and some of the automobile exhaust monitoring stations. Photochemical oxidant measurements were made using the neutral buffered automatic potassium iodide colorimetric method.

Figure 25.6 shows the annual mean concentrations of photochemical oxidants (monitored every day from 0500 hours to 2000 hours) at the ambient air monitoring sites operated by the TMG (TMG, 1989b). The monitoring data show that annual average levels of photochemical oxidants have decreased since the late 1960s. While annual mean concentrations of oxidants were as high as 0.035–0.045 ppm (70–90 µg m^{-3}, expressed as O_3) in 1968–1970, the oxidant levels dropped to about 0.02 ppm (40 µg m^{-3}) in 1978. Since then, levels have been stable until the present. Interestingly, annual average values at the suburban sites were about 10 µg m^{-3} higher than in the city centre (TMG, 1989b). However, photochemical oxidants are above the national standards at many monitoring sites.

On days when any one-hour O_3 concentration exceeds 0.12 ppm (240 µg m^{-3}), an "oxidant warning" is given. At this level adverse health effects may occur

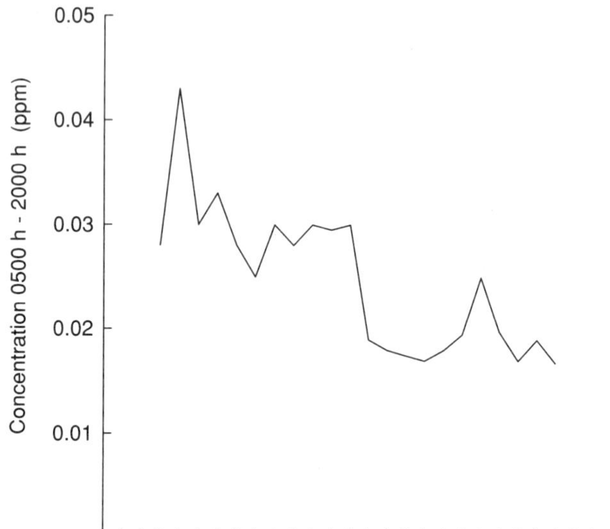

Figure 25.6 *Annual average photo-oxidants concentrations*

After TMG 1989b

in a significant proportion of the exposed population. The number of "oxidant warning days" per year is an index for the frequency of short-term high oxidant levels.

In the whole of Japan, the number of "oxidant warning days" per year varied between 59 and 171. The numbers vary greatly from year to year, and no trend has emerged since 1977. Nationally, there were 168 warning days in 1987 and 86 warning days in 1988. In Tokyo, a warning was issued on 45 days in 1988, (12 per cent of all days in 1988). As the national one-hour warning standard of 240 $\mu g\ m^{-3}$ is higher than the WHO one-hour average guidelines of 150–200 $\mu g\ m^{-3}$, it can be assumed that the WHO criteria are exceeded even more frequently.

25.4 Conclusions

Air Pollution Situation For a commercial city with a population of 12 million, the air pollution situation in Tokyo is very good. The levels of SO_2, SPM, CO, and NO_2 are below the applicable WHO guidelines. However, short-term peak levels of photochemical oxidants (particularly ozone) exceed national and WHO guidelines on many days each year. Ozone levels are even higher in suburban areas than in the city centre.

Main Problems Most air pollution sources have been controlled in the last 20 years. However, control has not been as effective for NO_x because the reduction of emissions at individual sources was offset by the increase in traffic. While NO_x emissions from petrol-powered vehicles can be reduced by the use of catalytic converters, control for heavy diesel vehicles is more difficult and is only about to start. Diesel vehicles are the main contributors to NO_x pollution.

Short-term high concentrations of photo-oxidants are regarded as the main air pollution problem in Tokyo. Oxides of nitrogen are necessary precursors for the formation of photo-oxidants (e.g., O_3). Thus a reduction of NO_2 (together with a reduction of VOC) will also result in a reduction of ozone levels.

Control Measures Tokyo is a recognized example of how air pollution in an industrial megacity can be controlled. Tokyo has experienced rapid growth in terms of its population and its economy in the past 30 years. Correspondingly with the rapid concentration of industries and rapid progress of motorization, environmental pollution in Tokyo continued to worsen until the late 1960s when pollution became a major social problem. Since then the TMG has promoted various measures, such as controlling emissions from factories and developing sewage and waste disposal plants (TMG, 1990).

In order to reduce emissions of air pollutants, the metropolitan government issued regulations and standards for industries to follow. Through that, emissions (and ambient levels) of SO_2, SPM and CO were reduced dramatically. In particular, the reduction of SO_2 through flue gas treatment technologies in stationary sources has served as an example of the technological emissions control potential.

In addition to flue gas treatment, the metropolitan government has been implementing controls on emissions depending on the height of individual chimneys (so that the concentration after coming out of the chimneys and reaching the ground should be below a certain value), fuel regulation and total emission regulation (Government of Japan, 1990).

The projected urban and industrial growth will limit the effectiveness of existing control programmes. An alternative might be to use different energy sources. Since cars account for the largest source of NO_2 and CO emissions, changes in fuel type might aid in reducing these levels. Additional solutions include car-pooling, increasing use of railways and reduction of unnecessary motor vehicle travel.

Health Effects The available monitoring data suggest that deleterious health effects from SO_2, SPM, Pb, CO, and NO_x pollution in Tokyo might be relatively low. It is, however, important to recognize that despite low mean concentrations there might be much higher levels at specific locations at specific times (e.g., at roadside locations during peak hours of traffic congestion). This can be especially important for CO pollution which might be much above the WHO guidelines (of 10 mg m^{-3} as an eight-hour mean) in certain conditions. Health effects may occur for people with high exposure, such as road-repair and tunnel workers.

With respect to NO_x pollution, epidemiological studies in Japan and the Tokyo area have not shown adverse effects. Ambient levels of NO_x are relatively low compared with those where effects have been demonstrated in humans. Nonetheless, concern is warranted since the effects of long-term, low-level exposures to NO_x may "promote" disease and increase susceptibility to illness.

Ozone is generally recognized as being the most irritating and toxic secondary air pollutant. Extensive research in clinical exposure facilities have demonstrated dose-response relationships between O_3 exposure (concentration and time factors) and adverse, reversible, respiratory health effects.

In Tokyo, health effects of air pollution have been

studied scientifically and statistically. According to the law for "Compensation for Health Damages due to Air Pollution", the number of people reporting "recognized" air pollution diseases has been recorded since 1975. However, although the air pollution situation in Tokyo greatly improved from 1975 to 1988, the number of cases reporting diseases rose from about 6,500 to 43,600 in the same time period (TMG, 1989b). Thus, in 1988 it was concluded that air pollution was not the only cause of the diseases and the law was changed accordingly.

25.5 References

Government of Japan 1988 *Quality of the Environment in Japan*, Environment Agency, Government of Japan, Tokyo.

Government of Japan 1990 *Japanese Performance of Energy Conservation and Air Pollution Control*, Environment Agency, Government of Japan, Tokyo.

Komeiji, T., Aoki, K., Koyama, I. and Okita, T. 1990 Trends of Air Quality and Atmospheric Deposition in Tokyo, *Atmospheric Environment* **24A**, 2099–2103.

OECD 1991 *OECD Environmental Data Compendium 1991*, Organisation for Economic Co-operation and Development, Paris.

TMG 1989a *Automobile Pollution Control Plan: Towards a Better Living Environment in Tokyo*, Tokyo Metropolitan Government, Tokyo.

TMG 1989b *Summary of the Results of Regular Monitoring for Air Pollution in the FY 1989*, Air Monitoring Division, Tokyo Metropolitan Government, Tokyo.

TMG 1990 *Protecting Tokyo's Environment*, Tokyo Metropolitan Government, Tokyo.

UN 1989 *Prospects of World Urbanization 1988*, Population Studies No. 112, United Nations, New York.

APPENDICES

Appendix I

Air quality guidelines and standards

In this report, urban air quality has been assessed by comparing national air quality data with established WHO air quality guidelines. Where air pollution levels exceed these guidelines, health and welfare effects may be encountered.

Most countries and some of the cities included in this report are required to comply with specific air quality standards which are usually incorporated in national legislation. These standards are often supplemented with alert or emergency levels which are concentrations at which specific remedial measures are taken. National standards are often subdivided into primary and secondary (and sometimes tertiary) standards. Such a hierarchy of standards may be designed to apply successively to protect health, welfare or specific land-use, such as residential or industrial zones.

Standards and guidelines have been adopted for different reasons. Annual mean guidelines and standards are designed to protect the population from regular exposure to high levels of pollution. By ensuring that the mean daily concentrations remain low, long-term chronic effects on health are minimized.

In this report annual data and guidelines and standards, unless stated otherwise, are presented as annual arithmetic means and are referred to as mean or average. This is the sum of all observations divided by the number of observations. In some cases reference is made to the median or 50 percentile which is the "middle" value when all observed values are ranked. In some cases the geometric mean is used (especially for SPM) which is the nth root of the product of n numbers (the antilog of the arithmetic mean of the logarithms of the data).

There is also a need to control acute effects which result when abnormally high levels of pollution persist for short periods. Therefore, short-term guidelines and standards place limits on pollutant concentrations over periods ranging from 10 minutes to 24 hours. Typically, 1-hour, 8-hour and 24-hour averaging times are used in short-term guidelines and standards. In some cases the term 98-percentile guideline or standard is used and this is defined as:

> A value (i.e., WHO daily guideline) below which 98 per cent of all daily means should fall. Where measurements are taken 365 days a year, values should not rise above the 98 percentile on more than seven days. Where fewer readings are made, the number of days in which levels may exceed this guideline falls proportionally.

The guidelines and standards applied and discussed within this report are given in the following tables.

All standards have been converted to SI units for comparative purposes. The relevant conversion factors (WHO, 1987) are given at the foot of each table. These conversion factors may not be the same as those applied to data throughout the report. National data may have been converted using different conversion factors (e.g. 0°C, 1 atm, 20°C, 1 atm or 25°C, 1 atm). In the following tables, concentrations originally given as parts per million (ppm) are given in parenthesis where appropriate.

Notes and sources for the tables are listed at the end of Appendix I.

Sulphur dioxide

Standard	Time-weighted average ($\mu g\ m^{-3}$)				Guidelines and standards with other averaging times ($\mu g\ m^{-3}$)	Source(s)
	1 hour	8 hours	24 hours	1 year		
WHO	350		125[a]	50[a]	500 (10 minutes)	1
			100–150	40–60		2
Argentina					70 (30 days)	3
Alert	(1)[b]	(0.3)[b]				
Alarm	(5)[b]					
Emergency	(10)[b]					
Brazil						
Primary			365	80		4
Secondary			100	40		
São Paulo						
Attention			800			
Alert			1,600			
Emergency			2,100			
China						
Class I[c]			50		150 (not once)	5
Class II[d]			150		500 (not once)	
Class III[e]			250		700 (not once)	
European Community (UK)						
98 percentile			350[f]		6	
98 percentile			250[g]			
50 percentile			120[h]			
50 percentile			80[i]			
50 percentile			180 (Winter)[j]			
50 percentile			130 (Winter)[k]			
Guide value			100–150	40–60		
India						
Sensitive area		30	20			7
Residential area		80	60			
Industrial area		120	80			
Indonesia	286 (0.1)					8
Japan	286 (0.1)					9
Korea	429 (0.15)			143 (0.05)		10
Mexico			350			11
Philippines	850		369			12
Russia			150			13
Thailand			300	100[l]		14
USA						
Federal			365	80	1,300 (3 hours)	15
California	715 (0.25)		143 (0.05)[m]			16

Numbers in parenthesis are original standard concentrations in ppm   Conversion factor: 1ppm = 2,860 $\mu g\ m^{-3}$

Suspended particulate matter

Standard	Measurement method	Standard	Time-weighted average (μg m^{-3}) 8 hours	24 hours	1 year	Guidelines and standards with other averaging times (μg m^{-3})	Source(s)
WHO							
	BS[n]			125[a]	50[a]		1
	BS[n]			100–150	40–60		2
	TSP			120[a]			1
	TSP			150–230	60–90		2
	TP			70[a]			1
Argentina	TSP					150 (1 month)	3
Brazil	BS	Primary		150	60		4
	BS	Secondary		100	40		
	TSP	Primary		240	80[l]		
	TSP	Secondary		150	60[l]		
	PM_{10}			150	50		
São Paulo							
	TSP	Attention		375			
	TSP	Alert		625			
	TSP	Emergency		875			
	IP	Attention		250			
	IP	Alert		420			
	IP	Emergency		500			
China							
	TSP	Class I[c]		150		300 (not once)	5
	TSP	Class II[d]	420	300		1,000 (not once)	
	TSP	Class III[e]	680	500		1,500 (not once)	
European Community							
(UK)	BS	98 percentile		250			6
	BS	50 percentile		130 (winter)			
	BS	50 percentile		80			
India							
	TSP	Sensitive area	100	70			7
	TSP	Residential area	200	140			
	TSP	Industrial area	500	360			
Indonesia	TSP			260			8
Japan	PM_{10}			100		200 (1 hour)	9
Korea	TSP			300[p]	150		10
Mexico	TSP			275			11
Philippines	TSP			180		250 (1 hour)	12
Russia	TSP			150			13
Thailand	TSP		330	100		14	
USA							
Federal	PM_{10}		150	50		15	
California	PM_{10}		50	30[l]		16	

Lead

Standard	Time-weighted average (μg m^{-3}) 24 hours	1 year	Guidelines and standards with other averaging times (μg m^{-3})	Source(s)
WHO		0.5–1		1,17
European Community (UK)		2		18
Indonesia	60			8
Thailand	10			14
USA				
Federal			1.5 (quarterly)	15
California			1.5 (30 days)	16

Carbon monoxide

Standard	Time-weighted average (mg m^{-3}) 1 hour	8 hours	24 hours	Guidelines and standards with other averaging times (mg m^{-3})	Source(s)
WHO	30	10		60 (30 minutes) 100 (15 minutes)	1
Argentina	57.2 (50)	11.4 (10)			3
Alert	114.5 (100)	17.2 (15)			
Alarm	137.4 (120)	34.3 (30)			
Emergency	171.8 (150)	57.2 (50)			
Brazil	40	10			4
São Paulo					
Attention		17.2(15)			
Alert		34.3 (30)			
Emergency		45.8 (40)			
China					
Class I[c]			4	10 (not once)	5
Class II[d]			4	10 (not once)	
Class III[e]			6	20 (not once)	
Indonesia		22.9 (20)			8
Japan	22.9 (20)		11.45 (10)		9
Korea	9.2 (8)	22.9 (20)			10
Mexico		14.8[q]		11	
Philippines	85	10			12
Russia			1		13
Thailand	50	20			14
USA					
Federal	40	10			15
California	22.9 (20)	10.3 (9)			16

Numbers in parenthesis are original standard concentrations in ppm. Conversion factor: 1ppm = 1.145 mg m^{-3}

Nitrogen dioxide

Standard	Time-weighted average (μg m^{-3}) 1 hour	8 hours	24 hours	1 year	Guidelines and standards with other averaging times (μg m^{-3})	Source(s)
WHO	400		150			1
	190–320[r]					20
Argentina	846 (0.45)[s]					3
Alert	1,128 (0.60)[s]		282 (0.15)[s]			
Alarm	2,256 (1.20)[s]		564 (0.30)[s]			
Emergency			752 (0.40)[s]			
Brazil	320			100		4
São Paulo						
Attention	1,130					
Alert	2,260					
Emergency	3,000					
China						
Class I[c]			50		100 (not once)	5
Class II[d]			100		150 (not once)	
Class III[e]			150		300 (not once)	
European Community (UK)						
98 percentile limit value	200					21
98 percentile guide value	135					
50 percentile guide value	50					
India						
Sensitive area		30	20			7
Residential area		80	60			
Industrial area		120	80			
Indonesia			94 (0.05)			8
Japan			75–113 (0.04–0.06)			9
Korea	282 (0.15)			94 (0.05)		10
Mexico	395					11
Philippines	190					12
Thailand	320					14
USA						
Federal				100		15
California	470 (0.25)					16

Note: Numbers in parentheses are original standard concentrations in ppm. Conversion factor: 1ppm = 1,880 μg m^{-3}

Ozone

Standard	Time-weighted average (µg m^{-3}) 1 hour	Guidelines and standards with other averaging times (µg m^{-3})	Source(s)
WHO	150–200	100–120 (8 hours)	1
	100–200		22
Argentina	200		3
Alert	300		
Alarm	400		
Emergency	800		
Brazil	160		4
São Paulo			
Attention	(200)[t]		
Alert	800		
Emergency	1,000		
China			
Class I	120		5
Class II	160		
Class III	200		
Indonesia	160		8
Japan	120		9
Korea	200[p]	40 (1 year)	10
Mexico	220[u]		11
Philippines	120		12
Thailand	200		14
USA			
Federal	235[v]		15
California	180[v]		16

Conversion factor: 1ppm = 2,000 µg m^{-3}

Notes

a Guideline values for combined exposure to sulphur dioxide and suspended particulate matter (they may not apply to situations where only one of the components is present)
b ppm volume/volume
c Class I = Tourist, historical and conservation areas
d Class II = Residential urban areas and rural areas
e Class III = Industrial areas and heavy traffic areas
f If black smoke < 150 μg m^{-3}
g If black smoke > 150 μg m^{-3}
h If black smoke < 40 μg m^{-3}
i If black smoke > 40 μg m^{-3}
j If black smoke < 60 μg m^{-3}
k If black smoke > 60 μg m^{-3}
l Geometric mean
m With ozone > 0.10 ppm, one-hour mean or TSP > 100 μg m^{-3}, 24-hour mean
n Application of the black smoke value is recommended only in areas where coal smoke from domestic fires is the dominant component of the particulates. It does not necessarily apply where diesel smoke is an important contributor
p Not to be exceeded more than two times in a year
q Eight-hour moving average
r Not to be exceeded more than once per month
s Oxides of nitrogen as nitrogen dioxide
t Attention level for São Paulo State only
u Maximum daily one-hour average
v Value not to be exceeded more than once in a year

Suspended particulate matter measurement methods

BS = Black smoke; a concentration of a standard smoke with an equivalent reflectance reduction to that of the atmospheric particles as collected on a filter paper.
TSP = Total suspended particulate matter; the mass of collected particulate matter by gravimetric analysis divided by total volume sampled.
PM_{10} = Particulate matter less than 10 μm in aerodynamic diameter; the mass of particulate matter collected by a sampler having an inlet with 50 per cent penetration at 10 μm aerodynamic diameter determined gravimetrically divided by the total volume sampled.
TP = Thoracic particles (as PM_{10})
IP = Inhalable particles (as PM_{10})

Sources

1 WHO 1987 *Air quality guidelines for Europe,* WHO Regional Publications, European Series No. 23, World Health Organization, Regional Office for Europe, Copenhagen.
2 WHO 1979 *Sulfur oxides and suspended particulate matter,* Environmental Health Criteria 8, World Health Organization, Geneva.
3 Law No. 20,284, Buenos Aires, 16 April 1973, Official Journal No. 22658. In: Murley, L. (Ed.) 1991 *Clean Air Around the World,* Second Edition, International Union of Air Pollution Prevention Associations, Brighton.
4 CONAMA Resolution No. 03 of 28 June 1990. In: Murley, L. (Ed.) 1991 *Clean Air Around the World,* Second Edition, International Union of Air Pollution Prevention Associations, Brighton.
5 Faiz, A., Sinha, K., Walsh, M. and Valma, A. 1990 *Automotive Air Pollution: Issues and Options for Developing Countries,* WPS 492, The World Bank, Washington DC.
6 80/779/EEC. In: Murley, L. (Ed.) 1991 *Clean Air Around the World,* Second Edition, International Union of Air Pollution Prevention Associations, Brighton.
7 NEERI 1988 *Air quality status in ten cities: India 1982-1985,* National Environmental Engineering Research Institute, Nagpur.
8 Achmadi U.F. 1988 Air Pollution in Indonesia: Current status and its control strategies. In: Proceedings of the Third Joint Conference of Air Pollution Studies in Asian Areas, Japan Society of Air Pollution, Tokyo, 387-408.
9 Basic Law for Environmental Pollution Control, 1969. In: Murley, L. (Ed.) 1991 *Clean Air Around the World,* Second Edition, International Union of Air Pollution Prevention Associations, Brighton.
10 Basic Law for National Environmental Policy. In: Murley, L. (Ed.) 1991 *Clean Air Around the World,* Second Edition, International Union of Air Pollution Prevention Associations, Brighton.
11 Gobierno de la Republica 1990 *Programa integral contra la contaminacion atmospherica,* Gobierno de la Republica, Mexico.
12 Santos-Ocfemia E. 1988 Air Pollution Situation in the Philippines. In: Proceedings of the Third Joint Conference of Air Pollution Studies in Asian Areas, Japan Society of Air Pollution, Tokyo, 424-436.
13 NEERI 1983 *Air quality in selected cities in India 1980-1981,* National Environmental Engineering Research Institute, Nagpur.
14 Pairoj-Boriboon S. 1988 Air Pollution Abatement Strategy in Thailand. In: Proceedings of the Third

Joint Conference of Air Pollution Studies in Asian Areas, Japan Society of Air Pollution, Tokyo, 437-445.

15 USEPA 1991 *National Air Quality and Emissions Trends Report,* Office of Air Quality Planning and Standards, US Environmental Protection Agency, Research Triangle Park.

16 SCAG 1991 *Air Quality Management Plan - South Coast Air Basin,* South Coast Air Quality Management District, Southern California Association of Governments, Los Angeles.

17 WHO 1977 *Lead,* Environmental Health Criteria 3, World Health Organization, Geneva.

18 82/884/EEC. In: Murley, L. (Ed.) 1991 *Clean Air Around the World,* Second Edition, International Union of Air Pollution Prevention Associations, Brighton.

19 WHO 1979 *Carbon monoxide,* Environmental Health Criteria 13, World Health Organization, Geneva.

20 WHO 1977 *Oxides of nitrogen,* Environmental Health Criteria 4, World Health Organization, Geneva.

21 85/203/EEC. In: Murley, L. (Ed.) 1991 *Clean Air Around the World,* Second Edition, International Union of Air Pollution Prevention Associations, Brighton.

22 WHO 1978 *Photochemical oxidants,* Environmental Health Criteria 7, World Health Organization, Geneva.

Appendix II

The description and classification of cities, urban agglomerations and countries in this study and the arrangement of the material do not imply the expression of any opinion whatsoever on the part of UNEP or WHO concerning the legal status of any country, territory, city or area or of its authorities, or concerning the delimitation of its frontiers or boundaries, or regarding its economic system or degree of development.

The names of cities/urban agglomerations and countries listed below include the abbreviated names used within the body of this report together with their full names.

List of city and country names

City/urban agglomeration name		Country name[1]	
Short name	Full name	Short name	Full name
Bangkok	Metropolitan Bangkok	Thailand	Kingdom of Thailand
Beijing	Beijing Municipality	China	People's Republic of China
Bombay	Greater Bombay	India	Republic of India
Buenos Aires	Metropolitan Area of Buenos Aires	Argentina	Argentine Republic
Cairo[a]	Greater Cairo	Egypt	Arab Republic of Egypt
Calcutta	Calcutta Metropolitan District	India	Republic of India
Delhi	Delhi	India	Republic of India
Jakarta	Metropolitan Jakarta	Indonesia	Republic of Indonesia
Karachi	Karachi	Pakistan	Islamic Republic of Pakistan
London	Greater London	UK	United Kingdom of Great Britain and Northern Ireland
Los Angeles	Los Angeles Metropolitan Area	USA	United States of America
Manila[b]	Metropolitan Manila	Philippines	Republic of the Philippines
Mexico City	Metropolitan Area of Mexico City	Mexico	United Mexican States
Moscow	Moscow	Russia	Russian Federation
New York	New York City Metropolitan Area	USA	United States of America
Rio de Janeiro	Metropolitan Area of Rio de Janeiro	Brazil	Federative Republic of Brazil
São Paulo	Greater São Paulo Area	Brazil	Federative Republic of Brazil
Seoul	Greater Seoul	Korea	Republic of Korea
Shanghai	Metropolitan Shanghai	China	People's Republic of China
Tokyo[c]	Metropolitan Tokyo	Japan	Japan

[a]Cairo/Giza in UN population estimates[2]
[b]Manila/Quezon in UN population estimates[2]
[c]Tokyo/Yokohama in UN population estimates[2]

Sources:
1. UNEP 1991 United Nations Environment Programme Environmental Data Report , 3rd edition, Basil Blackwell, Oxford.
2. UN 1989 Prospects of World Urbanization, 1988, Population Studies No. 112, United Nations, New York.

Appendix III

Units and abbreviations

1. Units

Mass:	gram (g)
	tonne or metric ton (t)
Length:	metre (m)
Area:	square metre (m^2)
Volume:	cubic metre (m^3)
Capacity:	litre (l)
Time:	second (s)
	hour (h)
	day (d)
	annum or year (a)
Energy:	watt (W)
Temperature:	degree Celsius (°C)
Pressure:	atmosphere (atm)

2. Fractions and multiples of units

10^{15}	peta	P
10^{12}	tera	T
10^{9}	giga	G
10^{6}	mega	M
10^{3}	kilo	k
10^{-2}	centi	c
10^{-3}	milli	m
10^{-6}	micro	µ
10^{-9}	nano	n

3. Common units used in the text

µg m^{-3}	=	micrograms per cubic metre
mg m^{-3}	=	milligrams per cubic metre
t a^{-1}	=	tonnes per annum
kWh	=	kilowatt-hour
ppm	=	parts per million (volume/volume)
ppb	=	parts per billion (volume/volume)

4. Conversion factors

		0°C, 1 atm	**25°C, 1 atm**
Sulphur dioxide (SO_2)			
1 ppm	=	2,856 µg m^{-3}	2,600 µg m^{-3}
Carbon monoxide (CO)			
1 ppm	=	1.250 mg m^{-3}	1.145 mg m^{-3}
Nitric oxide (NO)			
1 ppm	=	1,340 µg m^{-3}	1,230 µg m^{-3}
Nitrogen dioxide (NO_2)			
1 ppm	=	2,050 µg m^{-3}	1,880 µg m^{-3}
Ozone (O_3)			
1 ppm	=	2,140 µg m^{-3}	2,000 µg m^{-3}

5. GEMS/Air monitoring site classifications

Site location

City Centre = CC
Suburban = S

Dominant source

Commercial = C
Residential = R
Industrial = I
Mobile = M

Site classification not available = NA